Advances in Genetic Programming

Advances in Genetic Programming

edited by
Kenneth E. Kinnear, Jr.

A Bradford Book
The MIT Press
Cambridge, Massachusetts
London, England

This book was printed and bound in the United States of America.

Library of Congress Cataloging-in-Publication Data

Advances in genetic programming / edited by Kenneth E. Kinnear, Jr.
    p.   cm. — (Complex adaptive systems)
  "A Bradford book."
  Includes bibliographical references and index.
  ISBN 0-262-11188-8
  1. Electronic digital computers—Programming. I. Kinnear, Kenneth E. II. Series.
QA76.6.A3333 1994
006.3—dc20   93-47518

Contents

Contributors

Lee Altenberg
Institute of Statistics and Decision Sciences
Duke University
Durham, NC 27708-0251 USA
altenber@acpub.duke.edu

Martin Andrews
Cambridge University Engineering Department
Trumpington Street
Cambridge, CB2 1PZ UK
mdda@eng.cam.ac.uk

David Andre
Symbolic Systems Department
Stanford University
Building 60
Stanford, CA 94305 USA
(415) 497-9426
phred@leland.stanford.edu

Peter J. Angeline
Laboratory for Artificial Intelligence Research
Computer and Information Sciences Department
The Ohio State University
Columbus, OH 43210 USA
pja@cis.ohio-state.edu

Jason Bluming
Computer Science Department
Stanford University
Stanford, CA 94305 USA
jbluming@cs.stanford.edu

Aviram Carmi
Cal State University at Northridge and
Hughes Missile Systems Co.
1137 Mohawk
Topanga, CA 90290 USA
carmi@ipld01.hac.com
secsavc@secs.csun.edu

Patrik D'haeseleer
Computer Science Department
Stanford University
Stanford, CA 94305 USA
pdhaes@cs.stanford.edu

Hugo de Garis
ATR Human Information Processing
Research Laboratories
2-2 Hikari-dai
Seiko-cho
Soraku-gun
Kyoto, 619-02, Japan
+81-7749-5-1079
degaris@hip.atr.co.jp

Frédéric Gruau
C. E. N. G.
Département de Recherche Fondamentale
Matière Condensée
SP2M/PSC BP85X,38041
Grenoble Cedex
France
gruau@drfmc.ceng.cea.fr

Simon G. Handley
Computer Science Department
Stanford University
Stanford, CA 94305 USA
(415) 723-4096
shandley@cs.stanford.edu

Thomas S. Huang
Beckman Institute and
Coordinated Science Laboratory
University of Illinois
Urbana, IL 61801 USA
huang@uicsl.csl.uiuc.edu

Hitoshi Iba
Machine Inference Section
Electrotechnical Laboratory
1-1-4 Umezono
Tsukuba-city
Ibaraki, 305, Japan
+81-298-58-5918
iba@etl.go.jp

Jan Jannink
Stanford University
Computer Science Department
Stanford, CA 94305 USA
jan@cs.stanford.edu

Mike J. Keith
Allen Bradley Corporation
Heighland Heights, OH 44143 USA
(216) 646-3464
keithm@odin.icd.ab.com

Kenneth E. Kinnear, Jr.
Adaptive Computing Technology
62 Picnic Rd.
Boxboro, MA 01719 USA
(508) 263-7102
kim.kinnear@adapt.com

John R. Koza
Computer Science Department
Margaret Jacks Hall
Stanford University
Stanford, CA 94305-2140 USA
(415) 941-0336
koza@cs.stanford.edu

Martin C. Martin
The Robotics Institute
Carnegie Mellon University
5000 Forbes Ave.
Pittsburgh, PA 15213-3890 USA
(412) 268-1418
martin.martin@ri.cmu.edu

Brij Masand
Thinking Machines Corp
245 First Street
Cambridge, MA 02142 USA
(617) 234-1000
brij@think.com

Thang C. Nguyen
Beckman Institute and
Coordinated Science Laboratory
University of Illinois
Urbana, IL 61801 USA
cthang@uirvld.csl.uiuc.edu

Peter Nordin
Infologics/DaCapo AB
Torggatan 8
411 05 Göteborg
Sweden
+46 31 60 69 07
Peter.Nordin@dacapo.se

E. Howard N. Oakley
Institute of Naval Medicine
Alverstoke Gosport
Hants PO12 2DL UK
howard@quercus.demon.co.uk

Richard Prager
Cambridge University Engineering Department
Trumpington Street
Cambridge, CB2 1PZ UK
rwp@eng.cam.ac.uk

Craig W. Reynolds
Electronic Arts
1450 Fashion Island Boulevard
San Mateo, CA 94404 USA
(415) 513-7442
creynolds@ea.com
cwr@red.com

James P. Rice
Knowledge Systems Laboratory
Department of Computer Science
Stanford University
701 Welch Road, Bldg. C
Palo Alto, CA 94304 USA
(415) 723-8405
rice@ksl.stanford.edu

Conor Ryan
Computer Science Department
University College Cork
Ireland
conor@csvax1.ucc.ie

Taisuke Sato
Machine Inference Section
Electrotechnical Laboratory
1-1-4 Umezono
Tsukuba-city
Ibaraki, 305, Japan
+81-298-58-5918
sato@etl.go.jp

Eric V. Siegel
Computer Science Department
Columbia University
New York, NY 10027 USA
evs@cs.columbia.edu

Graham Spencer
Stanford University
3958 Sutherland
Palo Alto, CA 94303 USA
(415) 857-9313
graham@cs.stanford.edu

Walter Alden Tackett
University of Southern California and
Hughes Missile Systems Co.
Mail Stop X23
20307 Vose St.
Canoga Park, CA 91306 USA
tackett@enee.usc.edu
tackett@ipld01.hac.com

Astro Teller
Carnegie Mellon University
4019 Winterburn Ave.
Pittsburgh, PA 15207 USA
astro@cs.cmu.edu

Preface

Genetic programming, the actual *evolution* of computer programs using analogues of many of the same mechanisms used by natural biological evolution, has sparked the interest of a diverse audience. Professional software engineers, research scientists in machine learning, and even evolutionary biologists each find something fascinating and potentially explosive in these techniques for creating working computer programs untouched by human hands.

This book will:

- introduce you to genetic programming

- give you access to research results focused on improving the power of the genetic programming paradigm

- demonstrate innovative applications of genetic programming to diverse problem areas.

The ultimate goal of this book is to:

Assist you in successfully applying genetic programming, in research as well as in real-world applications.

The material is structured into three parts. Part I places genetic programming in the context of evolutionary computation as a whole and gives a brief but detailed introduction to genetic algorithms and genetic programming. It also gives some practical advice on applying genetic programming to real problems, and contains numerous pointers to additional information in the form of books, articles, email mailing lists, email digests, and ftp sites.

Part II contains contributions whose primary focus is improving the power of the genetic programming paradigm, although in some cases the problems solved in these contributions are themselves of significant interest.

Part III presents a wide variety of applications of genetic programming. Several of these applications are now or will soon be deployed on real-world problems. Others will require considerable scaling up to be of practical value. Many of these contributions also contain new techniques for improving the power of genetic programming.

Genetic programming as a field is still growing rapidly, and much remains to be discovered — and yet, it is being deployed on an increasing number of real-world problems. While there is room for new ideas and need for greater understanding, significant problems are being solved with genetic programming today.

Whatever your interest in genetic programming, be it theoretical or practical, there is something of interest for you in this volume.

Acknowledgments

This volume is the result of the work of many people. It grew directly from the genetic programming community, and our desire to share the work that we are doing with a larger audience than we could otherwise reach.

James Rice of the Knowledge Systems Laboratory at Stanford has created the genetic programming community almost by himself, through his tireless efforts as what someone called the "un-moderator" of the genetic programming electronic mailing list. Prior to this mailing list, most of the people doing genetic programming were the only people at their respective locations doing so, and frequently knew of only a few others around the world who were interested in similar endeavors. James set up the initial mailing list, and through his incisive technical contributions, openness to new ideas, and firm guidance has created a true electronic community.

It was this community that I asked for contributions for both this volume and the genetic programming workshop at ICGA in the summer of 1993. James also contributed many thorough and thoughtful chapter reviews as well as general advice and encouragement.

John Koza generously contributed the chapter which introduces genetic programming, to allow this book to reach a wider audience than would otherwise be possible, as well as participated as a contributor of a technical chapter. He provided highly valued advice, many reviews, and frequent encouragement throughout the process of editing the book. John also, of course, created genetic programming — without which we wouldn't have anything to write about.

The following individuals devoted extensive time and thought to reviews of submissions to the book: Peter J. Angeline, The Ohio State University; Robert J. Collins, UCLA; John R. Koza, Stanford University; Craig W. Reynolds, Electronic Arts; James P. Rice, Stanford University; Walter Alden Tackett, USC/Hughes Missile Systems.

Even the editor needs an editor, and my wife Karen Kinnear generously provided extensive technical and editorial reviews as well as exceptional support throughout the long hours required to bring this book to print.

The able assistance of the professionals at MIT Press helped make the process of editing this book a positive experience.

The enthusiasm and hard work of the individual authors of the chapters is, of course, what makes this book valuable. They deserve credit not only for performing the research and writing up the results reported here, but also for extensive review of each other's submissions. From six countries and as many time zones, the authors managed to meet tight deadlines with exceptional quality material.

Thanks to you all!

I INTRODUCTION

In this introductory section, the concepts necessary to understand the remainder of the book are explained. In Chapter 1, a quick sketch of the overall field of *evolutionary computation* is offered and *genetic programming* is defined and placed in context within the field of evolutionary computation. A brief overview of the work contained in the rest of the book is presented, along with some practical advice on applying genetic programming.

Chapter 2 contains an introduction to the technology of genetic algorithms and genetic programming. The background on genetic programming supplied in this chapter or its equivalent is required to fully comprehend the contributions in the rest of the book.

Both of the first two chapters contain extensive and distinct bibliographic references to additional information in the form of books, research papers, email mailing lists, email digests, and ftp sites.

1 A Perspective on the Work in this Book

Kenneth E. Kinnear, Jr.

This general introduction is designed to put the work in the rest of this book into perspective. In order to do that, it first briefly examines the field of evolutionary computation and where genetic programming fits within that framework. Then it looks at a brief overview of the kinds of work presented in this book. Following that, it presents some practical guidance toward applying genetic programming to a problem of interest to you, the reader, and concludes with suggestions of where to look for more information and inspiration.

1.1 What is Evolutionary Computation?

Over many years, biologists have identified many principles which govern the evolution of living things, at several levels of detail. At the highest level, the theory of natural selection or *survival of the fittest* governs the *evolutionary adaptation* of the biological world from the smallest virus to the most complicated mammal. Natural selection operates on the organism through its performance on a specific task — that of producing offspring. The more viable offspring produced by an organism, the more successful the organism.

Offspring usually result from the growth of a single cell which contains a single specific example of the genetic material for that species. It is through the creation of that specific genetic material that the parent(s) influence the inheritable structure and function of the offspring.

Different species create the genetic material for offspring in different ways, but in all species the parent(s) provide that material in some way or other. In some cases there are two parents, and these parents' genetic material is merged, some from one and some from the other parent, through sexual recombination. In other cases, mutation (random modification to the genetic material) is the only process that distinguishes the genetic endowment of the offspring from that of the parent.

However it is done, the evaluation of the individual through its reproductive performance (called the *fitness*), and the creation of a genetic plan for the offspring based on, but usually different than, the parent's, are central themes in the evolutionary adaptation of biological organisms.

These central themes can be applied to the evolutionary adaptation of computer structures, resulting in the field of *Evolutionary Computation*. The central focus of evolutionary computation is applying the concepts of selection based on fitness to a population of structures in the memory of a computer. These structures reproduce in such a way that the genetic content of their offspring is related in some way to the genetic content of the parents. It is this related but usually different genetic content which makes the offsprings' fitness related to but potentially different from that of their parents.

In most cases, the fitness functions used in the field of evolutionary computation are not oriented towards reproduction directly, but rather are created by the human overseer of the evolutionary process. The individuals in the population are allowed to reproduce in differing amounts based on how well they do on this human designed fitness function. Over time, the population will come to be comprised of individuals which do well as measured by the fitness function, and have been allowed to reproduce.

1.1.1   Why is Evolutionary Computation interesting?

The natural world abounds with structures of incredible complexity and apparent cleverness. Since the time of Darwin, it has been increasingly clear that these structures didn't just happen, but rather evolved. The concept of natural selection, or survival of the fittest has been seen to be a very powerful paradigm, given the existence proof of the biological world.

Prior to the advent of evolutionary computation as a field of study, few would have believed that computer structures could be operated on by evolutionary adaptation, and most of them would have expected this process to take thousands or even millions of years of computer time. Performing repeatable experiments on continental drift would have seemed more practical.

Unexpectedly, evolutionary computation has been shown to be efficient enough for practical use. One can conceptualize the solution of a problem as a search through the space of all potential solutions to that problem. Taking this view, evolutionary computation is considerably more powerful than alternative search techniques such as exhaustive or random search. Evolutionary computation implicitly utilizes a directed search, allowing it to search the space of possible computer structures and find solutions in hours on tasks that would take random search considerably longer than the life of the universe. The search is directed by the contents of the genetic material in the population to portions of the search space that are likely to contain solutions [Holland 1975, 1992, Chapter 1].

While computationally still challenging, evolutionary computation can produce useful and dramatic results in minutes and hours instead of millennia.

1.1.2   Styles of Evolutionary Computation

There are several main styles in the evolutionary computation field, distinguished first by the types of structures which comprise the individuals in the population. These differences determine the factors by which one individual may differ from another, and thus the allowable genetic variation. Equally important differences also exist in the genetic operators used to create offspring, as well as many of the selection procedures based on fitness and a host of other less significant things.

It is worth noting that these differing styles, while representing true differences in approach, each were developed by different and initially unrelated groups of people with little cross fertilization of ideas in the early days of development. Only in the last several years are individuals appearing that have successfully employed more than one of these approaches in solving problems, and even today the number of people who could be considered truly interdisciplinary is small. The relationships and individual strengths and weaknesses of each of these styles are just beginning to be understood.

The term *Evolutionary Computation* has itself been created specifically to be the overall description of all the differing styles of computing approaches based on evolutionary adaptation.

Below is a brief sketch of each of the predominant styles.

- **Evolution Strategy** defines a style of evolutionary computation frequently associated with engineering optimization problems. The structures which undergo adaptation are typically sets of real-valued *objective variables* which are associated with real-valued *strategy variables* in an individual. Fitness is determined by executing task specific routines and algorithms using objective variables as parameters. The strategy variables control the way in which mutation varies each objective variable during the production of new individuals. Recombination is usually employed for both objective variables as well as strategy variables. Rechenberg [1965] began the field. For further information, see [Bäck, Hoffmeister and Schwefel 1991] and [Bäck and Schwefel 1993].

- **Evolutionary Programming** operates on a variety of representational structures, frequently real-valued objective variables although more complex structures have been used (e.g. finite state machines). Again, the objective variables function as arguments to task specific routines and algorithms designed to solve the problem of interest. Mutation is the sole genetic operator employed, with significant strategy built into the overall algorithm to direct the mutation in a computationally fruitful direction. The field was begun by Fogel, Owens and Walsh [1966]. For further information, see [Fogel 1993].

- **Genetic Algorithms** frequently operate on fixed length character strings, often binary, as the structure undergoing adaptation. Fitness is determined by executing task specific routines and algorithms using an interpretation of the character string as the set of parameters. Sexual recombination, also called *crossover*, is the principal genetic operator employed, with mutation usually included as an operator of secondary importance. Other representational structures are possible, and are being used more frequently. See Chapter 2 for a detailed explanation of genetic algorithms. See also the seminal work by Holland [1975,1992], Goldberg's text [Goldberg 1989], and Section 2.6 for further references.

- **Genetic Programming** is an offshoot of genetic algorithms in which the computer structures which undergo adaptation are themselves computer programs. Specialized genetic operators are used which generalize sexual recombination and mutation for the tree structured computer programs undergoing adaptation. Chapter 2 includes a detailed explanation of genetic programming, and this entire volume addresses both the issue of increasing the power of the genetic programming approach, and that of demonstrating applications in a wide variety of areas. Genetic programming was developed by John Koza [1989], and his book, *Genetic Programming: On the Programming of Computers by Means of Natural Selection* [Koza 1992], remains the standard work in the field. *Genetic Programming: The Movie* [Koza and Rice 1992] provides a visual introduction to genetic programming. *Genetic Programming II* [Koza 1994] provides additional information concerning genetic programming. For additional references see Section 2.5.

1.2   What defines Genetic Programming?

Chapter 2 of this volume contains a thorough introduction to the technology of genetic algorithms and genetic programming. One of the questions that became relevant as this book was being put together concerns the definition of *genetic programming*. Genetic programming is an offshoot of genetic algorithms, but in some ways it is more of a generalization rather than a specialization of its parent discipline. What, then, distinguishes genetic programming from genetic algorithms?

Since the field of genetic programming is expanding rapidly, any definition given today will doubtless seem absurd in the near future. Still, the following are the distinguishing features of genetic programming at this time:

- **Nonlinear, and usually tree structured, genetic material** — while some genetic algorithms have genetic material which is nonlinear, linear genetic material remains the rule for genetic algorithms. However, genetic programming almost always operates on nonlinear genetic material, and it is usually explicitly tree structured.

- **Variable length genetic material** — genetic programming almost always operates on genetic material that can vary in size. For practical reasons, size limitations on its growth are usually implemented, but these usually allow considerable growth from the original randomly generated generation.

- **Executable genetic material** — genetic programming is the direct evolution of *programs* for computers. Thus, in almost all cases the genetic material that is evolved is in some sense executable. Executable is not a precise term, and covers a lot of gray areas. Usually the structures are interpreted by some interpreter — sometimes in a language identical or very similar to an existing computer language, sometimes in a very specialized task-oriented language designed for the problem at hand. However, in almost all cases, there is some concept of *executing* the genetic material by some sort of interpreter in order to perform the desired function from which the fitness is derived.

• **Syntax preserving crossover** — while many crossover operators for genetic programming have been reported, in most cases they are defined so as to preserve the syntactic correctness of the program which is the genetic material, as defined by whatever language has been chosen as the representation.

1.3   Current activity in Genetic Programming

Genetic programming is a new field, defined by John Koza's paper at IJCAI in 1989 [Koza 1989], his Stanford report in 1990 [Koza 1990], and his 800 page book in 1992 [Koza 1992]. These references have been the point of departure for an increasing number of individuals working on both the theory and applications of genetic programming.

This volume represents the first collection of technical advances devoted exclusively to genetic programming. Part I consists of this introduction and an introduction to genetic programming in Chapter 2. Part II reports on research designed to increase our understanding of fundamental aspects of genetic programming along with a wide variety of new techniques which increase the power of the genetic programming approach. Part III reports on applications of genetic programming to real world problems.

The next two sections give a very brief overview of parts II and III.

1.3.1   Part II: Increasing the Power of Genetic Programming

Part II reports on research that runs the gamut from theoretical to experimental, all designed to increase our understanding of genetic programming.

• Chapter 3 increases our theoretical understanding of genetic programming through the use of analysis techniques drawn from population genetics.

• Finding automatic ways of generating hierarchy in the programs evolved with genetic programming is one way to increase its power, and is discussed in Chapters 4, 5, and 6.

• Maintaining diversity in the population of a genetic programming system is very important. Chapters 7 and 8 each examine techniques drawn from evolutionary biology in order to maintain diversity in the population.

• Usually memory is held in the structure of the individuals in the population. Chapter 9 discusses a novel way in which memory can be made explicit in genetic programming.

- Evolving fully general algorithms capable of processing an unlimited number of fitness cases while using only a small number of fitness tests during the actual evolution is a challenging task. Several approaches to this problem are discussed in Chapters 7, 10, 11, and 12.

- A genetic programming run sometimes gets stuck on a local optimum, and fails to find either the global optimum or even a very good local optimum. Coevolution is increasingly used to avoid this problem, and is discussed in Chapters 4, 19, and 20.

- Genetic programming remains computationally intensive. Two very different techniques for improving its execution time performance are discussed in Chapters 13 and 14.

1.3.2   Part III: Innovative Applications of Genetic Programming

Genetic programming is just beginning to be applied to real world problems. Some of the applications reported in this section are to simplified, laboratory style problems, and require considerable scaling up to be practical. Some of the applications reported here are in production or will soon be used in a production environment (Chapters 17 and 21). Many are somewhere in between.

Part III demonstrates the wide variety of application areas to which genetic programming can be applied. The diversity of applications and the creativity of the authors in the ways that they have applied genetic programming to the various applications are exciting. In addition, there are a variety of broadly applicable techniques for increasing the power of genetic programming found in the contributions in this section.

The application areas for which solutions are evolved include:

- Image Classification (Chapter 7).
- Autonomous behavior (Chapters 15, 18, 8, 10).
- Double auction market strategies (Chapter 16).
- Design of stack filters (Chapter 17).
- Nonlinear equation fitting to chaotic data (Chapter 17).
- Decision tree induction to disambiguate natural language (Chapter 19).
- Analysis and construction of random number generators (Chapter 20).
- Determining confidence limits for text classification (Chapter 21).
- Evolving 3-D models to support model-based object recognition (Chapter 22).
- Feature detection for handwriting (Chapter 23).
- Evolving neural networks (Chapter 24).

1.4   Practical Guidance

While working with genetic programming over the past several years, and in talking with others working in the field as well, a number of common themes have become more and more evident. These themes deal in one way or another with practical issues that face the newcomer to genetic programming, and are rarely dealt with in research papers. The most common of these themes are listed below, and are presented in the hope that you will benefit from the lessons learned by others.

Properly, these issues should be read **after** you read the rest of the book. If you are new to genetic programming, you will probably find these issues very hard to understand if you haven't read much of the rest of the book. Perhaps the best strategy would be to remember that this section exists, and after reading the sections of the book that relate to your particular interest, come back to this section when you are in the midst of creating a genetic programming system of your own, or when you are using an existing genetic programming system to solve a problem of interest to you.

1.4.1   It isn't as easy as it looks — but it does work.

One of the frequently heard laments from people just beginning to do genetic programming is that "It didn't work!", spoken with a note of indignity in their voice. If you read much of the published work in the field, you would take away the impression that genetic programming can solve in a straightforward way almost any problem presented to it.

Certainly some problems are simply too complex for today's genetic programming systems using the computer resources available. However, genetic programming often **can** solve the problem, but the path to the solution is sometimes far from straightforward. This is rarely highlighted in the technical papers that you might read.

A newcomer to the field should pretty much assume that significant adjustment of the fitness function, representation, and other parameters will be required to solve any really interesting and thereby nontrivial problem. In many cases these adjustments will in fact yield a solution, so don't give up just because the first try didn't converge on a useful solution.

1.4.2   The fitness function is exceptionally important.

You simply cannot take too much care in crafting your fitness function. Any and all boundary conditions in your fitness function will be ruthlessly exploited by the individuals in your population. Outright though subtle bugs in your fitness function will almost certainly be recognized by the evolutionary process, and the only way in which this can be determined is to examine the individuals that evolve.

Remember that your fitness function is the only chance that you have to communicate your intentions to the powerful process that genetic programming represents. Make sure that it communicates precisely what you desire.

A central aspect of fitness functions is the concept of *partial credit*. A fitness function needs to score an individual on how well it performs on the problem to be solved. However, it needs to do more than this — it needs to score individuals in such a way that they can be compared and a more successful individual can be distinguished from a less successful individual. In order to do this, the fitness function needs to be able to distinguish a more successful solution from a less successful solution. This is a distinction which is not common in computer science. Usually, you write a program which determines if a solution was found or not, but rarely is there any interest in evaluating how well a problem was solved if it was not a complete solution. Mathematical programming has this concept and uses it widely, but many other areas do not.

Any fitness function for genetic programing must be able to evaluate partial solutions to the problem and produce a score which distinguishes a more complete solution from a less complete solution. Further, it is worth giving considerable thought to which intermediate partial solution is favored over another. The favoring of one partial solution over another defines the direction that the evolutionary processes in genetic programming encourage individuals in the population to move. If you have two partial solutions which are about the same, consider carefully which sorts of programs would produce each partial solution and make sure that the program that you believe is more correct gets the better score.

An aspect of partial credit is the granularity of the fitness function. A fitness function which has only a few values overall may not provide enough information about the problem to enable the evolutionary processes to proceed towards the solution.

In addition to partial credit, there are other factors to consider when designing a fitness function. One of these factors is how to aggregate the results of multiple fitness tests, in those cases where the overall fitness of an individual in the population is determined by its performance on several separate fitness tests. Simply adding the results of the individual fitness tests together may not be appropriate, since this suggests that you are trying to create a program which will minimize (or maximize) the fitness over the sum of several tests. That may well be what you desire, but you should determine **why** you have several fitness tests. If you have several fitness tests, but they are selected from a much larger and possibly infinite set of possible tests, your goal may be to evolve a program which is *general*. A program which is general should be able to process any test presented to it. Obviously, you can in practice only evolve any program by using a small set of fitness tests. In order to produce a general program, they must be constructed in such a way as to direct the evolutionary process towards the path of a fully general program. If your goal is to evolve a general program, you will want to combine the results of the individual fitness tests in such a way as to promote that goal. Simply summing the results of the individual fitness tests may do that, but sometimes it doesn't.

An illustrative example of this comes from work done on evolving a general algorithm for sorting an arbitrary sequence of numbers [Kinnear 1993]. The goal is to evolve an individual which will correctly sort **any** sequence presented to it. The individuals are presented with 15 sequences, and the disorder of each sequence is measured both before and after the sequence is presented to the individual for testing. Since the individuals are all presented with the same 15 sequences, it seems reasonable simply to add the disorder remaining in the 15 sequences and call it the raw fitness. However, this fitness test doesn't actually reward correct sorting behavior (which is defined for a single sequence) — it rewards an individual which can figure out how to minimize the total disorder of the 15 sequences. Ultimately, sorting the sequences does that, but this is a misleading fitness test. If a certain series of operations will remove a large amount of disorder from half the sequences while increasing the disorder of the other half by a lesser amount, that is considered behavior to be rewarded by this fitness test.

A more precise fitness test is to continue to sum the remaining disorder over the entire set of 15 fitness cases, but to examine the disorder before and after each sequence is presented to the individual under test and, in those cases where the disorder is increased, to add a penalty by multiplying the increase in disorder by a large factor. This penalizes any individual which increases the disorder in any sequence, thus removing the reinforcement for an individual which reduces the disorder in some sequences by increasing the disorder in others. This will avoid "leakage" from fitness test to fitness test, and create a fitness test that rewards sorting individual sequences instead of one that rewards minimizing the aggregate disorder in a set of sequences.

1.4.3   Representation is important too.

The problem, as it is seen by the genetic programming system, is defined by the representation and the fitness function. The representation consists of the *functions*, *terminals*, and the data on which they operate. Terminals are the constants and variables which take on the data values. The representation thus defines a language. The individuals in the population consist of computer programs in this language, and they are executed in order to solve the problem of interest.

Here are some things to consider when defining the representation for a problem:

- The boundary conditions for the functions are very important. What do the functions do when handed data that is illegal or in some way out of range? You should examine the values returned from **every** function and terminal, and make sure that every function will have a defined and reasonable value for **every** one of the possible data types and ranges that could possibly be presented to it.

- Even the number of terminals can make a large difference in the difficulty perceived by the genetic programming system. Remember, in defining the functions and terminals, you are creating a space of possible programs which you are then asking the genetic programming system to search. This space must be large enough to contain the solution you desire, of course, but the larger it is the less chance of encountering that solution despite the tremendous power of the search technique. There is a balance between having a rich enough set of functions and terminals and having too many — a balance which you will need to determine.

- While it is possible to allow individuals to be created or evolved which are syntactically incorrect or semantically invalid to the point of aborting the fitness test evaluation, generally this is not a good idea. Each time that an individual aborts, it is lowering the diversity of the population, in that the population is effectively just that much smaller than it would otherwise have been.

- You may find that preprocessing the data for the fitness tests, if it is real-world data, might make the problem much easier to solve. While genetic programming can certainly evolve structures as part of a solution which will remove a bias from some parameter, doing so may push the problem beyond the limits (e.g. intrinsic power, memory, or machine time) of your genetic programming system. If you are having problems, and you know that the data could be put in some more straightforward format, do so.

- If your goal is to apply genetic programming to problems of practical utility, as opposed to doing research into the fundamental properties of genetic programming, always pick the most powerful and useful seeming functions from the problem domain that you can think of. For example, high order parity functions are trivially soluble if the two-argument exclusive-or function is provided, whereas without it high-order parity functions are **very** hard to evolve. Providing the exclusive-or function may reduce high-order parity functions to such a degree of triviality that they cease to be of academic interest, but doing so is obviously the correct course of action from an engineering standpoint.

1.4.4   It all comes together in the transmission function.

One of the interesting observations from Chapter 3 is that a *transmission function* describes the production of a new individual from existing individuals. This transmission function is made up of the representation (the functions and terminals), and the genetic operators that operate on the individuals. Traditionally, the representation and the genetic operators have been seen to be linked, but usually they are considered separate areas of innovation. The concept of a transmission function shows clearly how certain classes of changes can be made to either the representation or the genetic operators with equivalent results.

You might find that you could change the representation in one way to make the problem easier for genetic programming, or you might change the genetic operators to effect an equivalent change. More creatively, you might do some of each, and thus find ways to solve a hitherto unsolvable problem.

1.4.5   Population size and diversity are also important.

Genetic programming typically operates with large population sizes compared to genetic algorithms and other evolutionary algorithms. Much work remains to be done to determine what makes a problem easy or hard for genetic programming, and what population sizes are best suited to solving each type of problem. However, it is fairly clear that a population must be larger than a critical minimum size in order to generate a solution reliably. This size is different for each problem, but it is probably true that each problem does have such a minimum size. It appears to be the case that, if the population is below this critical minimum size, running for more generations will usually not produce a correct solution.

There are two ways to see if the population size that you are using is preventing what would otherwise be the rapid solution to your problem. The first, obviously, is to use a larger population size. The second is to perform many runs with a given population size, looking for at least one run which successfully solves the problem. Frequently, when only the occasional run converges to a solution, and most do not, an increase (say a doubling) of the population size can disproportionately increase the frequency with which a solution appears. However, don't ignore the possibility of using multiple runs as an effective way to utilize a much larger virtual population. It isn't the same as running one much larger population, but it can be effective nonetheless.

Increasing the population size beyond some optimum will, in most cases, actually slow down the production of a solution as measured in real time, though the solution will probably still appear earlier as measured by the number of generations necessary to produce it. This is because as the population size grows, the number of individuals processed per generation increases as well, and so each generation takes longer to process. Usually, at some point the speedup caused by a solution appearing earlier as measured in generations is overcome by the additional time used for processing each generation.

It is worth noting, however, that for most interesting problems, genetic programming runs are likely to be population starved. This is simply because the complexity of problems tackled using genetic programming tends to increase to absorb the computational resources available. Thus, it is a generally good heuristic to use the largest population that your machine can handle, since until you exceed the optimum population size you typically get better than linear speedups from increasing the population size. On virtual memory machines, be careful to monitor the real-time behavior of your system as you increase the population size, since thrashing can result in abrupt degradation in the real-time performance of genetic programming systems.

1.4.6 Don't generalize from one run.

When trying to evolve a solution to a problem for the first time or when trying to tune or experiment with the parameters of a genetic programming system, it is tempting to do what you've always done before in computer science: set the parameters, make a run, and see how well it worked. Then, based on that knowledge, try some different parameters, make another run, and continue the process.

If the run is one sequence from generation 0 to the generation at which you usually stop, then you probably have not learned anything useful since **the normal variation on one run is very large**. Therefore, if you then try different parameters based on what you think you've learned and make another single run, and continue this process, you will, if you are lucky, eventually realize that you are going in circles because later information seems to contradict what you thought you had learned earlier.

The key point here is that with all of the same settings except a different random seed for the various random number streams in your system, you will get a **range** of results! This is a highly stochastic process, and an examination of multiple runs is necessary to see the results of **any** changes in parameters. There is no single number of runs that will work for all circumstances. For the kinds of problems that I operate with, I usually feel pretty confident that I've learned something when that learning is based on the average of 20 runs. For these problems, I find 10 runs to be sometimes misleading, and for close calls or a particularly significant result, I use many more than 20 runs to distinguish between two different approaches or parameter settings.

It is so tempting to generalize from too few runs. In the long run, though, you will waste far more time than you will save if you try to do so.

1.4.7   Genetic programming is robust.

Few of us have experience with truly robust computer algorithms or systems. Genetic programming is such a system, and through the same power that allows it to search out every loophole in a fitness test it can also sometimes manage to evolve correct solutions to problems despite amazingly severe bugs. Most systems we operate with every day are sufficiently brittle that tiny bugs will halt them in their tracks. Most of a genetic programming system is equally brittle, but a surprising amount of it is robust enough to survive fairly severe bugs. Problems in selection, bugs in genetic operators (as long as they produce **some** new individual), and similar areas can go unnoticed for a long time in a genetic programming system. These bugs will usually impair the efficiency of the production of a solution or perhaps be benign. (Otherwise, they are obviously not bugs but rather *discoveries*!) As always in debugging, be alert for occurrences that seem to violate your mental model of what the system should be doing. In addition, don't assume that since the system produces solutions that it is working entirely correctly. You must continue to be alert for behavior that violates your model of what should be happening.

Checking that your system works at least as well as the code in [Koza 1992] for benchmark problems such as the 6-multiplexer can go a long way towards establishing confidence in a genetic programming implementation.

1.4.8   Know your problem, know your data.

You will ultimately have to get intimate with the particular problem you are trying to solve. Perhaps you already are, but you may need to think about it in ways that are unusual, possibly so as to create an effective fitness test or a particularly expressive representation. Only in rare cases will your current understanding of your problem be sufficient to enable you to solve it with genetic programming.

1.5   Where to go for more information and inspiration.

Whether you are creating a genetic programming system from scratch to solve a problem, or employing one that already exists, you will almost certainly need to make changes to it in order to increase the power of the genetic programming system with respect to solving your particular problem. Certainly it is possible that the problem you are trying to solve can be solved in a straightforward way by genetic programming, but it is highly unlikely.

Chapter 2 of this volume contains pointers to useful material on genetic programming (Section 2.5) and genetic algorithms (Section 2.6), including mailing lists, conference proceedings, and ftp sites.

Some additional areas worthy of exploration are listed below.

1.5.1 Biology

Ultimately molecular biology and evolutionary biology are the source of most of the fundamental ideas in all areas of evolutionary computation. A very readable and detailed overview of molecular biology for the technical nonspecialist is *Dealing with Genes*, by Paul Berg and Maxine Singer [Berg and Singer 1992]. I like this book because in general it discusses the subject just above the level of laboratory procedure, allowing concentration on the essentials of the processes involved and not the procedures used to discover them. Another excellent book in this area is the *Molecular Design of Life* by Lubert Stryer [1989]. This is excerpted from his textbook *Biochemistry* [Stryer 1988].

If you are not already familiar with the kind of thinking done by evolutionary biologists, they can provide direct inspiration as well as a general mindset that is very helpful in solving practical problem with genetic programming. Richard Dawkins' *The Selfish Gene* [Dawkins 1989] and *The Extended Phenotype* [Dawkins 1982] are both classics of a sort and deservedly so. The second edition of *The Selfish Gene* contains a one chapter summary of *The Extended Phenotype*, and is a good place to start.

The work of Lynn Margulis on evolutionary innovation through symbiosis contains many things to think about as we look toward increasing the power of genetic programming. *Symbiosis in Cell Evolution* [Margulis 1993] is one place to start to examine this interesting area. Margulis has written a great deal, and I seldom fail to find interesting ideas applicable to genetic programming whenever I read her work.

1.5.2 Complex Adaptive Systems

Evolutionary computation is a part of the field of Complex Adaptive Systems. There are many books, both technical and popular, which deal with this emerging field, far more than can be mentioned here. A few are mentioned below, oriented toward the newcomer to the field.

Christopher Langton, of the Sante Fe Institute, more or less defined the field of artificial life (with a lot of help from many other people). There have been three Artificial Life conferences to date, and proceedings from these conferences are available as books called *Artificial Life* [Langton 1989a], *Artificial Life II* [Langton et. al. 1992], and *Artificial Life III* [Langton 1993]. The first of the series has an excellent overview chapter by Langton [1989b].

Emergent behavior is one of the themes of Complex Adaptive Systems, and two books worth looking into are *The Self Organizing Universe* by Erich Jantsch [1980] and *Emergent Computation* edited by Stephanie Forrest [1991].

There are three popular books on aspects of complex systems and artificial life, and the people working in these fields. They are *Artificial Life: The Quest for New Creation* by

Stephen Levy [1992], *Complexity: The Emerging Science at the Edge of Order and Chaos* by M. Mitchell Waldrop [1992], and *Complexity: Life at the Edge of Chaos* by Roger Lewin [1992]. Each is oriented differently, and all are widely available at the larger bookstores.

Finally, Stuart A. Kauffman's book *The Origins of Order: Self Organization and Selection in Evolution* [Kauffman 1993] is a summary of his work and that of others over the past several years in studying the fundamental processes that drive organic evolution no less than the evolution that takes place in genetic programming. Personally, I find that practically every section suggests something that could make genetic programming more powerful or could increase our understanding of what is going on in genetic programming.

1.5.3   Genetic Algorithms and other Evolutionary Computation paradigms

In many ways genetic programming is an offshoot of genetic algorithms, and the large body of work that exists in genetic algorithms can be applied to genetic programming, in some cases directly and in some cases less so. Chapter 2 (Section 2.6) lists many references for genetic algorithms, each of which has something worthwhile to offer. See Section 1.1.2 in this chapter for some beginning references to other styles of evolutionary computation, since the lessons learned elsewhere in evolutionary computation can be used for genetic programming as well.

Two useful areas to explore are the frequently asked questions digests (FAQ) from the genetic programming mailing list (see Section 2.5) and the FAQ from the comp.ai.genetic newsgroup [Heitkötter 1993].

One of the as yet unrealized strengths of genetic programming is that it is general enough to integrate many of the different techniques and approaches used in other styles of evolutionary computation into a more powerful whole.

1.6   Conclusion

The field of genetic programming is just beginning. Not a day goes by but someone, somewhere is observing something that has never been seen before, and there are many discoveries waiting to be made. At the same time, there is sufficient power today in genetic programming to solve some real-world problems, and one of the especially exciting aspects of this technology is the juxtaposition of these two events.

In few other places can you discover something that has never been known before in the morning, and by afternoon apply it to solving (perhaps more effectively than ever before) a real-world problem.

Whether your delight is in the discovery or the application, or both, enjoy yourself!

Bibliography

Bäck, T., Hoffmeister, F., Schwefel, H-P., (1991) "A Survey of Evolution Strategies", in *Proceedings of the Fourth International Conference on Genetic Algorithms*, R. K. Belew and L. B. Booker, Eds. San Mateo, CA: Morgan Kaufmann.

Bäck, T., Schwefel, H-P., (1993) "An Overview of Evolutionary Algorithms for Parameter Optimization", *Evolutionary Computation* 1(1), 1-23.

Berg, P., Singer, M. (1992) *Dealing with Genes: The Language of Heredity.* Mill Valley, CA: University Science Books.

Dawkins, R. (1982) *The Extended Phenotype.* Oxford: Oxford University Press.

Dawkins, R. (1989) *The Selfish Gene (new edition).* Oxford: Oxford University Press.

Fogel, D. B. (1993) "On the Philosophical Differences between Evolutionary Algorithms and Genetic Algorithms", in *Proceedings of the Second Annual Conference on Evolutionary Programming,* D. B. Fogel and W. Atmar, Eds., Evolutionary Programming Society, La Jolla, CA.

Fogel, L. J., Owens, A. J., and Walsh, M. J. (1966) *Artificial Intelligence through Simulated Evolution.* New York, NY: John Wiley.

Forrest, S. (1991) *Emergent Computation.* Cambridge, MA: MIT Press.

Goldberg, D. E. (1989) *Genetic Algorithms in Search, Optimization and Machine Learning.* Reading, MA: Addison-Wesley Publishing Company, Inc.

Heitkötter, J., Ed. (1993) "The Hitch-Hiker's Guide to Evolutionary Computation: A list of Frequently Asked Questions (FAQ)", Usenet: comp.ai.genetic. Available via anonymous ftp from **rtfm.mit.edu** in pub/usenet/news.answers/ai-faq/genetic/part?. ~60 pages. Postscript format from **lumpi.informatik.uni-dortmund.de** in /pub/EA/docs/hhgec.ps.Z.

Holland, J. H. (1975) *Adaptation in Natural and Artificial Systems.* Ann Arbor, MI: The University of Michigan Press.

Holland, J. H. (1992) *Adaptation in Natural and Artificial Systems (1992 edition).* Cambridge, MA: MIT Press.

Jantsch, E. (1980) *The Self-Organizing Universe: Scientific and Human Implications of the Emerging Paradigm of Evolution.* Oxford: Pergamon Press.

Kauffman, S. A., (1993) *The Origins of Order: Self-Organization and Selection in Evolution.* New York, NY: Oxford University Press.

Kinnear, K. E. Jr. (1993) "Evolving a Sort: Lessons in Genetic Programming," in *Proceedings of the 1993 International Conference on Neural Networks*, New York, NY: IEEE Press.

Koza, J. R. (1989) "Hierarchical Genetic Algorithms Operating on Populations of Computer Programs", in *Proceedings of the 11th International Joint Conference on Artificial Intelligence.* San Mateo, CA: Morgan Kaufmann.

Koza, J. R. (1990) "Genetic Programming: A Paradigm for Genetically Breeding Populations of Computer Programs to Solve Problems." Technical Report No. STAN-CS-90-1314, Computer Science Department, Stanford University.

Koza, J. R. (1992) *Genetic Programming.* Cambridge, MA: MIT Press.

Koza, J. R. (1994) *Genetic Programming II.* Cambridge, MA: MIT Press.

Langton, C. G., (1989a) *Artificial Life.* Volume VI, Santa Fe Institute Studies in the Sciences of Complexity. Redwood City, CA: Addison Wesley Publishing Company, Inc.

Langton, C. G., (1989b) "Artificial Life", in *Artificial Life*, C. G. Langton, Ed. Redwood City, CA: Addison-Wesley Publishing Company, Inc.

Langton, C. G., Taylor, C., Farmer, J. D., Rasmussen, S., (1992) *Artificial Life II.* Volume X, Santa Fe Institute Studies in the Sciences of Complexity. Redwood City, CA: Addison-Wesley Publishing Company.

Langton, C. G., (1993) *Artificial Life III.* Santa Fe Institute Studies in the Sciences of Complexity. Redwood City, CA: Addison-Wesley Publishing Company, Inc.

Levy, S. (1992) *Artificial Life: The Quest for a New Creation.* New York, NY: Pantheon Books.

Lewin, R. (1992) *Complexity: Life at the Edge of Chaos.* New York, NY: Macmillan Publishing Company.

Margulis, L. (1993) *Symbiosis in Cell Evolution: Microbial Communities in the Archean and Proterozoic Eons.* New York, NY: W. H. Freeman and Company.

Rechenberg, I. (1965) Cybernetic solution path of an experimental problem. Royal Aircraft Establishment Translation No. 1122, Farnborough Hants: Ministry of Aviation, Royal Aircraft Establishment.

Stryer, L. (1988) *Biochemistry (third edition).* New York, NY: W. H. Freeman and Company.

Stryer, L. (1989) *The Molecular Design of Life.* New York, NY: W. H. Freeman and Company.

Waldrop, M. M., (1992) *Complexity: The Emerging Science at the Edge of Order and Chaos.* New York, NY: Simon and Schuster.

2 Introduction to Genetic Programming

John R. Koza

This chapter provides an introduction to genetic algorithms, the LISP programming language, genetic programming, and automatic function definition. This chapter also outlines additional sources of information about genetic algorithms and genetic programming.

2.1 Introduction to Genetic Algorithms

John Holland's pioneering book *Adaptation in Natural and Artificial Systems* [1975, 1992] showed how the evolutionary process can be applied to solve a wide variety of problems using a highly parallel technique that is now called the *genetic algorithm*.

The *genetic algorithm* transforms a *population* of individual objects, each with an associated *fitness* value, into a new *generation* of the population using the Darwinian principle of reproduction and survival of the fittest and naturally occurring genetic operations such as *crossover (recombination)* and *mutation*. Each *individual* in the population represents a possible solution to a given problem. The genetic algorithm attempts to find a very good or best solution to the problem by genetically breeding the population of individuals.

In preparing to use the conventional genetic algorithm operating on fixed-length character strings to solve a problem, the user must

1. determine the representation scheme,
2. determine the fitness measure,
3. determine the parameters and variables for controlling the algorithm, and
4. determine a way of designating the result and a criterion for terminating a run.

In the conventional genetic algorithm, the individuals in the population are usually fixed-length character strings patterned after chromosome strings. Thus, specification of the *representation scheme* in the conventional genetic algorithm starts with a selection of the string length L and the alphabet size K. Often the alphabet is binary, so K equals 2. The most important part of the representation scheme is the mapping that expresses each possible point in the search space of the problem as a fixed-length character string (i.e., as a *chromosome*) and each chromosome as a point in the search space of the problem. Selecting a representation scheme that facilitates solution of the problem by the genetic algorithm often requires considerable insight into the problem and good judgment.

The evolutionary process is driven by the *fitness measure*. The fitness measure assigns a fitness value to each possible fixed-length character string in the population.

The primary parameters for controlling the genetic algorithm are the population size, *M*, and the maximum number of generations to be run, *G*. Populations can consist of hundreds, thousands, tens of thousands or more individuals. There can be dozens, hundreds, thousands, or more generations in a run of the genetic algorithm.

Each run of the genetic algorithm requires specification of a *termination criterion* for deciding when to terminate a run and a method of *result designation*. One frequently used method of result designation for a run of the genetic algorithm is to designate the best individual obtained in any generation of the population during the run (i.e., the *best-so-far individual*) as the result of the run.

Once the four preparatory steps for setting up the genetic algorithm have been completed, the genetic algorithm can be run.

The three steps in executing the genetic algorithm operating on fixed-length character strings are as follows:

I.   Randomly create an initial population of individual fixed-length character strings.
II. Iteratively perform the following substeps on the population of strings until the termination criterion has been satisfied:
     A. Assign a fitness value to each individual in the population using the fitness measure.
     B. Create a new population of strings by applying the following three genetic operations. The genetic operations are applied to individual string(s) in the population chosen with a probability based on fitness.
         1 Reproduce an existing individual string by copying it into the new population.
         2 Create two new strings from two existing strings by genetically recombining substrings using the crossover operation (described below) at a randomly chosen crossover point.
         3 Create a new string from an existing string by randomly mutating the character at one randomly chosen position in the string.
III. The string that is identified by the method of result designation (e.g., the best-so-far individual) is designated as the result of the genetic algorithm for the run. This result may represent a solution (or an approximate solution) to the problem.

The genetic algorithm involves probabilistic steps for at least three points in the algorithm, namely

• creating the initial population,

• selecting individuals from the population on which to perform each genetic operation (e.g., reproduction, crossover), and

• choosing a point (i.e., the crossover point or the mutation point) within the selected individual at which to perform the genetic operation.

Moreover, there is often additional randomness involved in the genetic algorithm. For example, the value of fitness may be measured for randomly chosen *fitness cases*.

As a result of the probabilistic nature of the genetic algorithm, it may be necessary to make multiple independent runs of the algorithm in order to obtain a satisfactory result for a given problem.   Thus, the above three steps are embedded in an outer loop representing the runs.

Figure 2.1 is a flowchart of these steps for the conventional genetic algorithm. The variable RUN refers to the current run number. The variable GEN refers to the current generation number. The index i refers to the current individual in a population.

The genetic operation of *reproduction* is based on the Darwinian principle of reproduction and survival of the fittest. In the reproduction operation, an individual is probabilistically selected from the population based on its fitness (with reselection allowed) and then the individual is copied, without change, into the next generation of the population. The selection is done in such a way that the better an individual's fitness, the more likely it is to be selected. An important aspect of this probabilistic selection is that every individual, however poor its fitness, has some probability of selection.

The genetic operation of *crossover* (sexual *recombination*) allows new individuals (i.e., new points in the search space) to be created and tested. The operation of crossover starts with two parents independently selected probabilistically from the population based on their fitness (with reselection allowed). As before, the selection is done in such a way that the better an individual's fitness, the more likely it is to be selected. The crossover operation produces two offspring. Each offspring contains some genetic material from each of its parents.

The table below illustrates the crossover operation being applied to the two parental strings 10110 and 01101 of length $L = 5$ over an alphabet of size $K = 2$.

Parent 1	Parent 2
10110	01101

The crossover operation begins by randomly selecting a number between 1 and $L-1$ using a uniform probability distribution. There are $L-1 = 4$ interstitial locations lying between the positions of a string of length five. Suppose that the third interstitial location is selected. This location becomes the *crossover point*. Each parent is then split at this crossover point into a crossover fragment and a remainder. The table below shows the *crossover fragments* of parents 1 and 2.

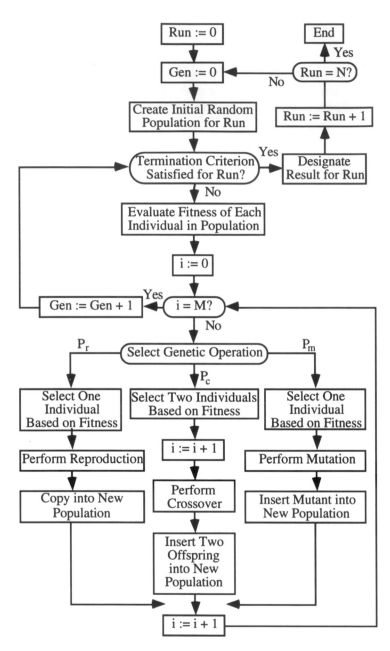

Figure 2.1
Flowchart of the conventional genetic algorithm.

Crossover fragment 1	Crossover fragment 2
101 – –	011 – –

After the crossover fragment is identified, something remains of each parent. The table below shows the *remainders* of parents 1 and 2.

Remainder 1	Remainder 2
– – – 10	– – – 01

The crossover operation combines remainder 1 (i.e., – – – 1 0) with crossover fragment 2 (i.e., 011 – –) to create offspring 2 (i.e., 01110). The crossover operation similarly combines remainder 2 (i.e., – – – 01) with crossover fragment 1 (i.e., 101 – –) to create offspring 1 (i.e., 10101). The following table shows the two offspring.

Offspring 1	Offspring 2
10101	01110

The two offspring are usually different from their two parents and different from each other.

The operation of mutation allows new individuals to be created. It begins by selecting an individual from the population based on its fitness (with reselection allowed). A point along the string is selected at random and the character at that point is randomly changed. The altered individual is then copied into the next generation of the population. Mutation is used very sparingly in genetic algorithm work.

The genetic algorithm works in a domain-independent way on the fixed-length character strings in the population. The genetic algorithm searches the space of possible character strings in an attempt to find high-fitness strings. The fitness landscape may be very rugged and nonlinear. To guide this search, the genetic algorithm uses only the numerical fitness values associated with the explicitly tested strings in the population. Regardless of the particular problem domain, the genetic algorithm carries out its search by performing the same disarmingly simple operations of copying, recombining, and occasionally randomly mutating the strings.

In practice, the genetic algorithm is surprisingly rapid in effectively searching complex, highly nonlinear, multidimensional search spaces. This is all the more surprising because the genetic algorithm does not know anything about the problem domain or the internal workings of the fitness measure being used.

2.2 Program Trees and the LISP Programming Language

All computer programs – whether they are written in FORTRAN, PASCAL, C, assembly code, or any other programming language – can be viewed as a sequence of applications of functions (operations) to arguments (values). Compilers use this fact by first internally translating a given program into a parse tree and then converting the parse tree into the more elementary assembly code instructions that actually run on the computer. However this important commonality underlying all computer programs is usually obscured by the large variety of different types of statements, operations, instructions, syntactic constructions, and grammatical restrictions found in most programming languages.

Genetic programming is most easily understood if one thinks about it in terms of a programming language that overtly and transparently views a computer program as a sequence of applications of functions to arguments.

Moreover, since genetic programming initially creates computer programs at random and then manipulates the programs by various genetically motivated operations, genetic programming may be implemented in a conceptually straightforward way in a programming language that permits a computer program to be easily manipulated as data and then permits the newly created data to be immediately executed as a program.

For these two reasons, the LISP (LISt Processing) programming language [Steele 1990] is especially well suited for genetic programming. However, it should be recognized that genetic programming does not require LISP for its implementation and is not in any way based on LISP or dependent on LISP. Indeed, the majority of the authors in this book implemented genetic programming in C, rather than LISP. Nonetheless, even when researchers do not actually use LISP for writing their programs to implement genetic programming, most (including most of the authors in this book) find it convenient to use the style of the LISP programming language for presenting and discussing the programs evolved by genetic programming.

LISP has only two main types of entities: atoms and lists. The constant 7 and the variable TIME are examples of atoms in LISP. A list in LISP is written as an ordered collection of items inside a pair of parentheses. (A B C D) and (+ 1 2) are examples of lists in LISP.

Both lists and atoms in LISP are called *symbolic expressions* (*S-expressions*). The S-expression is the only syntactic form in pure LISP. There is no syntactic distinction between programs and data in LISP. In particular, all data in LISP are S-expressions and all programs in LISP are S-expressions.

The LISP system works so as to evaluates whatever it sees. When seen by LISP, a constant atom (e.g., 7) evaluates to itself, and a variable atom (e.g., TIME) evaluates to the current value of the variable. When a list is seen by LISP, the list is evaluated by treating the first element of the list (i.e., whatever is just inside the opening parenthesis) as a function and then causing the application of that function to the results of evaluating the remaining elements of the list. That is, the remaining elements are treated as arguments to the function. If an argument is a constant atom or a variable atom, this evaluation is immediate; however, if an argument is a list, the evaluation of such an argument involves a recursive application of the above steps.

For example, in the LISP S-expression (+ 1 2), the addition function + appears just inside the opening parenthesis. The S-expression (+ 1 2) calls for the application of the addition function + to two arguments (i.e., the constant atoms 1 and 2). Since both arguments are atoms, they can be immediately evaluated to their values (i.e., 1 and 2). Thus, the value returned as a result of the evaluation of the entire S-expression (+ 1 2) is 3.

If any of the arguments in an S-expression are themselves lists (rather than constant atoms or variable atoms that can be immediately evaluated), LISP first evaluates these arguments. In Common LISP, this evaluation is done in a recursive, depth-first way, starting from the left). For example, the S-expression (+ (* 2 3) 4) calls for the application of the addition function + to two arguments, namely the sub-S-expression (* 2 3) and the constant atom 4. In order to evaluate the entire S-expression, LISP must first evaluate the sub-S-expression (* 2 3). This argument (* 2 3) calls for the application of the multiplication function * to the two constant atoms 2 and 3, so it evaluates to 6 and the entire S-expression evaluates to 10. LISP S-expressions are examples of prefix notation. FORTRAN, PASCAL, and C are similar to ordinary mathematical notation in that they use ordinary ("infix") notation, so the above program in LISP would be written as 2*3+4 in those languages.

For example, the LISP S-expression

```
(+ 1 2 (IF (> TIME 10) 3 4))
```

further illustrates how LISP views conditional and relational elements of computer programs as applications of functions to arguments. In the sub-S-expression (> TIME 10), the relation > is viewed as a function and > is then applied to the variable atom TIME and the constant atom 10. The subexpression (> TIME 10) then evaluates to either T (True) or NIL (False), depending on the current value of the variable atom TIME. The conditional operator IF is then viewed as a function and IF is then applied to three arguments: the logical value (T or NIL) returned by the subexpression (> TIME 10),

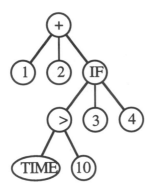

Figure 2.2
The LISP S-expression (+ 1 2 (IF (> TIME 10) 3 4)) depicted as a rooted, point-labeled tree
with ordered branches.

the constant atom 3, and the constant atom 4. If its first argument evaluates to T (more
precisely, anything other than NIL), the function IF returns the result of evaluating its
second argument (i.e., the constant atom 3), but if its first argument evaluates to NIL,
the function IF returns the result of evaluating its third argument (i.e., the constant atom
4). Thus, the S-expression evaluates to either 6 or 7, depending on whether the current
value of the variable atom TIME is or is not greater than 10. Most other programming
languages create differing syntactic forms and statement types for functions such as *, >,
and IF whereas LISP treats all of these functions in the same way.

   Any LISP S-expression can be graphically depicted as a rooted point-labeled tree with
ordered branches. Figure 2.2 shows the parse tree (program tree) corresponding to the
above S-expression.

   In this graphical depiction, the three internal points of the tree are labeled with
functions (i.e., +, IF, and >). The six external points (leaves) of the tree are labeled with
terminals (i.e., the variable atom TIME and the constant atoms 1, 2, 10, 3, and 4). The
root of the tree is labeled with the function appearing just inside the leftmost opening
parenthesis of the S-expression (i.e., the +).

   Note that this tree form of a LISP S-expression is equivalent to the parse tree that
many compilers construct internally to represent a given computer program.

   An important feature of LISP is that all LISP computer programs have just one
syntactic form (i.e., the S-expression). The programs of the LISP programming language
are S-expressions, and an S-expression is, in effect, the parse tree of the program.

2.3 Genetic Programming

Genetic programming is an extension of the conventional genetic algorithm in which each individual in the population is a computer program.

The search space in genetic programming is the space of all possible computer programs composed of functions and terminals appropriate to the problem domain. The functions may be standard arithmetic operations, standard programming operations, standard mathematical functions, logical functions, or domain-specific functions.

Genetic programming is an attempt to deal with one of the central questions in computer science, namely

How can computers learn to solve problems without being explicitly programmed? In other words, how can computers be made to do what needs to be done, without being told exactly how to do it?

The book *Genetic Programming: On the Programming of Computers by Means of Natural Selection* [Koza 1992] demonstrated a result that many found surprising and counterintuitive, namely that an automatic, domain-independent method can genetically breed computer programs capable of solving, or approximately solving, a wide variety of problems from a wide variety of fields.

In applying genetic programming to a problem, there are five major preparatory steps. These five steps involve determining

1.   the set of terminals,
2.   the set of primitive functions,
3.   the fitness measure,
4.   the parameters for controlling the run, and
5.   the method for designating a result and the criterion for terminating a run.

The first major step in preparing to use genetic programming is to identify the set of terminals. The terminals can be viewed as the inputs to the as-yet-undiscovered computer program. The set of terminals (along with the set of functions) are the ingredients from which genetic programming attempts to construct a computer program to solve, or approximately solve, the problem.

The second major step in preparing to use genetic programming is to identify the set of functions that are to be used to generate the mathematical expression that attempts to fit the given finite sample of data.

Each computer program (i.e., mathematical expression, LISP S-expression, parse tree) is a composition of functions from the function set $\mathcal{F}$ and terminals from the terminal set $\mathcal{T}$.

Each of the functions in the function set should be able to accept, as its arguments, any value and data type that may possibly be returned by any function in the function set and any value and data type that may possibly be assumed by any terminal in the terminal set. That is, the function set and terminal set selected should have the closure property.

These first two major steps correspond to the step of specifying the representation scheme for the conventional genetic algorithm. The remaining three major steps for genetic programming correspond to the last three major preparatory steps for the conventional genetic algorithm.

In genetic programming, populations of hundreds, thousands, tens of thousands or more computer programs are genetically bred. This breeding is done using the Darwinian principle of survival and reproduction of the fittest along with a genetic crossover operation appropriate for mating computer programs. As will be seen in the numerous chapters of this book, a computer program that solves (or approximately solves) a given problem may emerge from this combination of Darwinian natural selection and genetic operations.

Genetic programming starts with an initial population of randomly generated computer programs composed of functions and terminals appropriate to the problem domain. The creation of this initial random population is, in effect, a blind random search of the search space of the problem represented as computer programs.

Each individual computer program in the population is measured in terms of how well it performs in the particular problem environment. This measure is called the *fitness measure*. The nature of the fitness measure varies with the problem.

For many problems, fitness is naturally measured by the error produced by the computer program. The closer this error is to zero, the better the computer program. In a problem of optimal control, the fitness of a computer program may be the amount of time (or fuel, or money, etc.) it takes to bring the system to a desired target state. The smaller the amount of time (or fuel, or money, etc.), the better. If one is trying to recognize patterns or classify examples, the fitness of a particular program may be measured by some combination of the number of instances handled correctly (i.e., true positive and true negatives) and the number of instances handled incorrectly (i.e., false positives and false negatives). On the other hand, if one is trying to find a good randomizer, the fitness of a given computer program might be measured by means of entropy, satisfaction of the gap test, satisfaction of the run test, or some combination of these factors. For some problems, it may be appropriate to use a multiobjective fitness measure incorporating a combination of factors such as correctness, parsimony, or efficiency.

Typically, each computer program in the population is run over a number of different *fitness cases* so that its fitness is measured as a sum or an average over a variety of representative different situations. These fitness cases sometimes represent a sampling of different values of an independent variable or a sampling of different initial conditions of a system. For example, the fitness of an individual computer program in the population may be measured in terms of the sum of the absolute value of the differences between the output produced by the program and the correct answer to the problem (i.e., the Minkowski distance) or the square root of the sum of the squares (i.e., Euclidean distance). These sums are taken over a sampling of different inputs (fitness cases) to the program. The fitness cases may be chosen at random or may be chosen in some structured way (e.g., at regular intervals or over a regular grid).

The computer programs in the initial generation (i.e., generation 0) of the process will generally have exceedingly poor fitness. Nonetheless, some individuals in the population will turn out to be somewhat more fit than others. These differences in performance are then exploited.

The Darwinian principle of reproduction and survival of the fittest and the genetic operation of crossover are used to create a new offspring population of individual computer programs from the current population of programs.

The reproduction operation involves selecting a computer program from the current population of programs based on fitness (i.e., the better the fitness, the more likely the individual is to be selected) and allowing it to survive by copying it into the new population.

The crossover operation is used to create new offspring computer programs from two parental programs selected based on fitness. The parental programs in genetic programming are typically of different sizes and shapes. The offspring programs are composed of subexpressions (subtrees, subprograms, subroutines, building blocks) from their parents. These offspring programs are typically of different sizes and shapes than their parents.

Intuitively, if two computer programs are somewhat effective in solving a problem, then some of their parts probably have some merit. By recombining randomly chosen parts of somewhat effective programs, we sometimes produce new computer programs that are even more fit at solving the given problem than either parent.

The mutation operation may also be used in genetic programming.

After the genetic operations are performed on the current population, the population of offspring (i.e., the new generation) replaces the old population (i.e., the old generation).

Each individual in the new population of computer programs is then measured for fitness, and the process is repeated over many generations.

At each stage of this highly parallel, locally controlled, decentralized process, the state of the process will consist only of the current population of individuals.

The force driving this process consists only of the observed fitness of the individuals in the current population in grappling with the problem environment.

As will be seen, this algorithm will produce populations of computer programs which, over many generations, tend to exhibit increasing average fitness in dealing with their environment. In addition, these populations of computer programs can rapidly and effectively adapt to changes in the environment.

The best individual appearing in any generation of a run (i.e., the best-so-far individual) is typically designated as the result produced by the run of genetic programming.

The hierarchical character of the computer programs that are produced is an important feature of genetic programming. The results of genetic programming are inherently hierarchical. In many cases the results produced by genetic programming are default hierarchies, prioritized hierarchies of tasks, or hierarchies in which one behavior subsumes or suppresses another.

The dynamic variability of the computer programs that are developed along the way to a solution is also an important feature of genetic programming. It is often difficult and unnatural to try to specify or restrict the size and shape of the eventual solution in advance. Moreover, advance specification or restriction of the size and shape of the solution to a problem narrows the window by which the system views the world and might well preclude finding the solution to the problem at all.

Another important feature of genetic programming is the absence or relatively minor role of preprocessing of inputs and postprocessing of outputs. The inputs, intermediate results, and outputs are typically expressed directly in terms of the natural terminology of the problem domain. The computer programs produced by genetic programming consist of functions that are natural for the problem domain. The postprocessing of the output of a program, if any, is done by a *wrapper* (*output interface*).

Finally, another important feature of genetic programming is that the structures undergoing adaptation in genetic programming are active. They are not passive encodings (i.e., chromosomes) of the solution to the problem. Instead, given a computer on which to run, the structures in genetic programming are active structures that are capable of being executed in their current form.

In summary, genetic programming breeds computer programs to solve problems by executing the following three steps:

1. Generate an initial population of random compositions of the functions and terminals of the problem (i.e., computer programs).

2. Iteratively perform the following substeps until the termination criterion has been satisfied:

    a. Execute each program in the population and assign it a fitness value using the fitness measure.

    b. Create a new population of computer programs by applying the following two primary operations. The operations are applied to computer program(s) in the population chosen with a probability based on fitness.

        i   Reproduce an existing program by copying it into the new population.

        ii Create two new computer programs from two existing programs by genetically recombining randomly chosen parts of two existing programs using the crossover operation (described below) applied at a randomly chosen crossover point within each program.

3. The program that is identified by the method of result designation (e.g., the best-so-far individual) is designated as the result of the genetic algorithm for the run. This result may represent a solution (or an approximate solution) to the problem.

The genetic crossover (sexual recombination) operation operates on two parental computer programs selected with a probability based on fitness and produces two new offspring programs consisting of parts of each parent.

For example, consider the following computer program (LISP S-expression):

`(+ (* 0.234 Z) (- X 0.789)),`

which we would ordinarily write as

$$0.234 \ Z + X - 0.789.$$

This program takes two inputs (X and Z) and produces a floating point output.

Also, consider a second program:

`(* (* Z Y) (+ Y (* 0.314 Z))),`

which is equivalent to

$$Z \ Y \ (Y + 0.314 \ Z).$$

In figure 2.3, these two programs are depicted as rooted, point-labeled trees with ordered branches. Internal points (i.e., nodes) of the tree correspond to functions (i.e., operations) and external points (i.e., leaves, endpoints) correspond to terminals (i.e., input data). The numbers beside the function and terminal points of the tree appear for reference only.

The crossover operation creates new offspring by exchanging sub-trees (i.e., sub-lists, subroutines, subprocedures) between the two parents.

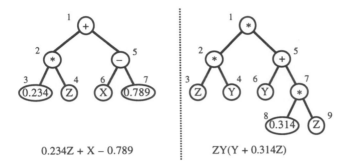

Figure 2.3
Two Parental computer programs.

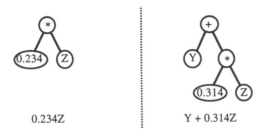

Figure 2.4
Two Crossover Fragments.

Assume that the points of both trees are numbered in a depth-first way starting at the left. Suppose that the point number 2 (out of 7 points of the first parent) is randomly selected as the crossover point for the first parent and that the point number 5 (out of 9 points of the second parent) is randomly selected as the crossover point of the second parent. The crossover points in the trees above are therefore the * in the first parent and the + in the second parent. The two crossover fragments are the two sub-trees shown in figure 2.4.

These two crossover fragments correspond to the underlined sub-programs (sub-lists) in the two parental computer programs.

The two offspring resulting from crossover are

```
(+ (+ Y (* 0.314 Z)) (- X 0.789))
```

and

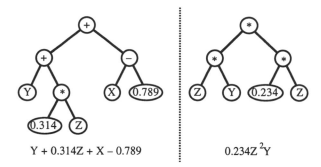

$$Y + 0.314Z + X - 0.789 \qquad\qquad 0.234Z^2Y$$

Figure 2.5
Two Offspring.

`(* (* Z Y) (* 0.234 Z)).`

The two offspring are shown in figure 2.5.

Thus, crossover creates new computer programs using parts of existing parental programs. Because entire sub-trees are swapped, the crossover operation always produces syntactically and semantically valid programs as offspring regardless of the choice of the two crossover points. Because programs are selected to participate in the crossover operation with a probability based on fitness, crossover allocates future trials to regions of the search space whose programs contains parts from promising programs.

The videotape *Genetic Programming: The Movie* [Koza and Rice 1992] provides a visualization of the genetic programming process and of solutions to various problems.

2.4 Automatic Function Definition in Genetic Programming

Several of the papers in this book deal with the issue of scaling up genetic programming to more difficult problems by means of the automatic discovery of functional subunits. This section discusses one such approach, namely automatically defined functions (ADFs).

A human programmer writing a program containing a common calculation involving the exponential of two numbers would probably write a subroutine (defined function, subprogram, procedure) for the common calculation and then call the subroutine twice from the main program. The six lines of code below show a one-line main LISP program for finding the difference $e^3 - e^2$ and a three-line defined function called exp-approx for calculating an approximation to the value of the exponential function:

```
1.  ;;;---main  program---
2.  (values (- (exp-approx 3.0) (exp-approx 2.0)))
3.  ;;;---definition of the function "exp-approx"---
4.  (defun exp-approx (arg0)
5.     (values (+ 1.0   arg0   (* 0.5 arg0 arg0)
6.                     (* 0.1667 arg0 arg0 arg0))))
```

Lines 1 and 3 contain comments to identify the main program in line 2 and the definition of the function in lines 4, 5, and 6.

Line 2 contains the main program that calls the function exp-approx twice and then computes the difference (–) between the two. The single value to be returned by the main program in line 2 is highlighted with an explicit invocation of the values function (which ordinarily would not be used).

Lines 4, 5 and 6 contain the definition of the function exp-approx. This function definition (called a defun in LISP) does four things.

First, this defun (line 4) assigns a name, exp-approx, to the function being defined. The name permits subsequent reference to this function by the main (calling) program (line 2).

Second, this defun identifies the argument list (line 4) of the function being defined. In this defun, the argument list is (arg0) containing one dummy variable (formal parameter) arg0.

Third, this defun contains a body (lines 5 and 6) that performs the work of the function. In this defun, the work consists of the addition of the first three terms of the Taylor series approximation for e^x using the four-argument primitive function of addition (+) and the three- and four-argument primitive function of multiplication (*). Note that the body of the function uses the dummy variable arg0 that is localized within the defun. This dummy variable does not appear in the main program.

Fourth, this defun identifies the value to be returned by the function. In this defun, the single value to be returned is emphasized and highlighted with an explicit invocation of the values function (line 5).

This particular illustrative defun has one local dummy variable, returns only a single value, has no side effects, and refers only to its one local dummy variable (i.e., it does not refer to any of the actual or "global" variables of the overall problem). However, in general, defined functions may have any number of arguments (including no arguments), may return multiple values (or no values), may or may not perform side effects, and may or may not explicitly refer to the actual (global) variables of the overall problem.

Automatic function definition can be implemented within the context of genetic programming by establishing a constrained syntactic structure for the individual programs

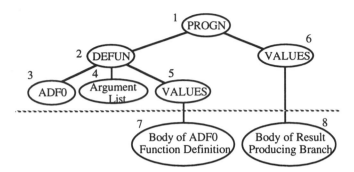

Figure 2.6
An overall computer program consisting of one function-defining branch and one result-producing branch.

in the population as described in [Koza 1992]. Each program in the population contains one (or more) function-defining branches and one (or more) main result-producing branches. The result-producing branch usually has the ability to call one or more of the defined functions. A function-defining branch may have the ability to refer hierarchically to another already-defined function (and potentially even itself). When the result-producing branch returns only a single value, we sometimes refer to it as the *value-returning branch*.

Figure 2.6 shows the overall structure of a program consisting of one function-defining branch and one main result-producing branch. The function-defining branch appears in the left part of this figure and the result-producing branch appears on the right.

There are eight different "types" of points in this overall program. The first six types are invariant and appear above the horizontal dotted line in this figure. The eight types are as follows:

1. the root of the tree (which consists of the place-holding PROGN connective function),
2. the top point, DEFUN, of the function-defining branch,
3. the name, ADF0, of the automatically defined function,
4. the argument list of the automatically defined function,
5. the VALUES function of the function-defining branch identifying, for emphasis, the value(s) to be returned by the automatically defined function,
6. the VALUES function of the result-producing branch identifying, for emphasis, the value(s) to be returned by the result-producing branch,
7. the body (i.e., work) of the automatically defined function ADF0, and
8. the body of the result-producing branch.

When the overall program is evaluated, the PROGN causes the sequential evaluation of the two branches. The PROGN starts by evaluating the first branch, namely the function-defining branch. The function-defining branch merely defines the automatically defined function ADF0. When a defun is evaluated by the PROGN, the result returned (which happens in LISP to be the name of the function being defined) is irrelevant. Since the PROGN returns only the result of the evaluation of its last argument, the evaluation of the function-defining branch does not cause the return of any value for the PROGN as a whole. The PROGN now evaluates its second branch, namely the result-producing branch. The body of the result-producing branch may refer to the automatically defined function ADF0. The value returned by the overall program consists only of the value returned by the VALUES function associated with the result-producing branch. Note that in this formulation all references to an automatically defined function are to the automatically defined function within the same overall program (and not to the automatically defined functions of other programs in the population).

Genetic programming will evolve a population of programs, each consisting of a function definition in the function-defining branch and a result-producing branch. The structures of both the function-defining branches and the result-producing branch are determined by the combined effect, over many generations, of the selective pressure exerted by the fitness measure and by the effects of the operations of Darwinian fitness proportionate reproduction and crossover. The function defined by the function-defining branch is available for use by the result-producing branch. Whether or not the defined function will be actually called is not predetermined, but instead, determined by the evolutionary process.

Since each individual program in the population of this example consists of one function-defining branch and one result-producing branch, the initial random generation must be created so that every individual program in the population has this particular constrained syntactic structure. Specifically, every individual program in generation 0 must have the invariant structure represented by the six points of types 1 through 6 described above. Each function and terminal in the function-defining branch is of type 7. The function-defining branch is a random composition of functions from the function set $\mathcal{F}_{fd}$ for the function-defining branch and terminals from the terminal set $\mathcal{T}_{fd}$ for the function-defining branch. The terminal set $\mathcal{T}_{fd}$ for the function-defining branch typically contains the localized dummy variables (e.g., ARG0). Each function and terminal in the result-producing branch is of type 8. The result-producing branch is a random composition of functions from the function set $\mathcal{F}_{rp}$ and terminals from the terminal set $\mathcal{T}_{rp}$. The function set for the result-producing branch typically contains the available defined functions (e.g., ADF0) and does not contain the dummy variables of the defined function (e.g., ARG0). The result-producing branch typically contains the actual variables of the problem. The actual variables of the problem usually do not appear in the function-defining branch, although they may.

Since a constrained syntactic structure is involved, crossover must be performed so as to preserve this syntactic structure in all offspring. Since each program must have the invariant structure represented by the six points of types 1 through 6, crossover is limited to points of types 7 and 8. Structure-preserving crossover is implemented by allowing any point of type 7 or 8 to be the crossover point in the first parent. However, once the crossover point in the first parent has been selected, the crossover point of the second parent must be of the same type (i.e. type 7 or 8). In the context of this example, this means that crossover will only exchange a sub-tree from the function-defining branch of one parent with a sub-tree from the function-defining branch of the other parent or that crossover will exchange a sub-tree from the result-producing branch of one parent with a sub-tree from the result-producing branch of the other parent. This restriction on the selection of the crossover point of the second parent ensures the preservation of the constrained syntactic structure (originally created in generation 0) for all offspring in all subsequent generations.

Automatic function definition is the main subject of the forthcoming book *Genetic Programming II* [Koza 1994]. A visualization of automatic function definition and numerous example problems can be found in the forthcoming videotape *Genetic Programming II Videotape: The Next Generation* [Koza and Rice 1994]. There is one example of automatic function definition in *Genetic Programming: The Movie* [Koza and Rice 1992].

Angeline and Pollack [1993a, 1993b] have developed a "module acquisition" strategy which is discussed elsewhere in this book.

2.5 Sources of Additional Information about Genetic Programming

An electronic mail mailing list on genetic programming has been established and is currently maintained by James P. Rice of the Knowledge Systems Laboratory of Stanford University. You may subscribe to this on-line mailing list at no charge by sending a subscription request consisting of the message subscribe genetic-programming to genetic-programming-request@cs.stanford.edu by electronic mail.

An on-line public repository and FTP (file transfer protocol) site containing computer code, papers on genetic programming, and frequently asked questions (FAQs) has been established and is currently maintained by James McCoy of the Computation Center at the University of Texas at Austin. This FTP site may be accessed by electronic mail by anonymous FTP from the `pub/genetic-programming` directory from the site `ftp.cc.utexas.edu`. This FTP site contains the original "Little LISP" computer code written in Common LISP for genetic programming as described in *Genetic Programming: On the Programming of Computers by Means of Natural Selection* [Koza 1992], the SGPC ("Simple Genetic Programming in C") computer code written in the C programming language by Walter Alden Tackett and Aviram Carmi of Hughes Aircraft Corporation for genetic programming, and computer code appearing in *Genetic Programming II* [Koza 1994] in Common LISP for implementing automatic function definition and constrained syntactic structures within genetic programming.

The proceedings of the Fifth International Conference on Genetic Algorithms contains recent work on genetic programming by Angeline and Pollack [1993a, 1993b], Banzhaf [1993], Gruau [1993], Handley [1993], Iba et. al [1993], Kinnear [1993b], Koza [1993], Spencer [1993], and Tackett [1993]. The proceedings of the Second International Conference on Simulation of Adaptive Behavior contains work by Reynolds [1993]. The proceedings of the *1993 IEEE International Conference on Neural Networks* contains work by Kinnear [1993a]. The proceedings of the Third Workshop on Artificial Life contains work by Reynolds [1994].

In addition, the proceedings of the IEEE World Congress on Computational Intelligence to be held in Florida in June 1994 will contain additional papers on genetic programming.

2.6 Sources of Additional Information about Genetic Algorithms

Additional information on genetic algorithms can be found in Goldberg [1989], Davis [1987, 1991], Michalewicz [1992], Maenner and Manderick [1992], and Buckles and Petry [1992]. Conference proceedings include Grefenstette [1985, 1987], Schaffer [1989], Belew and Booker [1991], Forrest [1993], Rawlins [1991], and Whitley [1992]. Stender [1993] describes parallelization of genetic algorithms. Davidor [1992] describes application of genetic algorithms to robotics. Schaffer and Whitley [1992] and Albrecht, Reeves, and Steele [1993] describe work on combinations of genetic algorithms and neural networks. Forrest [1991] describes application of genetic classifier systems to semantic nets. Schwefel and Maenner [1991] and Maenner and Manderick [1992] contain recent work on evolutionary strategies (ES). Fogel and Atmar [1992, 1993] contain recent work on evolutionary programming (EP).

Additional information about genetic algorithms may be obtained from the GA-LIST electronic mailing list to which you may subscribe at no charge by sending a subscription request to `GA-List-Request@AIC.NRL.NAVY.MIL`. Additional information about evolutionary programming may be obtained from the EP-LIST electronic mailing list to which you may subscribe at no charge by sending a subscription request to EP-List-Request@Magenta.Me.Fau.Edu.

Bibliography

Albrecht, R. F. (1993), C. R. Reeves, and N. C. Steele, *Artificial Neural Nets and Genetic Algorithms*. Springer-Verlag.

Angeline, P. J. (1993a) and J. B. Pollack. *Coevolving high-level representations*. Technical report 92-PA-COEVOLVE. Laboratory for Artificial Intelligence. Ohio State University. July 1993.

Angeline, Peter J. and Pollack, Jordan B. Coevolving high-level representations. In C. G. Langton, Ed. *Artificial Life III*. In Press.

Angeline, P. J. (1993b) and J. B. Pollack, Competitive environments evolve better solutions for complex tasks. In S. Forrest, Ed. *Proceedings of the Fifth International Conference on Genetic Algorithms*. San Mateo, CA: Morgan Kaufmann Publishers Inc. Pages 264-270.

Banzhaf, W (1993) Genetic programming for pedestrians. In S. Forrest, Ed. *Proceedings of the Fifth International Conference on Genetic Algorithms*. San Mateo, CA: Morgan Kaufmann Publishers Inc.. Page 628.

Belew, R. (1991) and L. Booker, Eds., *Proceedings of the Fourth International Conference on Genetic Algorithms*. San Mateo, CA: Morgan Kaufmann.

Buckles, B. P. (1992) and F. E. Petry, *Genetic Algorithms*. Los Alamitos, CA: The IEEE Computer Society Press.

Davidor, Y. (1991) *Genetic Algorithms and Robotics*. Singapore: World Scientific.

Davis, L., Ed. (1987) *Genetic Algorithms and Simulated Annealing*. London: Pittman.

Davis, L. (1991), *Handbook of Genetic Algorithms*. New York: Van Nostrand Reinhold.

Fogel, D. B. (1992) and W. Atmar, Eds., *Proceedings of the First Annual Conference on Evolutionary Programming*. San Diego, CA: Evolutionary Programming Society .

Fogel, D. B. (1993) and W. Atmar, Eds., *Proceedings of the Second Annual Conference on Evolutionary Programming*. San Diego, CA: Evolutionary Programming Society.

Forrest, S. (1991) *Parallelism and Programming in Classifier Systems*. London: Pittman.

Forrest, S. (1993) *Proceedings of the Fifth International Conference on Genetic Algorithms*. San Mateo, CA: Morgan Kaufmann Publishers Inc.

Goldberg, D. E. (1989) *Genetic Algorithms in Search, Optimization, and Machine Learning*. Reading, MA: Addison-Wesley.

Grefenstette, J. J., Ed. (1985) *Proceedings of an International Conference on Genetic Algorithms and Their Applications*. Hillsdale, NJ: Lawrence Erlbaum Associates.

Grefenstette, J. J., Ed. (1987) *Genetic Algorithms and Their Applications: Proceedings of the Second International Conference on Genetic Algorithms*. Hillsdale, NJ: Lawrence Erlbaum Associates.

Gruau, F. (1993) Genetic synthesis of modular neural networks. In S. Forrest, Ed. *Proceedings of the Fifth International Conference on Genetic Algorithms*. San Mateo, CA: Morgan Kaufmann Publishers Inc. Pages 318–325.

Handley, S. (1993) Automated learning of a detector for a-helices in protein sequences via genetic programming. In S. Forrest, Ed. *Proceedings of the Fifth International Conference on Genetic Algorithms*. San Mateo, CA: Morgan Kaufmann Publishers Inc. Pages 271-278.

Holland, J. H. (1975) *Adaptation in Natural and Artificial Systems: An Introductory Analysis with Applications to Biology, Control, and Artificial Intelligence.* Ann Arbor, MI: University of Michigan Press 1975. Also Cambridge, MA: The MIT Press 1992.

Iba, H. (1993), T. Jurita, H. de Garis, and T. Sat, System identification using structured genetic algorithms. In S. Forrest, Ed. *Proceedings of the Fifth International Conference on Genetic Algorithms.* San Mateo, CA: Morgan Kaufmann Publishers Inc. Pages 279-286.

Kinnear, K. E. Jr. (1993a) Evolving a sort: Lessons in genetic programming. *1993 IEEE International Conference on Neural Networks, San Francisco.* Piscataway, NJ: IEEE 1993. Volume 2. Pages 881-888.

Kinnear, K. E. Jr. (1993b) Generality and difficulty in genetic programming: Evolving a sort. In S. Forrest. Ed. *Proceedings of the Fifth International Conference on Genetic Algorithms.* San Mateo, CA: Morgan Kaufmann Publishers Inc. Pages 287–294.

Koza, J. R. (1992) *Genetic Programming: On the Programming of Computers by Means of Natural Selection.* Cambridge, MA: The MIT Press.

Koza, J. R. (1993) Simultaneous discovery of reusable detectors and subroutines using genetic programming. In S. Forrest, Ed. *Proceedings of the Fifth International Conference on Genetic Algorithms.* San Mateo, CA: Morgan Kaufmann Publishers Inc. Pages 295–302.

Koza, J. R. (1994) *Genetic Programming II.* Cambridge, MA: The MIT Press. In Press.

Koza, J. R.(1992), and J. P. Rice *Genetic Programming: The Movie.* Cambridge, MA: The MIT Press.

Koza, J. R. (1994), and J. P. Rice *Genetic Programming II Videotape: The Next Generation* Cambridge, MA: The MIT Press. In Press.

Maenner, R. (1992), and B. Manderick, *Proceedings of the Second International Conference on Parallel Problem Solving from Nature.* North Holland.

Michalewicz, Z. (1992). *Genetic Algorithms + Data Structures = Evolution Programs.* Berlin: Springer-Verlag.

Rawlins, G., Ed. (1991) *Foundations of Genetic Algorithms.* San Mateo, CA: Morgan Kaufmann Publishers Inc.

Reynolds, C. W. (1993) An evolved vision-based behavioral model of coordinated group motion. In Meyer, Jean-Arcady, Roitblat, Herbert L. and S. Wilson, Ed. *From Animals to Animats 2: Proceedings of the Second International Conference on Simulation of Adaptive Behavior.* Cambridge, MA: The MIT Press. Pages 384-392..

Reynolds, C. W. (1994) An evolved vision-based model of obstacle avoidance behavior. In C. G. Langton, Ed. *Artificial Life III.* In Press.

Schaffer, J. D., Ed. (1989). *Proceedings of the Third International Conference on Genetic Algorithms.* San Mateo, CA: Morgan Kaufmann Publishers Inc.

Schaffer, J. D. (1992) and D. Whitley, Eds., *Proceedings of the Workshop on Combinations of Genetic Algorithms and Neural Networks 1992.* Los Alamitos, CA: The IEEE Computer Society Press.

Schwefel, H.-P. (1991) and R. Maenner, Eds., *Parallel Problem Solving from Nature.* Berlin: Springer-Verlag.

Spencer, G. (1993) Automatic generation of programs for crawling and walking. In S. Forrest, Ed. *Proceedings of the Fifth International Conference on Genetic Algorithms.* San Mateo, CA: Morgan Kaufmann Publishers Inc. Page 654.

Steele, Guy L. Jr. (1990) *Common LISP.* Digital Press. Second Edition.

Stender, J., Ed. *Parallel Genetic Algorithms.* IOS Publishing.

Tackett, W. A. (1993) Genetic programming for feature discovery and image discrimination. In S. Forrest, Ed. *Proceedings of the Fifth International Conference on Genetic Algorithms.* San Mateo, CA: Morgan Kaufmann Publishers Inc. Pages 303–309.

Whitley, D. (editor) (1992) *Foundations of Genetic Algorithms and Classifier Systems 2,* Vail, Colorado 1992. San Mateo, CA: Morgan Kaufmann Publishers Inc.

II INCREASING THE POWER OF GENETIC PROGRAMMING

This section contains contributions whose principal purpose is to increase the power of the genetic programming paradigm. It leads off in Chapter 3 with a theoretical analysis of *evolvability* — the ability of a population to produce variants fitter than any yet existing. It explores the relationship between the fitness function, representation, and genetic operators that produce evolvability. It analyzes the relationship of evolvability to the observed proliferation of common blocks of code within programs evolved using genetic programming. This theoretical analysis points the way to several intriguing and quite practical ways to improve the power of a genetic programming system.

Chapter 4 argues that evolutionary computations, including genetic programming, belong to a unique class of problem solvers that display *emergent intelligence*, a method of problem solving that allows task specific knowledge to emerge from the interaction between a problem solver and a fitness function. It examines what is distinct about evolutionary computations and emergent intelligence, why they represent a novel approach to computational problem solving and how that difference is manifested in genetic programming.

The evolution of scalable learning is the subject of Chapter 5, where three differently sized variations of a particular problem are solved using genetic programming. It is demonstrated that the problem difficulty as perceived by genetic programming grows very rapidly as the scale of the problem is increased. The use of *automatically defined functions*, a technique which allows subprograms to evolve along with main programs in genetic programming, dramatically reduces the computational effort required to solve the problem and also reduces the rate of increase in difficulty as the scale of the problem is increased.

Chapter 6 begins with a comparison of performance between the automatically defined functions from Chapter 5 and an alternative approach to the evolution of subprograms called *modules*, defined in Chapter 4. On the problem used for comparison, automatically defined functions show a large performance gain and modules do not. A variety of hypotheses to explain these differences are tested with experiments, yielding the conclusion that a particular form of *structural regularity* created by automatically defined functions is largely responsible for their increase in performance.

A classification problem with scalable difficulty is presented in Chapter 7. This problem is designed to be analogous to real world problems where the solution is frequently unknown and an exact solution is often not possible. This problem is used to test the performance of a genetic programming system on evolving a classification algorithm by varying the size and degree of representation of the training data set. The effects of *locality*, taking into account the spatial relationship of members of the population as part of the breeding scheme, on performance are reported, as are the effects of using steady state genetic programming.

Chapter 8 focuses on the effect of taking locality into account in breeding scheme design as well as during fitness evaluation. In this contribution the dynamics of the population as a whole are examined instead of the usual focus on the best individual in order to analyze and understand the effects of introducing locality into a genetic programming system. A variety of population analysis tools are demonstrated which could be applied to a range of similar problems.

Most problems to not have solutions that are simple mappings from the inputs to the correct outputs; some kind of internal state or memory is needed to operate effectively on these problems. Chapter 9 proposes a simple addition to the genetic programming paradigm that seamlessly incorporates the evolution of the effective gathering, storage, and retrieval of arbitrarily complex state information. Experimental results show that the effective production and use of complex memory structures can be evolved, and that functions that do so perform better than those that do not.

Chapter 10 reports on investigations into the evolution of reactive control programs for obstacle avoidance in a "robot like" simulated vehicle for a corridor–following task. Typically, controllers developed for tasks such as these are quite brittle and overly specific, and the central focus of this chapter is the addition of noise to the fitness evaluation process in order to increase the robustness of the evolved solutions. Noise is shown to aid in the evolution of robustness, but it is not without associated costs, which are examined.

Premature convergence to a less optimal solution than is desired is sometimes a cause of difficulty in genetic programming systems. Chapter 11 proposes a new selection and breeding scheme which is similar to *disassortive mating*, taken directly from population genetics, to reduce premature convergence. This scheme, called the Pygmy algorithm, is shown to allow the use of small populations and strong selective pressure with little chance of losing genetic diversity.

The *minimum description length* principle is offered in Chapter 12 as a way to avoid the relentless growth of genetic programming programs. The use of the minimum description length principle in the fitness function of a genetic programming system not only keeps the size of the evolving individuals within bounds, but also increases the capability of the system to evolve general individuals. MDL based fitness is not appropriate for every problem, and criteria are presented for its effective use.

In Chapter 13 the design of a high performance genetic programming system is discussed. A large number of implementation issues are examined, with the focus on the C++ programming language, yet the concepts are applicable to any high-performance language. Both execution time performance as well as memory performance are examined, and proposals for the optimization of each are illustrated by experiments. For anyone about to write a genetic programming system, this chapter is a tremendous help.

Most genetic programming systems use a technique where a problem specific language is executed by an interpreter, essentially a virtual machine. This is highly flexible, but creates significant overhead. In a successful attempt to dramatically improve performance, Chapter 14 discusses a system which uses as its representation actual machine code. Each individual is a segment of machine code, and these segments are evolved directly using genetic operators. Fitness evaluation consists of giving control to the machine code which is the individual. The speedups available are very exciting, and the approach deserves serious consideration by anyone with a particularly complex problem.

3 The Evolution of Evolvability in Genetic Programming

Lee Altenberg

The notion of "evolvability" — the ability of a population to produce variants fitter than any yet existing — is developed as it applies to genetic algorithms. A theoretical analysis of the dynamics of genetic programming predicts the existence of a novel, emergent selection phenomenon: the evolution of evolvability. This is produced by the proliferation, within programs, of blocks of code that have a higher chance of increasing fitness when added to programs. Selection can then come to mold the *variational* aspects of the way evolved programs are represented. A model of code proliferation within programs is analyzed to illustrate this effect. The mathematical and conceptual framework includes: the definition of evolvability as a measure of performance for genetic algorithms; application of Price's *Covariance and Selection Theorem* to show how the fitness function, representation, and genetic operators must interact to produce evolvability — namely, that genetic operators produce offspring with fitnesses specifically correlated with their parent's fitnesses; how blocks of code emerge as a new level of replicator, proliferating as a function of their "constructional fitness", which is distinct from their schema fitness; and how programs may change from innovative code to conservative code as the populations mature. Several new selection techniques and genetic operators are proposed in order to give better control over the evolution of evolvability and improved evolutionary performance.

3.1 Introduction

The choice of genetic operators and representations has proven critical to the performance of genetic algorithms (GAs), because they comprise dual aspects of the same process: the creation of new elements of the search space from old. One hope in genetic algorithm research has been that the representation/operator problem could itself be solved through an evolutionary approach, by creating GAs in which the representations and/or operators can themselves evolve. What I discuss in this chapter is how genetic programming (GP) exhibits the evolution of representations as an inherent property. Moreover, I investigate how the direction of evolution of representations in genetic programming may be toward increasing the *evolvability* of the programs that evolve, and suggest ways that the evolution of evolvability can be controlled.

3.1.1 Evolvability

"Evolvability" is a concept I wish to develop as a performance measure for genetic algorithms. By evolvability I mean the ability of the genetic operator/representation scheme to produce offspring that are fitter than their parents. Clearly this is necessary for adaptation to progress. Because adaptation depends not only on how *often* offspring are fitter than their parents, but on how *much* fitter they are, the property ultimately under consideration, when speaking of evolvability, is the entire distribution of fitnesses among the offspring produced by a population. Since there is a chance that offspring are fitter than parents even in random search, good GA performance requires that the upper tail of the offspring fitness distribution be fatter than that for random search.

But this excess of fitter offspring as compared with random search needn't occur with all parents. It need only occur with parents that are fitter than average, because selection is biasing the population in that direction. In other words, the action of the genetic operator on the representation needs to produce high correlations between the performance of parents and the fitness distribution of their offspring. This is shown in Section 3.2.3 through an application of Price's *Covariance and Selection Theorem* [Price 1970], which in my mind serves as a fundamental theorem for genetic algorithms.

This correlation property is exemplified by the Building Blocks hypothesis [Goldberg 1989]. A building block is a schema [Holland 1975] from a fitter-than-average individual which, when recombined into another individual, is likely to increase its fitness. Thus a building block is *defined* by a correlation between parent and offspring fitnesses under recombination. A recombination operator which is able to pick out this schema from one individual and substitute it into another is then leaving intact the building blocks but producing variation in a way for which there is adaptive opportunity, namely, combining the proper building blocks. But for recombination to be of use, however, the representation must be such that building blocks — i.e. schemata which give correlated fitnesses when substituted into a chromosome by recombination — exist.

Evolvability can be understood to be the most "local" or finest-grained scale for measuring performance in a genetic algorithm, whereas the production of fitter offspring over the course of a run of a GA, or many runs, are more "global", large scale performance measures. As a population evolves, the distribution of offspring fitnesses is likely to change. The global performance of a genetic algorithm depends on it maintaining the evolvability of the population as the population evolves toward the global optimum. I will explore how genetic programming, through its ability to evolve its representations, may be able to maintain or increase the evolvability of the programs as a population evolves.

3.1.2 Representations

How is evolvability achieved in genetic algorithms? It comes from the genetic operators being able to transform the representation in ways that leave intact those aspects of the individual that are already adapted, while perturbing those aspects which are not yet highly adapted. Variation should be channeled toward those "dimensions" for which there is selective opportunity.

As an illustrative domestic example, consider the ways that water flow and temperature in a bathroom shower are "represented". There are two common control mechanisms:

1. One handle for hot water flow and one handle for cold water flow; and

2. A single handle which varies water flow by tilting up and down, and varies temperature by twisting clockwise and counterclockwise.

Adjusting the shower consists of perturbing the two degrees of freedom in either representation until flow and temperature are satisfactory. Representation 2 was invented to deal with the problem that 1 has, which is that bathers usually want to change the flow rate without changing the temperature, or vice versa. Under 1, that would require a sophisticated "operator" which changes both handles simultaneously in the right amounts, while in 2 it requires changing only the tilt of the handle or the twist. It generally takes many more adjustments under 1 to converge to satisfaction, with greater risks of scalding or freezing. The representation under 2 is more "attuned" to the dimensions of variation typical of the environment [Barwise and Perry 1983]. Another way of saying this is that the two degrees of freedom in 1 are more epistatic in their relation to the objective function of the bather.

The basic design task for genetic algorithms is to create a system whose "dimensions of variability" match the dimensions of variability of the task domain. The existence of sub-tasks or sub-goals in a problem correspond to dimensions of variability for the task domain. Representations and operators that are attuned to these dimensions can therefore exhibit greater evolvability.

When speaking about the evolution of representations in genetic programming, I mean not the fine-scale aspects of a genetic program, such as the choice of functions or terminals, the data-structure implementation, or LISP versus some other programming language; these are fixed at the outset. Rather, I mean the large-scale structure of the program.

Any particular program is a representation of the *behavior* that the program exhibits. Often the same program behavior can be represented by a variety of programs. These programs may exhibit different *evolvability*, however, because some representations may make it more likely that modification by genetic operators can produce still fitter programs. A representation must therefore be understood to have *variational* properties — how changes in the representation map to changes in behavior of the code. Object oriented programming, for example — although not usually described in this manner — is fundamentally a method of engineering desirable variational properties into code. So it is important to understand that the variational properties of a program are distinct from its fitness. In classical fixed-length GAs, selection can act only on the fitness of the population, not on the variational properties of the representation. In genetic programming, selection can act on both, as will be described.

3.1.3 Evolving evolvability

A number of modifications of genetic algorithms have been implemented to evolve their evolvability by adapting the genetic operators or representations. These include:

• the addition of modifier genes, whose allelic values control the genetic operators (e.g. loci controlling recombination or mutation rates and distributions) [Bäck, Hoffmeister and Schwefel 1991, Bergman and Feldman 1992, Shaffer and Morishima 1987];

• running a meta-level GA on parameters of the operators, in which each run of the primary GA produces a performance measure used as the fitness of the operator parameter [Grefenstette 1986];

• "messy GAs", a form of representation which allows more complex transformations from parent to offspring [Goldberg, Deb, and Korb 1990];

• focusing the probabilities of operator events to those that have had a history of generating fitter individuals [Davis 1989];

• adjusting the mutation distribution to maintain certain properties of the fitness distribution, e.g. the "one-fifth" rule of Evolution Strategies [Bäck, Hoffmeister and Schwefel 1991];

• dynamic parameter encoding [O'Neil and Shaefer 1989, Pitney, Smith, and Greenwood 1990, Schraudolph and Belew 1992, Shaeffer 1987, Szarkowicz 1991].

Genetic programming allows another means by which evolvability can evolve: the differential proliferation of blocks of code *within* programs. Such differential proliferation of blocks of code within programs has already been observed in practice (see [Angeline and Pollack 1992, Tackett 1993], and the chapters in this volume by Angeline (Ch. 4), Kinnear (Ch. 6), Teller (Ch. 9), Oakley (Ch. 17), and Handley (Ch. 18)).

3.1.4 Constructional selection

Proliferation of copies of code within programs is possible in genetic programming because of the syntactic closure of programs to such duplications. The recombination operators carry out these duplications. The exchange of whole subtrees has been the main genetic operator used in GP, but in general, recombination can involve any block of code (e.g. Automatically Defined Functions [Koza 1993]).

Because blocks of code can be duplicated indefinitely within programs, they constitute an emergent level of replication, and hence, selection. Here I employ the concept of "replicator": any entity that can spawn multiple copies of itself based on its interactions with an environment [Dawkins 1976, Brandon 1988]. Programs are the primary level of replicator in GP, since the selection operator explicitly duplicates them based on their fitness. Blocks of code can emerge as another level of replicator because copies of them can proliferate within programs. The rate at which they proliferate is not imposed by the selection operator as in the case of the programs themselves.

Suppose that different blocks of code had different likelihoods of improving fitness when substituted at random locations in randomly chosen programs in the population. If the genetic operator could increase the frequency with which it chooses those blocks of code bearing the highest likelihood of producing fitter offspring, then evolvability could increase. This frequency depends on the average number of copies of that block of code in the population, which can increase in two ways:

The first way is for programs that contain the block to increase their number in the population. This shows up in the "schema" or marginal fitness of a block of code, and is accounted for by applying the Schema Theorem [Holland 1975] to genetic programming [Koza 1992]. But this pure selection effect does not produce more copies of a block of code within programs. That requires the creation of novel programs through insertion of additional copies of the code from donor programs to recipient programs. If the offspring programs survive long enough, they will be recipients of furhter copies of the code, or will be used as donors for contributing additional copies of the code to other recipient programs.

There are thus two numbers that can evolve: the number of programs carrying a block of code, and the number of copies of the code within the programs. Change in the former is determined by selection as specified by the fitness function of the GP algorithm. But change in the latter is an emergent selection phenomenon, involving the effects of constructing new programs by adding copies of the block of code. What I am proposing is that the proliferation of copies of code within programs constitutes a secondary kind of selection process, which I will refer to as "constructional selection" [Altenberg 1985]. The Schema Theorem does not account for this latter process.

Increases in the number of copies of a block of code within programs requires multiple sampling by the genetic operator; it can't be the result of a one-time lucky event. Therefore, differential proliferation of code depends on repeatable *statistical* differences in the fitness effects of code substitutions by the different blocks. (In contrast, in a fixed-length GA, a single substitution of a schema into a chromosome can produce an individual so fit that the schema goes to fixation without further recombination events; at that point no further increase in the schema frequency is possible).

If a block of code has a relatively stationary distribution in its effects on the fitness of programs to which it is added, its rate of proliferation *within* programs can be referred to as its "constructional" fitness. The constructional fitness is distinct from the current average fitness in the population of a block of code (i.e. its marginal or "schema" fitness). It is an emergent selection phenomenon in that the constructional fitness is never evaluated by the GP algorithm, nor is it obtained by averaging current program fitnesses.

What determines the constructional fitness of a block of code? That is, how does the fitness distribution, for recombinants that carry a new copy of a block of code, determine whether that block proliferates or not? How rapidly a particular block of code proliferates within programs depends in nontrivial ways on the relationship between the population dynamics, the fitness distribution of the new programs carrying additional copies, and how stationary this distribution is as the population evolves.

The population dynamics which are most easy to analyze would be a GP version of "hill-climbing". A single program is modified by internally duplicating a block of code and substituting it at another location within the program. If the new program is fitter, keep it; if not, try again. GP hill-climbing will be the effective dynamics if very low recombination rates are used so that selection fixes the fittest program before the next recombination event occurs. GP hill-climbing resembles the dynamics of gene duplication in natural populations.

Under these dynamics, deleterious additions of a block of code go extinct before the code can be passed on again through recombination. Only advantageous additions are ever resampled. Therefore blocks of code proliferate based solely on their probability of increasing fitness when added to a program. As a consequence, in time, a growing proportion of code additions made by the genetic operator should involve blocks of code with a high probability of producing a fitness increase. In this way, the operator becomes focused in its action on those dimensions of variability with greater potential for increasing fitness. This is not simply an increase in the adaptedness of the programs in the population, but a change in their potential for further adaptation, i.e. a change in the evolvability of the gene pool.

Under the high recombination rates typical of GP, the picture is not so simple. Let me refer to a block of code's probability of increasing fitness when added to a program as its "evolvability value". With high recombination rates, blocks of code with the highest evolvability value need not have the highest constructional fitness. Even a program on its way to extinction under selection can be sampled several times by the genetic operator as a donor of its code to new programs. Thus a block of code that caused a deleterious effect when added to a program has a chance of being sampled again for further donation before its program goes extinct.

Blocks of code will proliferate depending on the survival time of the programs resulting from their addition. So code that produces a near neutral effect could have a constructional fitness advantage over "Babe Ruth"[1] code — i.e. code that occasionally produces very fit new programs, but which on the average has a deleterious effect when added to a program.

---

[1] Babe Ruth, the legendary American baseball player, had a mediocre batting average, but when he got a hit, it was very likely to be a home run.

In the early stages of the evolution of a population, for a block of code to survive (distinct from the question of proliferation within programs), it must be able to increase its fitness through recombination events with other blocks of code at least as fast as the mean fitness of the population is increasing. Code is then in competition for high evolvability, and so one would expect code with high evolvability to survive and/or proliferate within programs at that stage.

In the later stages of the population's evolution, the population reaches a recombination-selection balance, and the mean fitness approaches a quasi-stationary state. At that point the main selective opportunity is for a fitter-than-average program to preserve what fitness it's got when producing offspring [Altenberg and Feldman 1987]. Code is then in competition for the robustness of its behavior in the face of the actions of the genetic operators.

The evolution of populations of programs whose fitness is robust to crossover in the later stages of GP runs has been reported [Koza 1992], and Andrew Singleton and Nick Keenan (personal communication) have identified this phenomenon as "defense against crossover".

The possibilities in genetic programming for such emergent evolutionary phenomena indicate that it may prove to be a rich area for research. But it shows that a better understanding of GP dynamics is needed to control the direction of such emergent phenomena toward making GP a useful tool. Some suggestions are provided in the Discussion for how to maintain the evolution of evolvability through special selection regimes and genetic operators.

3.1.5   Synopsis of the models

In the next two sections, I provide some mathematical underpinnings to the concepts discussed so far. In section 3.2 I investigate the relations between selection, genetic operators, representations and evolvability. The main results are that, first, there is a duality between representations and genetic operators in producing a *transmission function*; second, it is the relationship between the transmission function and the selection function that determines the performance of the genetic algorithm; and third, the relationship between selection and transmission which produces good GA performance is that there be correlations between parental fitness and offspring fitness distributions. This is expressed mathematically by elaborating on Price's Covariance and Selection Theorem [Price 1970]. I propose that this theorem provides a more adequate basis to account for genetic algorithm performance than does the Schema Theorem [Holland 1975].

In section 3.3, I offer a simple model of genetic programming dynamics to illustrate how constructional selection can produce the evolution of evolvability. The model includes simplifying assumptions made to provide a tractable illustration of the mechanism of evolutionary increase in evolvability, but these simplifications undoubtedly limit the model's applicability to other emergent phenomena in genetic programming. These assumptions include that the genetic operator can add blocks of code to programs from other programs, but cannot delete blocks of code from programs or disrupt them, and that the population undergoes GP hill-climbing. But this model comprises a "generation 0" model from which other more complex models are being developed. It demonstrates how the different distributions of fitness effects of blocks of code lead to the exponential proliferation of those which confer greater evolvability to the programs that contain them. I also emphasize the distinction between the schema fitness of a block of code and its "constructional" fitness, which is what determines its proliferation rate within programs.

3.2   Selection, Transmission, and Evolvability

The conventional theoretical approach to genetic algorithms focuses on schemata and the Schema Theorem. This approach, however, relies on hidden assumptions which have obscured and even misled GA design (e.g. the principle of minimal alphabets [Goldberg 1989, Antonisse 1989]). The main hidden assumption I wish to bring out is that of "implicit correlations". This is approached by considering a general model of genetic algorithms with arbitrary representations, operators, and selection. With this approach, it is possible to extract features that are essential for the performance of GAs which have nothing to do with schemata.

The action of genetic operators on the representation produces a *transmission function* [Slatkin 1970, Cavalli-Sforza and Feldman 1976, Altenberg and Feldman 1987], which is simply the probability distribution of offspring from every possible mating. With two parents, for example, the transmission function would simply be

$$T(i \leftarrow j, k),$$

where j and k are labels for the parents and i the label for an offspring. To be precise, let S be the search space; then $T : S^3 \mapsto [0, 1]$, $T(i \leftarrow j, k) = T(i \leftarrow k, j)$, and $\sum_i T(i \leftarrow j, k) = 1$ for all $i, j, k \in S$.

Clearly, the transmission function can represent not only the action of genetic operators in binary, fixed length GAs, but also messy GAs, the parse trees of GP, and real-valued GAs as well (where $S \subset \Re^n$) or even functionals. Multi-parent transmission could simply be represented as $T(i \leftarrow j_1, \ldots, j_m)$.

Principle 1 *It is the relationship between the transmission function and the fitness function that determines GA performance. The transmission function "screens off" [Salmon 1971,*

Brandon 1990] the effect of the choice of representation and operators, in that changes in either affect the dynamics of the GA only through their effect on the transmission function.

For example, the effect of inserting "introns" (loci that are neutral to performance but are targets of crossover) [Levenick 1991, Forrest and Mitchell 1993], which is a change in the representation, can equivalently be achieved by an adjustment in the crossover operator. In the Discussion some implications of Principle 1 for genetic programming design will be considered.

What aspect of the relationship between transmission function and fitness function is crucial to GA performance? It is the correlation between parental performance and offspring performance under the action of the genetic operators. Without such a correlation, selection on the parents has no means of influencing the distribution of offspring fitnesses. I make explicit this condition in Section 3.2.3, where I utilize Price's *Selection and Covariance* theorem [Price 1970, 1972] to show how the change in the fitness distribution of the population depends on this correlation.

To develop this I first present the form of the recursion for the general "canonical genetic algorithm". Then I define the idea of "measurement functions" to extract information from a population, which allows the statement of Price's theorem. Use of the proper measurement function allows one to use Price's theorem to show how the distribution of fitnesses in the population changes over one generation, including the probability of producing individuals fitter than any in the population. This will be seen to depend on the covariance between parent and offspring fitness distributions, and a "search bias" indicating how much better in the current population the genetic operator is than random search.

3.2.1   A general model of the canonical genetic algorithm

For the purpose of mathematical exposition, a "canonical" model of genetic algorithms has been generally used since its formulation by Holland [1975], which incorporates assumptions common to many evolutionary models in population genetics: discrete, non-overlapping generations, frequency-independent selection, and infinite population size. The algorithm iterates three steps: selection, random mating, and production of offspring to constitute the population in the next generation.

Definition: Canonical Genetic Algorithm
The dynamical system representing the "canonical" genetic algorithm is:

$$x'_i = \sum_{jk} T(i \leftarrow j, k) \, \frac{w_j w_k}{\overline{w}^2} \, x_j x_k, \tag{3.1}$$

where

• x_i *is the frequency of type* i *in the population,* $i = 1 \ldots n$, *and* x_i' *is the frequency in the next generation;*

• w_i *is the fitness of type* i;[2]

• $\overline{w} = \sum_i w_i x_i$ *is the mean fitness of the population; and*

• $T(i \leftarrow j, k)$ *is the probability that offspring type* i *is produced by parental types* j *and* k *as a result of the action of genetic operators on the representation.*

This general form of the transmission-selection recursion was used at least as early as 1970 by Slatkin [1970], and has been used subsequently for a variety of quantitative genetic and complex transmission systems [Cavalli-Sforza and Feldman 1976, Karlin 1979, Altenberg and Feldman 1987]..

The quadratic structure of the recursion can be seen by displaying it in vector form as:

$$x' = \frac{1}{\overline{w}^2} \, T \, (W \otimes W) \, (x \otimes x) \tag{3.2}$$

where

$$T = \| \, T(i \leftarrow j_1, j_2) \, \|_{i, j_1, j_2 = 1}^{n}, \tag{3.3}$$

is the *transmission matrix*, the n by n^2 matrix of transmission function probabilities, n is the number of possible types (assuming n is finite), W is the diagonal matrix of fitness values, x is the frequency vector of the different types in the population, and $\otimes$ is the Kronecker (tensor) product. The Kronecker product gives all products of pairs of matrix elements, e.g.:

$$\| \, x \, y \, z \, \| \otimes \begin{Vmatrix} a & b \\ c & d \\ e & f \end{Vmatrix} = \begin{Vmatrix} xa & xb & ya & yb & za & zb \\ xc & xd & yc & yd & zc & zd \\ xe & xf & ye & yf & ze & zf \end{Vmatrix}.$$

The mathematics of (3.2) for $W = I$ (the identity matrix) have been explored in depth by Lyubich [1992].

3.2.2   Measurement functions

Let $F_i : S \mapsto \mathcal{V}$ be a function of the properties of individuals of type $i \in S$ (or let $F_i(p) : S \mapsto \mathcal{V}$ be a parameterized family of such functions, for some parameter p). Here

---

[2]Using the letter "w" for fitness is traditional in population genetics, and derives from "worth" (Sewall Wright, informal comment).

Table 3.1
Measurement functions, F_i (some taking arguments), and the population properties measured by their mean $\overline{F}$.

Population Property:	Measurement Function:
• Mean fitness:	$F_i = w_i$
• Fitness distribution:	$F_i(w) = \begin{cases} 1 & w_i \leq w \\ 0 & w_i > w \end{cases}$
• Subtree frequency:	$F_i(S) = \Pr[\text{Operator picks subtree } S \text{ for crossover}]$
• Schema $\mathcal{H}$ frequency:	$F_i(\mathcal{H}) = \begin{cases} 0 & i \notin \mathcal{H} \\ 1 & i \in \mathcal{H} \end{cases}$
• Mean phenotype:	$F_i \in \Re^n = \{\text{vector-valued phenotypes}\}$

we are interested in properties that can be averaged over the population, so $\mathcal{V}$ is a vector space over the real numbers (e.g. $\Re^n$ or $[0, 1]^n$), allowing averages $\overline{F} = \sum_i x_i \, F_i \in \mathcal{V}$ for $x_i \geq 0$ with $\sum_i x_i = 1$.

Different measurement functions allow one to extract different kinds of information from the population, by examining the change in the population mean for the measurement function:

$$\overline{F} = \sum_i F_i \, x_i, \quad \overline{F}' = \sum_i F_i \, x_i' \tag{3.4}$$

Some examples of different measurement functions and the population properties measured by $\overline{F}$ are shown in Table 3.1.

3.2.3 Price's theorem applied to evolvability

Price's theorem gives the one-generation change in the population mean value of F:

Theorem 1 (Covariance and Selection, Price, 1970)
For any parental pair $\{j, k\}$, let ϕ_{jk} represent the expected value of F among their offspring. Thus:

$$\phi_{jk} = \sum_i F_i \, T(i \leftarrow j, k).$$

Then the population average of the measurement function in the next generation is

$$\overline{F}' = \overline{\phi}_u + \text{Cov}[\phi_{jk},\ w_j w_k / \overline{w}^2] \tag{3.5}$$

where:

$$\overline{\phi}_u = \sum_{jk} \phi_{jk}\, x_j x_k$$

is the average offspring value in a population reproducing without selection, and

$$\text{Cov}[\phi_{jk},\ w_j w_k / \overline{w}^2] = \sum_{jk} \phi_{jk}\, \frac{w_j w_k}{\overline{w}^2}\, x_j x_k - \overline{\phi}_u$$

is the population covariance between the parental fitness values and the measured values of their offspring.

Proof. This is obtained directly from substituting (3.1) into (3.4). ■

Price's theorem shows that the covariance between parental fitness and offspring traits is the means by which selection directs the evolution of the population.

Now, Price's theorem can be used to extract the change in the *distribution* of fitness values in the population by using the measurement function

$$F_i(w) = \begin{cases} 0 & w_i \leq w \\ 1 & w_i > w \end{cases}.$$

Then

$$\overline{F}(w) = \sum_i F_i(w)\, x_i = \sum_{i:w_i > w} x_i$$

is the proportion of the population that has fitness greater than w (The standard cumulative distribution function would be $1 - \overline{F}(w)$).

First several definitions need to be made.

- Let $\mathcal{R}(w)$ be the probability that random search produces an individual fitter than w.

- For a parental pair $\{j, k\}$, define the *search bias*, $\beta_{jk}(w)$, to be the excess in their chance of producing offspring fitter than w, compared with random search:

$$\beta_{jk}(w) = \sum_i F_i(w)\, T(i \leftarrow j, k) - \mathcal{R}(w).$$

- The average search bias for a population before selection is

$$\overline{\beta}_u(w) = \sum_{jk} \beta_{jk}(w)\, x_j\, x_k.$$

- The variance in the (relative) fitness of parental pairs is

$$\text{Var}[w_j\, w_k/\overline{w}^2] = \sum_{jk} \frac{w_j\, w_k}{\overline{w}^2}\, x_j\, x_k - 1 = \text{Var}[w_i/\overline{w}](2 + \text{Var}[w_i/\overline{w}])$$

- The coefficient of regression of $\beta_{jk}(w)$ on $w_j\, w_k/\overline{w}^2$ for the population, which measures the magnitude of how $\beta_{j,k}(w)$ varies with $w_j w_k/\overline{w}^2$, is

$$\text{Reg}[\beta_{jk}(w) \to w_j\, w_k/\overline{w}^2] = \text{Cov}[\beta_{jk}(w),\; w_j\, w_k/\overline{w}^2] \,/\, \text{Var}[w_j\, w_k/\overline{w}^2].$$

Theorem 2 (Evolution of the Fitness Distribution)
The probability distribution of fitnesses in the next generation is

$$\overline{F}(w)' = \mathcal{R}(w) + \overline{\beta}_u(w) + \text{Var}[w_j\, w_k/\overline{w}^2]\, \text{Reg}[\beta_{jk}(w) \to w_j\, w_k/\overline{w}^2]. \qquad (3.6)$$

Theorem 2 shows that in order for the GA to perform better than random search in producing individuals fitter than w, the search bias, plus the parent-offspring regression scaled by the fitness variance,

$$\overline{\beta}_u(w) + \text{Reg}[\beta_{jk}(w) \to w_j\, w_k/\overline{w}^2]\, \text{Var}[w_j\, w_k/\overline{w}^2], \qquad (3.7)$$

must be positive.

We wish to know the frequency with which the population produces individuals that are fitter than any that exist. Here one simply uses:

$$w = w_{\text{max}} = \max_{i:x_i>0} (w_i)$$

in Theorem 2. But this can be further refined by quantifying the degree to which the transmission function is *exploring* the search space versus *exploiting* the current population. Holland [1975] introduced the idea that GAs must find a balance between exploration and exploitation for optimal search. For any transmission function, a global upper bound can be placed on its degree of exploration [Altenberg and Feldman 1987]:

Lemma 1 (Exploration Rate Limit)
Any transmission function $T(i \leftarrow j, k)$ can be parameterized uniquely as

$$T(i \leftarrow j, k) = (1 - \alpha)\,(\delta_{ij} + \delta_{ik})/2 + \alpha P(i \leftarrow j, k), \qquad (3.8)$$

with

$$\delta_{ij} = \begin{cases} 1 & \textit{if } i = j \\ 0 & \textit{if } i \neq j \end{cases},$$

where

$$\alpha = 1 - \min_{i,j \neq i}\{2\,T(i \leftarrow i, j),\, T(i \leftarrow i, i)\} \in [0, 1],$$

and $P(i \leftarrow j, k)$ is a transmission function in which $P(i \leftarrow i, j) = 0$ for some i, j.

The value α is defined as the global exploration rate limit for the transmission function, since it sets a maximum on the rate at which novel genotypes can be produced by transmission for all possible population compositions. The exploratory part of the transmission function, $P(i \leftarrow j, k)$, is referred to as the search kernel.

In classical GAs, for example, where the transmission function consists of recombination followed by mutation, the transmission matrix T from (3.3) can be decomposed into the product

$$T = [(1 - \mu)I + \mu M]\,[(1 - \rho)\,(\mathbf{1}^{\mathrm{T}} \otimes I) + \rho R],$$

where $1 - \rho$ is the chance of no crossover and $1 - \mu$ is the chance of no mutation during reproduction, M is an n by n Markov matrix representing mutation probabilities, R is an n by n^2 matrix representing recombinant offspring probabilities, and $\mathbf{1}$ is the n-long vector of 1s, yielding

$$\alpha = \mu + \rho - \mu\rho.$$

Using (3.8), recursion (3.1) can be rewritten as:

$$x_i' = (1 - \alpha)\,\frac{w_i}{\overline{w}}\,x_i + \alpha \sum_{jk} P(i \leftarrow j, k)\,\frac{w_j w_k}{\overline{w}^2}\,x_j x_k, \qquad (3.9)$$

This gives:

Theorem 3 (Evolvability)
The probability that a population generates individuals fitter than any existing is

$$\overline{F}(w_{\max})' =$$
$$\alpha\left\{\mathcal{R}(w_{\max}) + \overline{\beta}_u(w_{\max}) + \mathrm{Var}[w_j\,w_k/\overline{w}^2]\,\mathrm{Reg}[\beta_{jk}(w_{\max}) \to w_j\,w_k/\overline{w}^2]\right\},$$

where now the search bias $\beta_{jk}(w_{\max})$ is defined in terms of the search kernel:

$$\beta_{jk}(w) = \sum_i F_i(w) \, P(i \leftarrow j, k) - \mathcal{R}(w).$$

Proof. This result is direct, noting that for the terms times $1 - \alpha$ in (3.9), $\sum_i F_i(w_{\max}) \, x_i \, w_i / \overline{w} = 0$. ∎

Both the regression and the search bias terms require the transmission function to have "knowledge" about the fitness function. Under random search, both these terms would be zero. It is this implicit knowledge that is the source of power in genetic algorithms.

It would appear that maximum evolvability would be achieved by setting $\alpha = 1$. However, with $\alpha = 1$, cumulative increases in the frequency of the fittest types from selection cannot occur, and the average search bias would not be able to increase. The average search bias of the population reflects the accumulation over time of the covariance between the search bias and fitness, as seen by how $\overline{\beta}_u(w)$ itself changes in one generation:

$$
\begin{aligned}
\overline{\beta}_u(w)' &= \sum_{j,k} \beta_{jk}(w) \, x'_j \, x'_k \\
&= (1-\alpha)^2 \left[\overline{\beta}_u(w) + \mathrm{Cov}[\beta_{jk}(w), \frac{w_j w_k}{\overline{w}^2}] \right] + O(\alpha)
\end{aligned}
\tag{3.10}
$$

The optimal value of α is therefore a complex question, but has received a good amount of theoretical and empirical attention for special cases [Hesser and Männer 1991, Bäck 1992, Schaffer *et al.* 1989, Davis 1989, Grefenstette 1986].

3.2.4 Price's theorem and the Schema Theorem

None of these theorems on GA performance needed to invoke schemata. The Schema Theorem [Holland 1975] has been viewed as the theoretical foundation explaining the power of genetic algorithms. However, there is nothing in the Schema Theorem that can distinguish a "GA-easy" problem from a "GA-hard" problem; random and needle-in-a-haystack fitness functions will still obey the Schema Theorem, in that short schemata with above average fitness will increase in frequency, even though schemata are unrelated in any way to the fitness function. Mühlenbein [1991] and Radcliffe [1992] have also pointed out that the Schema Theorem does not explain the sources of power in genetic algorithms.

In expositions of how "schema processing" provides GAs with their power, and in the framing of the Building Blocks hypothesis [Goldberg 1989], it is implicit that the regression and search bias terms given in the theorems here will be positive under the recombination operator. In "deceptive" and other GA hard problems, at some point in the population's evolution before the optimum is found, the evolvability vanishes, i.e. the terms in (3.7) go to zero or below. Investigations on the performance of different representations and operators are implicitly studying how to keep (3.7) positive until the optimum is attained.

For example, Hamming cliffs – where single bit changes in classical binary string GAs cause large changes in the real valued parameter encoded by the string – are a problem because large jumps in the parameter space will have lower correlations in fitness for typical problems. Gray code increases the correlation resulting from single bit changes on average.

The fact that theorems 1 to 3 apply to any representation scheme and genetic operator is important, because they give the measure by which representations and operators should be judged. Under some representations, mutation may give greater evolvability over the search space; under other representations, recombination or some other operator may give greater evolvability. The operators that gives the greater evolvability may change as the population evolves.

Moreover, the question of whether non-binary representation or non-classical operators, (i.e. not recombination and mutation) are "legitimate" because they do not fit into the framework of "schema processing", should be seen as a misconception. Any combination of operators and representations that achieve the same evolvability properties over the course of the run are equally good and legitimate.

At least two studies [Manderick, de Weger, and Spiessens 1991, Menczer and Parisi 1992] have attempted to estimate the performance of different genetic operators by examining the correlation in mean fitness between parent and offspring. What Price's theorem shows is that this correlation (as embedded in the covariance expression) is an exact term in the equation for the change of the mean fitness of the population. In a model of a real-valued GA, Qi and Palmieri [Palmieri and Qi 1992, Qi and Palmieri 1992] go further and actually derive a version of Price's theorem in order to give the change in the mean phenotype of the population, using the vector-valued phenotype as the measurement function.

3.3 Dynamics of Genetic Programming

In this section, I look into how the search bias and parent-offspring regression themselves can evolve through the differential proliferation of blocks of code within programs. The dynamics of code proliferation in genetic programming are, in their exact behavior, a formidable analytical problem. So I will describe what I believe are their qualitative aspects, and analyze a simplified model for the dynamics of GP hill-climbing.

My basic conjecture is that blocks of code proliferate within programs based on the shape of the *distribution* of their fitness effects when added to the programs in the population.

The differential proliferation of code can then come to change the distribution of fitness effects resulting from the genetic operator so that the evolvability itself will evolve.

3.3.1 A model of genetic programming dynamics

Here I assume that the only genetic operator is recombination, and no mutation is acting. Let $\mathcal{B}$ be the space of different blocks of code. Let $C(i|j)$ be the probability that the recombination operator picks out a particular block i from a program j. So $\sum_{i \in \mathcal{B}} C(i|j) = 1$, for all $j \in \mathcal{S}$. Then the frequency, p_i, that the operator picks block i from a random program in the population is

$$p_i = \sum_{j \in \mathcal{S}} C(i|j)\, x_j, \ i \in \mathcal{B}.$$

The *marginal fitness*, u_i, (or if one prefers, "schema fitness") of block i is the average of the fitnesses of the programs that contain it, weighted by the frequency of the block within the programs:

$$u_i = \frac{1}{p_i} \sum_{j \in \mathcal{S}} C(i|j)\, w_j x_j, \ i \in \mathcal{B}.$$

With these definitions, one can give (3.9) in a GP form:

$$x_i' = (1 - \alpha)\frac{w_i}{w}\, x_i + \alpha \sum_{j \in \mathcal{S}} \sum_{k \in \mathcal{B}} P(i \leftarrow j, k)\frac{w_j}{w}\, x_j\, \frac{u_k}{w}\, p_k. \tag{3.11}$$

Note that $P(i \leftarrow j, k)$ is no longer symmetrical in arguments j and k. When block k is added to program j to make program i (thus $P(i \leftarrow j, k) > 0$), then as long as $j \neq i$, we know that $C(k|i) > C(k|j)$.

3.3.2 The "constructional" fitness of a block of code

As described in section 3.1.4, under high recombination rates, a block of code in a program that is less fit than average, and on its way to extinction, may nonetheless get sampled and added to another program, so blocks with a high probability of producing a near neutral effect on fitness may be able to proliferate when competing against blocks with large portions of deleterious effects.

Under low recombination rates, however, the population would tend to fix on the single fittest program between recombination events. Deleterious recombinants would usually go extinct before they could be sampled for further recombination. So blocks of code could proliferate only by producing fitness increases when added to the program fixed in the population. In this much more simple situation, which is effectively "GP hill-climbing",

I will show how the program can come to be composed of code that has a high chance of producing fitness increases when duplicated within the program.

To model differential code proliferation in genetic programming, I need to make additional simplifying assumptions. First, I assume that blocks of code are transmitted discretely and not nested within one another. Second, I shall consider only recombination events that add blocks of code, without deleting others. Third, there must be some sort of "regularity" assumption about the properties of a block of code in order for there to be any systematic differential proliferation of different blocks. Here I shall assume that when a given block of code is added to a program, the factor by which it changes the program's fitness has a relatively stationary distribution. This is necessary for it to have any consistent constructional fitness as the population evolves. In reality, this distribution doubtless depends on the stage of evolution of the population, the diversity of programs, and the task domain. These complications are, of course, areas for further investigation.

When a new block of code of type i is created, suppose that, with probability density $f_i(\omega)$, it multiplies the fitness of the organism by a factor ω. This probability density would be the result of the phenotypic properties of the block of code and the nature of the programs it is added to as well as the task domain.

Under small α, the population evolves only when the addition of a block of code increases the fitness of the program. Then, before other code duplications occur, the population will have already fixed on the new, fitter program. The probability, ξ_i, that a copy of a block of code, i, increases the fitness of the program is

$$\xi_k = \int_1^\infty f_k(\omega)\, d\omega = \sum_i F_i(w_j)\, P(i \leftarrow j, k) = \beta_{jk}(w_j) + \mathcal{R}(w_j), \qquad (3.12)$$

for block k being added to program j to produce program i. My major simplifying assumption is that ξ_k is a constant for all programs j it is added to. The value ξ_k is the rate that a block of code gives rise to new copies of itself that are successfully incorporated in the program, and one can therefore think of ξ_k as block k's "constructional fitness".

Let $n_i(t)$ be the number of blocks of code of type i in the program at time t, and set $\alpha \approx 0$ to be the rate of duplication per unit time under the genetic operator, for all blocks of code. The probability that the operator picks block i for duplication is

$$p_i(t) = n_i(t) \,/\, N(t), \text{ where } N(t) = \sum_i n_i(t).$$

Using the probability, ξ_i, that a copy of block i is successfully added to the program, one obtains this differential equation for the change in the composition of the program (approximating the number of blocks of code with a continuum):

$$\frac{d}{dt}n_i(t) = \alpha\,\xi_i\,n_i(t)\,/\,N(t). \tag{3.13}$$

Within the evolving program, the ratio between the frequencies of blocks of code grows exponentially with the difference between their constructional fitnesses:

$$\frac{n_i(t)}{n_j(t)} = \frac{n_i(0)}{n_j(0)}\,e^{\alpha\,(\xi_i-\xi_j)\,\int_0^t d\tau/N(\tau)}.$$

Theorem 4 (Evolution of Evolvability)

The evolvability of the program, which is the average constructional fitness of its blocks of code,

$$\bar{\xi}(t) = \sum_i \xi_i\,p_i(t),$$

— i.e., the chance that a duplicated random block of code increases the fitness of the program — increases at rate

$$\frac{\alpha}{N(t)}\,\mathrm{Var}(\xi) > 0.$$

Proof.

$$
\begin{aligned}
\frac{d}{dt}\bar{\xi}(t) &= \sum_i \xi_i \frac{d}{dt}\left(\frac{n_i(t)}{N(t)}\right) = \frac{\alpha}{N(t)^2}\sum_i \xi_i^2\,n_i(t) - \frac{\alpha}{N(t)^3}\left(\sum_i \xi_i\,n_i(t)\right)^2 \\
&= \frac{\alpha}{N(t)}\left[\sum_i \xi_i^2 p_i(t) - \bar{\xi}(t)^2\right] = \frac{\alpha}{N(t)}\mathrm{Var}(\xi).
\end{aligned}
$$

∎

Thus, the rate of increase in the evolvability of a program is proportional to the variance in the constructional fitnesses of its blocks of code. This result is Fisher's fundamental theorem of Natural Selection [Fisher 1930], but applied not to the *fitness* of the evolving programs, but to their *evolvability*, i.e. the probability that genetic operators acting on the program gives rise to new fitter programs.

The distribution of constructional fitnesses of the blocks of code of the program can be followed as the program grows in size through evolution according to (3.13). An example is plotted in figure 3.1, where the distribution of constructional fitness values among the blocks of the program in generation 0 is compared with the distribution when the program

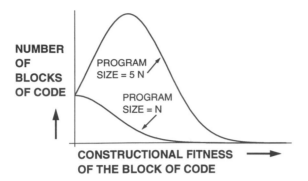

Figure 3.1
A comparison of the distribution of constructional fitnesses of blocks of code within the evolving program. Beginning with an initial program of size N at generation 0, a truncated normal distribution is assumed as an illustration. Under the dynamics of constructional selection, by the time the program has evolved to size $5N$, the distribution has shifted in favor of more constructionally fit blocks of code.

has evolved to 5 times its original size. The blocks with high likelihood of increasing program fitness when duplicated have increased their proportion enormously.

3.3.3   The constructional fitness is distinct from the marginal fitness

This is illustrated with the following example. Consider two programs of equal fitness, each of which has only one subtree which has any chance of increasing its fitness when duplicated. Call the subtree in the first program A, and the subtree in the second program B. All the rest of the code in both programs produces only deleterious results when duplicated.

Suppose that only 0.001 of the possible ways subtree A is duplicated within the program increases its fitness, and suppose that for subtree B the chance is 0.0001 in its program. The values 0.001 and 0.0001 are the constructional fitnesses of the two subtrees. But suppose further that when duplications of subtree A produce a fitness increase, the increase is by a factor of 1.1, while for subtree B that factor is 10. One can then follow the expected number of copies of the two subtrees, and the expected fitnesses of the programs carrying them over time, using the model (3.13). This is plotted in figure 3.2.

As one can see, the program whose subtree, A, has the higher constructional fitness but the smaller fitness effect, falls behind in its expected fitness at first, but later shoots ahead of the program with subtree B, as exponential growth in A takes off. This assumes that the proportion of the programs that are block A or B is still small. Therefore, the constructional fitness is not the same as the marginal fitness of a block of code, and the Schema Theorem does not account for the proliferation of blocks of code within programs.

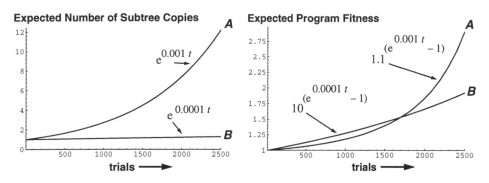

Figure 3.2
Example showing that the "constructional" fitness of a subtree is not the same as its marginal fitness in the population. The equations are given for each plot. On the left, $n_i(t) = e^{\xi_i t}$, $i \in \{A, B\}$. On the right, $w(t) = \sigma_i^{[\exp(\xi_i t) - 1]}$, where σ is the fitness effect factor for the subtree.

3.4   Discussion

3.4.1   Conservative versus exploratory code

In mature GP populations with typical high recombination rates, the blocks of code that proliferate may not be those that have the greatest chance of increasing fitness when added to programs, but may be those that play a more conservative game and trade-off their ability to increase fitness with an ability not to be deleterious when added, e.g nearly neutral code. Thus the constructional fitness of a block of code may not correspond to its evolvability value. This is because high recombination rates induce selection on the central part of a block's fitness distribution as well the upper tail. One can expect that the upper tail (i.e. the probability of a block increasing fitness when added) would shrink as the population becomes more highly adapted, so that the average fitness effect would become a more important statistic for a block of code. It should be possible to measure this in GP runs.

Low recombination rates, while probably not optimal for GP performance, have the benefit that constructional selection will favor blocks of code with the fattest upper tails for their fitness effects, yielding code with greater evolvability. The following ways to limit constructional selection to the upper tails of the distribution under high recombination rates suggest themselves:

"Upward-mobility" selection. When a recombinant offspring is produced, place it in the population only if it is fitter than its parents. This could be implemented with the requirement that it be fitter that both parents, one parent, a mid-parental value, or a supra-parental value. This regime could be exaggerated by adding to the population more than one copy of individuals who exceed their parents' fitnesses by a certain amount. Upward-mobility selection would put deleterious and near-neutral recombinations on an equal footing, thus preventing the proliferation of code that trades-off its evolvability value for an ability not to produce deleterious effects.

Soft brood selection. Suppose one were to produce a number of recombinant offspring from each mating, and use only the fittest of the "brood" as the result of the mating (effectively holding a tournament within the brood). This would be called soft brood selection in population biology — "soft" because selection within the brood does not affect how many offspring the mated pair contribute to the next generation.

If the total probability in the upper tail of the fitness effect distribution for block i were ξ_i, as in (3.12), then with a brood size of n, the probability that the offspring is fitter than its parent (the recipient of the block of code) would be $1 - (1 - \xi_i)^n \approx n\xi_i$ for ξ_i small. Therefore, scaling $n \approx 1/\xi_i$ as the population evolves should mimic the dynamics of constructional selection under small recombination rates. Thus one has the following:

Conjecture: Upward-mobility Selection, Brood Selection, and Evolvability
Under soft brood selection or upward-mobility selection, the distribution of fitness effects from recombination events in genetic programming should increase in both the upper and lower tails compared with conventional selection.

Tests of this conjecture are currently under way. Whether brood selection would be of practical benefit depends on whether the computational overhead of producing larger broods is not better expended on larger population sizes.

"Genetic engineering". A more direct way to counter constructional selection for conservative code is create intelligent recombination operators that seek out code with high evolvability value. This can be done by estimating the evolvability value of different blocks of code based on the sampling history of that code, creating a library of the code with highest evolvability values, and having the operator draw from that library as the source of donor code. This approach is also under development by Walter Tackett (personal communication). Constructing the library requires keeping tallies of the distribution of fitness changes produced by each genetic operation in order to estimate a block of code's evolvability value. It may be profitable as well to make the genetic operator more intelligent by adding context sensitivity. It could examine the arguments and calling nodes of a block of code to see which of them it increased fitness most often with.

3.4.2   Principle 1 applied to genetic programming

In the design of genetic programming algorithms, the point of Principle 1 may be helpful: changes in representations or operators are relevant to the evolutionary dynamics only through how they change the transmission function. This point can be obscured because representations serve a dual function in GP: the representation is used as a computer program to perform the computations upon which fitness will be based, and the representation is operated upon by the genetic operators to produce variants. These two functions of the representation I refer to as *generative* and *variational* aspects, respectively, and it is the variational aspect that is relevant to the evolutionary dynamics.

Two modifications to GP representation provide an illustration: "modules" [Angeline and Pollack 1992] and "automatically defined functions" (ADFs) [Koza 1993].

Angeline and Pollack introduce an operator which replaces a subtree with a single referent (possibly taking arguments), making the subtree into an atom, which they call a "module". The same evolutionary dynamics would result from adding markers that told the genetic operator to avoid crossover within the delimited region. The compression operator therefore changes the variational aspect of the representation.

Koza [1993] has introduced ADFs as a modification to GP representation which has been shown to improve its evolvability by orders of magnitude in a number of task domains. How can we understand why ADFs work? A way to see what ADFs are doing to the variational properties of the representation is to expand the "main" result-producing branch of the program by recursively replacing all ADF calls with the code being executed. Let us then consider what the operator acting on the ADF form of the representation is doing to the expanded form of the representation.

First off, in generation 0, where one has only new, randomly generated programs, the expanded form would show internally repeated structures of code. This would be vanishingly unlikely for random programs where each node and argument were sampled independently from the pool of functions and terminals. But this effect could be achieved by non-independent generation of the branches of the initial population.

In the ADF form of the representation, when the ADF branch undergoes recombination, in the expanded form one would see parallel substitutions in each instance of the repeated code. Thus an operator which recognized repeated instances of a block of code and changed them in parallel would have this same variational property as ADFs. One can translate between two equivalent forms — simple operators acting on more complex representations and more complex operators acting on simple representations.

The success of ADFs indicates that, for a variety of cases, the dimensions of variation produced using ADFs are highly attuned to the dimensions of variation of the task domain. A main research task is to further explore the variational possibilities of representations and operators to find those that are attuned to their task domains. A few suggestions are offered below:

Divergent ADFs. Allow different calls to an ADF to be replaced with their own instantiation of the ADF code, which can evolutionarily diverge from the original ADF. This mimics the process of gene duplication and divergence in biological evolution. The ADF that may be optimal for one calling function may not be optimal for another, so relaxing the constraint may allow better optimization. Alternatively, each ADF could be instantiated by its own code, but a gene-conversion-like operator could produce a tunable degree of information exchange among the family of related ADFs.

"Cassette" recombination[3]. Crossover occurs at both the root and branch boundaries of an internal part of a subtree, so that the set of inputs feeding into a root are unchanged, but the intervening code varies (This genetic operator grew out of a discussion with Kenneth Kinnear where we jointly defined the operator; he calls this operator "modular crossover" in Chapter 6 of this volume). Maintaining the relationship between distal arguments and proximal functions while varying the processing in between might be an advantageous dimension of variation.

Branching and junctions. Included with the set of terminals and functions could be primitives that facilitate the creation of programs with reticulated topologies. A branch would allow the value returned by a subtree to be piped as arguments to several different functions in the program.

Scaling-up. The problem of scaling-up from small "toy" problems to large ones is ubiquitous in artificial intelligence. The ability to scale-up may emerge from genetic programming if one can manage to make scaling-up an evolvability property. This could be attempted by maintaining a constant directional selection for performing at an increasing scale: when the population reached a certain performance level for fitness cases at one scale, fitness cases at the next higher scale could be introduced. Then blocks of code may proliferate which are able to increase, through simple actions of the genetic operator, the scale which the program handles.

3.4.3 Conclusions

In this chapter I have described how an emergent selection phenomenon, the evolution of evolvability, can occur in genetic programming through differential proliferation of blocks of code within programs. In preparing the conceptual framework for this phenomenon, a mathematical analysis of evolvability was given for genetic algorithms in general. This analysis employed Price's Covariance and Selection Theorem to demonstrate how genetic

---

[3]This term is borrowed from genetics, where it refers to the substitution of one block of genes by another through transposition; yeast use it in mating-type switching, for example [Egel, Willer, and Nielsen 1989].

algorithm performance requires that the genetic operator, acting on the representation, produce offspring whose fitness distributions are correlated with their parents' fitnesses. It was discussed how this theorem yields the necessary connection between representations, operators, and fitness that is missing from the Schema Theorem in order to tell how well the genetic algorithm will work.

With this general framework as a backdrop, I illustrated the evolution of evolvability in a simple mathematical model of code proliferation in genetic programming. Out of consideration of the dynamics of genetic programming, a number of suggestions were offered for ways to control the evolution of evolvability. These included GP hill-climbing, "upward-mobility" selection, soft brood selection, and "genetic engineering". Furthermore, other techniques were suggested to enhance the performance of genetic programming, including divergent ADFs, "cassette" recombination, branching and junctions, and scaling-up selection.

3.4.4 Further work

Certainly the various conjectures and suggestions made in this chapter provide directions for further work. But what is really needed are mathematical methods of analysis and simulation models to explore more thoroughly the population genetics of genetic programming. In this chapter I have been mainly concerned with population dynamics and not with the how the algorithmic properties of blocks of code may determine their evolvability value. This is an area that needs investigation. Some phenomena in GP will most likely be found to be primarily population-dynamic phenomena, while others will be found to be more deeply derived from the algorithmic nature of these evolving programs. Further simulations of genetic programming that implement the population dynamics, while controlling the fitness function, would be instrumental in distinguishing them.

Genetic programming, because of the open-ended complexity of what it can evolve, is potentially a rich new area for emergent evolutionary phenomena. Novel approaches will need to be developed before these phenomena, and their implications for genetic programming design, will be well understood.

Acknowledgements

The search for an alternative to the Schema Theorem began at the Santa Fe Institute's 1991 Complex Systems Summer School. Thanks to David Hall for pointing out the precedence of Price's work. This chapter benefited greatly from discussions held at the 1993 International Conference on Genetic Algorithms, and reviews of earlier versions. Thanks especially to Roger Altenberg, Kenneth Kinnear, Walter Tackett, Peter Angeline, Andrew Singleton, and John Koza. Cassette recombination was jointly invented in a discussion with Kenneth Kinnear.

Notice

Any application of the material in this chapter toward military use is expressly against the wishes of the author.

Bibliography

Altenberg, L. (1985), Knowledge representation in the genome: new genes, exons, and pleiotropy. *Genetics* 110, supplement: s41. Abstract of paper presented at the 1985 Meeting of the Genetics Society of America.

Altenberg, L. and M. W. Feldman (1987), Selection, generalized transmission, and the evolution of modifier genes. I. The reduction principle. *Genetics* 117: 559–572.

Angeline, P. J. (1994), Genetic programming and emergent intelligence. In K. E. Kinnear, editor, *Advances in Genetic Programming*, Cambridge, MA. MIT Press.

Angeline, P. J. and J. B. Pollack (1994), Coevolving high-level representations. In C. G. Langton, editor, *Artificial Life III*, Menlo Park, CA. Addison-Wesley. In press.

Antonisse, J. (1989), A new interpretation of schema notation that overturns the binary coding constraint. In J. D. Schaffer, editor, *Proceedings of the Third International Conference on Genetic Algorithms*, pages 86–91, San Mateo, CA. Morgan Kaufmann.

Bäck, T. (1992), Self-adaptation in genetic algorithms. In F. Varela and P. Bourgine, editors, *Toward a Practice of Autonomous Systems. Proceedings of the First European Conference on Artificial Life*, pages 263–271, Cambridge, MA. MIT Press.

Bäck, T., F. Hoffmeister, and H.-P. Schwefel (1991), A survey of evolution strategies. In R. K. Belew and L. B. Booker, editors, *Proceedings of the Fourth International Conference on Genetic Algorithms*, pages 2–9, San Mateo, CA. Morgan Kaufmann.

Barwise, J. and J. Perry (1983), *Situations and Attitudes*. M.I.T. Press, Boston, pages 292–295.

Bergman, A. and M. W. Feldman (1992), Recombination dynamics and the fitness landscape. *Physica D* 56(1): 57–67.

Brandon, R. N. (1988), The levels of selection a hierarchy of interactors. In H. C. Plotkin, editor, *The Role Of Behavior In Evolution*, pages 51–72. M.I.T. Press, Cambridge, Massachusetts.

Brandon, R. N. (1990), *Adaptation and Environment*. Princeton University Press, Princeton, pages 83–84.

Cavalli-Sforza, L. L. and M. W. Feldman (1976), Evolution of continuous variation: direct approach through joint distribution of genotypes and phenotypes. *Proceedings of the National Academy of Science U.S.A.* 73: 1689–1692.

Davis, L. (1989), Adapting operator probabilities in genetic algorithms. In J. D. Schaffer, editor, *Proceedings of the Third International Conference on Genetic Algorithms*, pages 61–69, San Mateo, CA. Morgan Kaufmann.

Dawkins, R. (1976), *The Selfish Gene*. Oxford University Press, Oxford.

Egel, R., M. Willer, and O. Nielsen (1989), Unblocking of meiotic crossing-over between the silent mating type cassettes of fission yeast, conditioned by the recessive, pleiotropic mutant rik1. *Current Genetics* 15(6): 407–410.

Fisher, R. A. (1930), *The Genetical Theory of Natural Selection*. Clarendon Press, Oxford, pages 30–37.

Forrest, S. and M. Mitchell (1993), Relative building-block fitness and the building-block hypothesis. In L. D. Whitley, editor, *Foundations of Genetic Algorithms 2*, pages 109–126. Morgan Kaufmann, San Mateo, CA.

Goldberg, D. (1989), *Genetic Algorithms in Search, Optimization and Machine Learning*. Addison Wesley.

Goldberg, D., K. Deb, and B. Korb (1990), Messy genetic algorithms revisited: studies in mixed size and scale. *Complex Systems* 4(4): 415–444.

Grefenstette, J. J. (1986), Optimization of control parameters for genetic algorithms. *IEEE Transactions on Systems, Man and Cybernetics* SMC-16(1): 122–128.

Handley, S. G. (1994), The automatic generation of plans for a mobile robot via genetic programming with automatically defined functions. In K. E. Kinnear, editor, *Advances in Genetic Programming*, Cambridge, MA. MIT Press.

Hesser, J. and R. Männer (1991), Towards an optimal mutation probability for genetic algorithms. In H.-P. Schwefel and R. Männer, editors, *Parallel Problem Solving from Nature*, pages 23–32, Berlin. Springer-Verlag.

Holland, J. H. (1975), *Adaptation in Natural and Artificial Systems*. University of Michigan Press, Ann Arbor.

Karlin, S. (1979), Models of multifactorial inheritance: I, Multivariate formulations and basic convergence results. *Theoretical Population Biology* 15: 308–355.

Kinnear, K. E. (1994), Alternatives in automatic function definition: a comparison of performance. In K. E. Kinnear, editor, *Advances in Genetic Programming*, Cambridge, MA. MIT Press.

Koza, J. R. (1992), *Genetic Programming: On the Programming of Computers by Means of Natural Selection*. MIT Press, Cambridge, MA.

Koza, J. R. (1993), Hierarchical automatic function definition in genetic programming. In L. D. Whitley, editor, *Foundations of Genetic Algorithms 2*, pages 297–318. Morgan Kaufmann, San Mateo, CA.

Levenick, J. R. (1991), Inserting introns improves genetic algorithm success rate: Taking a cue from biology. In R. K. Belew and L. B. Booker, editors, *Proceedings of the Fourth International Conference on Genetic Algorithms*, pages 123–127, San Mateo, CA. Morgan Kaufmann.

Lyubich, Y. I. (1992), *Mathematical Structures in Population Genetics*. Springer-Verlag, New York, pages 291–306.

Manderick, B., M. de Weger, and P. Spiessens (1991), The genetic algorithm and the structure of the fitness landscape. In R. K. Belew and L. B. Booker, editors, *Proceedings of the Fourth International Conference on Genetic Algorithms*, pages 143–150, San Mateo, CA. Morgan Kaufmann Publishers.

Menczer, F. and D. Parisi (1992), Evidence of hyperplanes in the genetic learning of neural networks. *Biological Cybernetics* 66(3): 283–289.

Mülenbein, H. (1991), Evolution in time and space — the parallel genetic algorithm. In G. Rawlins, editor, *Foundations of Genetic Algorithms*, pages 316–338, San Mateo, CA. Morgan Kaufmann.

Oakley, H. (1994), Two scientific applications of genetic programming: stack filters and non-linear equation fitting to chaotic data. In K. E. Kinnear, editor, *Advances in Genetic Programming*, Cambridge, MA. MIT Press.

O'Neil, E. and C. Shaefer (1989), The ARGOT strategy III: the BBN Butterfly multiprocessor. In J. Martin and S. Lundstrom, editors, *Proceedings of Supercomputing, Vol.II: Science and Applications*, pages 214–227. IEEE Computer Society Press.

Palmieri, F. and X. Qi (1992), Analyses of Darwinian optimization algorithms in continuous space. Technical Report EE-92-01, Department of Electrical and Systems Engineering U-157, University of Connecticut, Storrs, CT 06269-3157, Available by ftp from ftp@roma.eng2.uconn.edu.

Pitney, G., T. Smith, and D. Greenwood (1990), Genetic design of processing elements for path planning networks. In *IJCNN International Joint Conference on Neural Networks*, volume 3, pages 925–932, New York.

Price, G. R. (1970), Selection and covariance. *Nature* 227: 520–521.

Price, G. R. (1972), Extension of covariance selection mathematics. *Annals of Human Genetics* 35: 485–489.

Qi, X. and F. Palmieri (1992), General properties of genetic algorithms in the euclidean space with adaptive mutation and crossover. Technical Report EE-92-04, Department of Electrical and Systems Engineering U-157, University of Connecticut, Storrs, CT 06269-3157, Available by ftp from ftp@roma.eng2.uconn.edu.

Radcliffe, N. J. (1992), Non-linear genetic representations. In R. Männer and B. Manderick, editors, *Parallel Problem Solving from Nature, 2*, pages 259–268, Amsterdam. North-Holland.

Salmon, W. C. (1971), *Statistical Explanation and Statistical Relevance*. University of Pittsburgh Press, Pittsburgh.

Schaffer, J. and A. Morishima (1987), An adaptive crossover distribution mechanism for genetic algorithms. In J. Grefenstette, editor, *Genetic Algorithms and their Applications: Proceedings of the Second International Conference on Genetic Algorithms*, pages 36–40, Hillsdale, NJ. Lawrence Erlbaum Associates.

Schaffer, J. D., R. A. Caruana, L. J. Eshelman, and R. Das (1989), A study of control parameters affecting online performance of genetic algorithms for function optimization. In J. D. Schaffer, editor, *Proceedings of the Third International Conference on Genetic Algorithms*, pages 51–60, San Mateo, CA. Morgan Kaufmann.

Schraudolph, N. and R. Belew (1992), Dynamic parameter encoding for genetic algorithms. *Machine Learning* 9(1): 9–21.

Shaeffer, C. (1987), The argot strategy: adaptive representation genetic optimizer technique. In J. Grefenstette, editor, *Genetic Algorithms and their Applications: Proceedings of the Second International Conference on Genetic Algorithms*, pages 50–58, Hillsdale, NJ. Lawrence Erlbaum Associates.

Slatkin, M. (1970), Selection and polygenic characters. *Proceedings of the National Academy of Sciences U.S.A.* 66: 87–93.

Szarkowicz, D. (1991), A multi-stage adaptive-coding genetic algorithm for design applications. In D. Pace, editor, *Proceedings of the 1991 Summer Computer Simulation Conference*, pages 138–144, San Diego, CA.

Tackett, W. A. (1993), Genetic programming for feature discovery and image discrimination. In S. Forrest, editor, *Proceedings of the Fifth International Conference on Genetic Algorithms*, pages 303–309, San Mateo, CA. Morgan Kaufmann.

Teller, A. (1994), The evolution of mental models. In K. E. Kinnear, editor, *Advances in Genetic Programming*, Cambridge, MA. MIT Press.

4 Genetic Programming and Emergent Intelligence

Peter J. Angeline

Genetic programming is but one of several problem solving methods based on a computational analogy to natural evolution. Such algorithms, collectively titled *evolutionary computations*, embody dynamics that permit task specific knowledge to *emerge* while solving the problem. In contrast to the traditional knowledge representations of artificial intelligence, this method of problem solving is termed *emergent intelligence*. This chapter describes some of the basics of emergent intelligence, its implementation in evolutionary computations, and its contributions to genetic programming. Demonstrations and guidelines on how to exploit emergent intelligence to extend the problem solving capabilities of genetic programming and other evolutionary computations are also presented.

4.1 Prelude

Evolutionary computations, including *genetic algorithms* [Holland 1975; Goldberg 1989a], *evolution strategies* [Bäck, Hoffmiester and Schwefel 1992] and *evolutionary programming* [Fogel, Owens and Walsh 1966; Fogel 1992] are popular methods for problem solving and function optimization. Each of these techniques employ models of natural evolution at the core of their computational machinery. *Genetic programming* (GP) [Koza 1992], a variant of genetic algorithms, is an especially interesting form of computational problem solving. In GP, a population of programs, represented by expression trees in a task specific language, evolve to accomplish the desired task. But rather than rating the content of the trees, as is normally done in genetic algorithms, genetic programs are graded on their behavior when run as a program. Genetic programs often acquire elegant solutions with a degree of subtlety not anticipated by the programmer.

While using evolution as a model for computational problem solving is in itself interesting, it does nothing to illuminate the advantages of these techniques over standard problem solving methods. Understanding exactly how evolutionary computations differ from standard techniques should provide insight into the nature of evolutionary computations and may enlighten us about human problem solving.

This chapter argues that evolutionary computations belong to a unique class of problem solvers that display *emergent intelligence*, a method of problem solving that allows task specific knowledge to emerge from the interaction of the problem solver and a task description, here represented by the fitness function. This chapter examines what is distinct about evolutionary computations and emergent intelligence, why they represent a novel approach to computational problem solving and how that difference is manifested in genetic programming. Demonstrations and guidelines showing how to exploit emergent intelligence to extend the problem solving capabilities in genetic programming and other evolutionary computations are also presented. The next section begins with a review of traditional computational problem solving philosophy as represented by symbolic artificial intelligence (AI).

4.2 General Computational Problem Solving

General Computational problem solving is usually considered the jurisdiction of classical artificial intelligence (AI), a field which is as amorphous as the symbolic representations that dominate its techniques. Problem solving in AI takes one of two forms: computational models of humans solving problems and the construction of programs that solve problems as well or better than humans. Clearly, the techniques of genetic programming and all evolutionary computations fit into the latter methodology of building intelligent programs.

4.2.1 Weak and Strong Methods

At a very high level, AI problem solving methods divides into two distinct categories: *weak* and *strong*. A strong method is one that contains a significant amount of task specific knowledge in order to solve it. In contrast, a weak method requires little or no knowledge about a task. As the name implies, strong methods are more powerful algorithms than weak methods.

Strong methods can take the form of an analytic solution or an expert system using a knowledge base of a billion rules. Because their content is directed toward solving a particular problem, they are only narrowly applicable. However, their task specific knowledge often allows these techniques to find solutions that are more accurate with less computational effort. Analytic methods and rulebases with a billion rules supply the same sort of strength for solving a particular task since they contain the same problem solving information at the knowledge level [Newell 1982]. However, general analytic methods and billion element rulebases are not easily created with current techniques. Standard symbolic AI requires a *knowledge engineer* to extract the task specific knowledge out of a domain or an expert so it can be inserted into the knowledge base or generalized into an analytical solution. In this chapter, any information that is extracted from the task domain prior to solving the task, including rules or analytical relations, are referred to as *explicit* knowledge.

Weak methods, which include *best-first search*, *heuristic search*, *hill climbing*, *means-ends analysis* and others, are broadly applicable and extendable [Rich 1982; Winston 1983]. AI uses weak methods as general algorithmic shells in which to place knowledge as it is identified and codified. Interactive refinement and/or analytic generalization of the knowledge base is intended to lead to a strong method for the task.

4.2.2 Credit Assignment

The problem solving ability of an algorithm is chiefly determined by how "intelligently" the algorithm traverses the possible solutions in the space. One possibility is to perform *credit assignment* on the representational components of each solution encountered and bias future solutions towards organizations of representational components that perform well together. Credit assignment typically refers to the rating of a structure's components as being beneficial or harmful to solving the problem. The term is most associated with machine learning programs, however, a search algorithm also uses an implicit form of credit assignment when selecting which element in the search space to visit next. Poor search algorithms, such as random search, use inaccurate credit assignment, if any, to guide them through the search space. A more "intelligent" search mechanism embodies an accurate credit assignment mechanism and manipulates only problematic representational components to traverse the space.

Exactly how to assign credit to components of a particular representation is often a difficult problem. Occasionally, if enough assumptions are made, analytical solutions to credit assignment for a particular problem type and representation can be developed, as in the case of neural networks and back-propagation [Rumelhart, Hinton and Williams 1986].

In standard AI techniques, credit assignment is assumed to be task specific and must be engineered as explicit knowledge. In order to be general, weak methods make no commitments to specific operators that manipulate a specific representation [Winston 1983; Rich 1982] and thus make no commitment to credit assignment. At most, they supply a general algorithm for traversing a set of possible problem solutions, often called the *problem space*, in a orderly manner. Task specific knowledge specifies how to traverse the problem space "intelligently" [Winston 1983; Rich 1982].

AI's emphasis on explicit knowledge and strong methods as the central techniques of the field creates several problems. First, explicit knowledge becomes the commodity that determines the success of the technique. If the amount of knowledge required to adequately solve a problem is large and difficult to come by, then only a Herculean effort by the knowledge engineer will lead to a successful solution. Additionally, because knowledge is usually represented in discrete, situation specific chunks, knowledge bases often require significant testing and updating. It is common for a knowledge engineer to create a system that solves only the specific problem instances used to build the system and is too brittle to be applied to situations with even minimal differences. In fact, a complete knowledge base for some problems may be too large to write down. Lastly, if and when the knowledge base is filled out, it may be unmanageable both to maintain and to effectively search for appropriate knowledge to apply to a particular problem. Large quantities of explicit knowledge leads to scaling problems that limit its use. Consequently, AI's concentration on strong methods as its main technique has lead to the realization that task specific knowledge is not a cheap commodity.

Traditional AI's reliance on explicit knowledge is a consequence of its strong method emphasis and restricts its problem solving to be highly task specific. In a dynamic world, each task, even each instance of solving a particular task, has characteristics that are unique and changing. It may be that the only manner to decipher and explicitly encode the variety of combinations in a knowledge base is to solve a significant portion of the possible problems beforehand and encode all the various characteristics with information on how to address each individually. While the program performs intelligently on the task, it does so in a purely static manner.

The key to intelligent problem solving lies in supplying an ability for the program to react dynamically as information about solving the task is obtained. This allows the method to opportunistically adapt to the idiosyncracies of the current problem and its variants. If such a dynamic can be identified, then a successful program will not require explicit knowledge for handling the minute differences between problem instances and the knowledge associated with these differences can be removed.

4.2.3 Evolutionary Weak Methods and Empirical Credit Assignment

The credit assignment mechanism in evolutionary computations is representation specific rather than task specific as in AI's strong methods. As a result, evolutionary computations represent a distinct class of problem solving algorithms called *evolutionary weak methods*. By incorporating representation specific operators and *empirical credit assignment*, the form of credit assignment unique to these algorithms, evolutionary computations begin "weak" but gather knowledge about the task as they investigate additional points in the problem space. This modification of search behavior allows a "stronger" search as more elements of the problem space are visited. Such is the essence of an *adaptive algorithm* [Holland 1975]. The adaptive character of evolutionary weak methods distinguish these algorithms from traditional AI weak methods that never augment their traversal of the problem space with information gained from the intermediate search points.

The advantage of evolutionary computations over standard AI weak methods is that rather than relying on strong task specific knowledge to assign credit, a *weak* form of credit assignment, empirical credit assignment, is built into the dynamics of the algorithm.

This built-in mechanism for credit assignment in evolutionary computations provides support for descriptions of evolutionary computations as "strong weak methods."

At a high level, all evolutionary computations perform a simple SELECT-COPY-MODIFY loop over the population each generation. SELECT identifies the number of copies each member of the current population produces in the next population based on a relative ranking by the fitness function. COPY makes identical copies of the specified numbers for each member. Finally, MODIFY applies the reproductive operators. In order for the evolutionary weak method to be task independent, the operators must encode representation specific modifications rather than task specific modifications.

Given that the only information evolutionary computations receive are the relative rankings of the population, components of the representation are never explicitly assigned credit. Instead, evolutionary computations take an empirical approach to inducing a particular features's applicability to the problem. Probabilistic applications of the representation specific operators create new points in the problem space from the current population members. Each new population member can be considered an experiment testing the features preserved by the applied operator against those that are manipulated. Successful experiments, i.e. ones that construct comparatively fit individuals, are added to the population and used as the basis of future experimentation. Such an empirical approach leads to the selection of features that are both beneficial for solving the problem and conducive to manipulation by the operators being used.

Because this form of credit assignment is holistic, meaning that credit is assigned to only the complete individual and not its representational components, empirical credit assignment selects for any abstract feature of an individual that is preserved by the reproduction operators. This is distinct from Holland's schema theory [Holland 1975] for simple genetic algorithms. Holland's theory assumes that specific representational components are selected by the genetic algorithm and preserved in the population. But there is nothing in the dynamics of any evolutionary computation that permits the algorithm to distinguish between two individuals with distinct structures but identical fitness ratings. Thus it cannot be representational components that are manipulated by the algorithm but the more abstract *phenotypic* features that are being preserved. Interestingly enough, when a feature based encoding is used, such as is often the case in simple genetic algorithms, there is no distinction between genotypic and phenotypic features. But when the interpretation of the representation is not tied to a bit's position, such as in genetic programming, the mapping between structure and behavior is more complex. Selection for genotypic components in evolutionary computations using these more complex interpretation functions can not occur because the genotypic features are representationally inaccessible. In other words, it is improper to speak of "schema" [Holland 1975] in evolutionary weak methods such as genetic programming since there is no simple correspondence between structures in the genotype and the features of the phenotype.

It is important to note that the form of credit assigned in empirical credit assignment is the amount of the population the feature occupies. This makes a population based search algorithm a necessity. However, this does not imply that standard weak methods that use populations, such as *best-first search* and *beam search*, embody empirical credit assignment: a population is necessary but not sufficient. In empirical credit assignment, the operators must map between members of the problem space so that appropriate features are preserved in the subsequent populations. To the extent that best-first or beam search algorithms use representation specific operators and not task specific operators, then these weak methods would also be using a form of empirical credit assignment. However, if the operators that create new population members from old employ task specific knowledge, then the assignment of credit is strong.

4.2.4 Emergent Intelligence

Emergent intelligence (EI) is a method of reducing or removing the reliance on explicit knowledge in a problem solver by favoring instead the *emergence* of problem specific constraints as a direct consequence of the action of problem solving. Emergent phenomena are defined as local interactions creating global properties [Forrest 1991]. Rather than providing explicit knowledge to the problem solver, EI algorithms allow task specific knowledge to remain as natural constraints in the description of the task. These techniques work "as if" they had explicit knowledge because the original source of the constraints, the problem specification itself, is an integral part of the problem solving algorithm. Using empirical credit assignment, evolutionary weak methods opportunistically discover solutions that fill the various constraints of a task without the need for explicit knowledge.

The interaction of a problem solver with the problem environment allows appropriate knowledge to "emerge" as a by-product. From the outside, a program that has a reserve of a billion rules and knows how to locate those pertinent to the current problem is no different than one that allows the task environment itself to limit the computation in a similar manner. Both methods are equally intelligent, since both solve the problem. Both are also equivalent at the knowledge level since they have access to exactly the same knowledge. The difference lies only in the representation of the pertinent task specific knowledge.

The distinction between standard AI techniques and emergent intelligence are best illustrated by an example. Consider the pattern of a snowflake. Each snowflake has at its center a crumb of dust that forms the snowflake's skeleton. Ice crystals begin forming on the dust particle and successive layers are deposited with certain symmetrical patterns. The fluctuation of humidity, pressure, and temperature at the exposed surfaces of the flake at the instant of the next layer's crystallization determine the exact shape of the patterns. The uniqueness of a snowflake's shape is due to the sensitivity of the process constructing the flake to these variations. The snowflake's symmetry emerges as a natural constraint of the current structure interacting with the environment, i.e. snowflakes form "as if" they had knowledge of symmetry. This is the sense that emergent intelligence allows task specific knowledge to emerge "as if" the algorithm had the knowledge explicitly represented. Because the constraint is a natural consequence of the interaction of the snowflake's structure with the environment there is no need to represent it explicitly. It emerges naturally as a consequence of the interaction. In contrast, the standard AI strong method for determining the shape of a snowflake would be to place a rule in the knowledge base that all snowflakes must be symmetrical along with other rules for deducing a snowflake's final shape directly from the myriad of variations of humidity, pressure and temperature.

In short, using emergent intelligence allows the removal of explicit knowledge that is a natural consequence of the problem solving process interacting with the task environment. By allowing the task environment to be an integral component of the problem solving algorithm, all the natural constraints, including those too subtle for a knowledge engineer to extract, are available to the algorithm and emerge at appropriate moments while solving problems.

4.3 Genetic Programming: The Inside Story

Genetic programs, as defined in [Koza 1992], have a number of significant properties, both attributed and innate. Some of the attributed properties, especially in comparison to simple genetic algorithms, are not valid while others have been misdiagnosed or gone unnoticed completely. The following sections attempts to place the properties of GP into a more consistent perspective.

4.3.1 Comparing Genetic Algorithms and Genetic Programming

Genetic programs are often said to be distinct from simple genetic algorithms due to the dynamic nature of their tree representations and their interpretation as programs. Simple genetic algorithms [Holland 1975; Goldberg 1989a], which use fixed length bit strings, n-point crossover and point mutation are in principle both representationally and operationally general. In fact, there is no theoretical reason why simple genetic algorithms could not construct genetic programs with similar properties. However, as so often happens in science, the practice of genetic algorithms focuses on only a small portion of what is possible As we will see, genetic programming fills a previously unoccupied niche in genetic algorithms, which accounts for much of its interest as a technique.

4.3.1.1 Representation of Genotypes

One of the claimed advantages of genetic programs over genetic algorithms is its dynamic tree representation [Koza 1992; Angeline and Pollack 1993a]. However, this is not strictly accurate. Consider that a tree with known maximal depth and branching factor is easily represented within a fixed-length bit string. In genetic programs, there is always a practical limitation to tree size in order to keep the problem computationally tractable. Size limitations either come from a defined maximum depth, as in [Koza 1992] and [Angeline and Pollack 1993a], or from the virtual size of the computer's memory. The representation of the "dynamic" tree is done by simply using only the amount of the fixed-length bit string necessary to represent the tree. As the tree grows, more of the bit string is used. Jefferson's group [Jefferson et al. 1992] use such an approach in their genetic algorithm that evolves finite state machines. The designation of the expression tree representation as being dynamic is then only a logical convenience and offers no actual representational advantage. However, dynamic allocation of computational resources on an as needed basis does allow an implementational advantage over extremely large fixed-length strings.

4.3.1.2 Complexity of Interpretation

The fitness function in a genetic algorithm uses an *interpretation function* to convert the bit string representation into the evaluated form. For instance, in the classic experiments of DeJong [DeJong 1975], the bits were interpreted as binary numbers and passed as parameters to the objective functions being optimized. What makes genetic algorithms general is that they manipulate the bit strings without any regard to their eventual interpretation, i.e., their operators are representation specific rather than task specific.

Interpretation functions are typically very simple, usually a conversion from a bit string to a real number as in [DeJong 1975]. The typical interpretation function uses a *positional encoding* [Angeline and Pollack 1993a; Angeline and Pollack 1992] where the semantics of the interpretation are tied to the position of the bit in the genotype rather than its content. As a result, the interpretation of a bit string is likely to be a simple combination of the various positions, similar to a union of independent features. Positional encodings also force the programmer to organize the representation so that bit positions with related semantics are reasonably adjacent and have a lower probability of being separated by a crossover operation. In so called "messy" genetic algorithms [Goldberg, Korb and Deb

1989] where the organization of the genotype is allowed to evolve, a marker designates the "original" position of the allele and thus its semantics.

In genetic programming, the interpretation of a particular expression tree is performed without consideration to the position of the allele. Thus, as in computer programs, the interpretation of the allele "if" in a genetic program is the same regardless of its position in the expression tree. The arbitrarily complex association between an expression and its interpreted behavior allows for an unlimited variety of dynamics to emerge naturally from the evolutionary process. Again, while there is in principle no reason why a standard genetic algorithm could not use a similar interpretation functions, this has not been the practice.

4.3.1.3 Syntax Preserving Crossover

While it is always true that manipulating a bit string representation with operators that disregard its interpretation is the most general method of evolving solutions, to assume it is invariably the most accurate and the most efficient method is counter to both intuition and experimental results. Specialized genetic algorithms such as [Grefenstette 1989], [Michalewicz 1993] and other evolutionary computations such as evolutionary programs [Fogel 1992; Angeline, Saunders and Pollack 1994] make great use of operators that are more friendly to the representation's eventual interpretation. In essence, these evolutionary computations manipulate the evaluated representation more directly. This approach either eliminates the interpretation function or compensates for it with appropriate operators.

Genetic programming employs a similar strategy: a crossover operator that is sensitive to the expression tree structure it is manipulating. This simple crossover scheme preserves the syntactic constraints of the representation by swapping only complete subtrees between parents. Adding the ability to identify complete subtrees to GP's crossover operator ensures that only syntactically valid programs are constructed, removing all syntactically invalid programs from the search.

In actuality, GP's use of a crossover operator that preserves the syntactic constraints of the representation language is the only true distinction between standard genetic algorithms and genetic programming. More importantly, the conceptual conveniences of a dynamic representation, non-positional interpretations, and direct manipulation of representation, highlights the narrow focus of previous genetic algorithms research. Genetic programming's most significant contribution to the field may be to re-emphasize the variety of options not yet explored.

4.3.2 Innate Emergent Intelligence in GP

The interaction of the dynamic representation and non-positional interpretation with the dynamics of empirical credit assignment in genetic programming provide some innate emergent properties that assist in the acquisition of solutions. These properties emerge not because they were designed into the genetic programming process but because the dynamics of the genetic program determine them to be useful or necessary for success.

4.3.2.1 Emergence of Introns

One of the most misinterpreted aspects of genetic programming, and one that shows very clearly the tension between human programming preconceptions and the dynamics of GP, is the more than occasional appearance of redundancy in evolved genetic programs. Several solutions evolved by Koza [Koza 1992] contain semantically extraneous components. These components are unnecessary since they can be removed from the program without altering the solution the program represents. Such superfluous constructions appear in the resulting expression trees so often that Koza uses an automatic editor to remove them at the conclusion of a run.

The ubiquity of redundant constructs suggests a more basic truth about their nature. There is an advantage from the viewpoint of the evolving genetic program for syntactically redundant constructions that are semantically transparent. Because the redundant functions do not effect the semantics of the sub-expression, if a crossover operation is performed such that it splits the redundant components, the semantics of the removed subtree are still preserved.

These redundant functions serve as *introns* for the evolving subexpression giving the blind crossover operator a higher probability of capturing the complete subexpression intact. An intron is a portion of a chromosome that is never expressed and provides spacing between the genes. Levenick [Levenick 1991] shows that introducing introns into a genetic algorithm's representation improves the success rate as much as a factor of 10. Work by Forrest and Mitchell [Forrest and Mitchell 1992], investigating a class of fitness functions called the *royal road functions* showed the standard genetic algorithm could only solve their problem when introns were introduced into the representation. When padding separates expressed components of the bit string, crossover has a better chance of transferring a substring intact. The situation is the same in genetic programs. By introducing redundant structures that are semantically unexpressed, crossover in GP has a better chance of transferring a complete subtree The advantage of GP introns is that they, they emerge naturally from the dynamics of the algorithm rather than being designed into the representation. It is important then to not impede this emergent property as it may be crucial to the successful development of genetic programs.

4.3.2.2 Emergent Diploidy and Dominance

Another interesting form of emergence occurs when the primitive language for a genetic program includes a conditional statement with semantics of the form *if <condition> then <action1> else <action2>*. When this statement is available, the otherwise haploid structure of the genetic program create a task dependent diploidic representation with some interesting dominance characteristics. Consider that when the conditional is executed, the *<condition>* is first evaluated and based on its result either *<action1>* or *<action2>* is evaluated. In the event that the *<condition>* is independent of the state of the environment, then the conditional represents a static form of diploidy with the dominant action determined by the *<condition>*. For instance, with a *<condition>* that always computes to false, *<action2>* would be the dominant subtree. The "recessive" subtree is never executed but, in the matter of a few generations, could become dominate in an offspring if required. This is no different than the form of diploidy and dominance explicitly designed into some genetic algorithm representations [Goldberg 1989a].

Unlike standard forms of dominance computation, dominance in these genetic programs is not designed into the representation but emerges dynamically. The *<condition>* is local to the conditional statement and thus is itself mutable by the evolutionary process. Consequently, the dominance of the actions is specific to the individual. Beneficial dominance relations emerge from the interaction of fit individuals with the task. In addition, the dominance calculation need not be static but can be dependent on the context of the situation. When a *<condition>* uses elements of the environment's state, then the dominance of the various actions is determined by the context of the environment. This permits an environment specific form of dominance.

4.4 Exploiting Emergent Intelligence

The last section described emergent phenomena that are implicit in the dynamics of genetic programs. However, if emergent intelligence is to be useful as a technique then it must be possible to extend the dynamics of evolutionary weak methods such as genetic programming to allow additional features to emerge. The following sections describe some examples of modifications to genetic programs that harness inherent dynamics for emergent intelligence in problem solving.

4.4.1 Emergent Problem Decomposition in Genetic Programs

One of the most important aspects of problem solving is its hierarchical nature. Often, it is prudent to decompose a difficult problem into several more simple subproblems, each of which may be more easily solved. Means-ends analysis [Rich 1982; Winston 1983], a standard AI weak method, embodies this approach to solving problems. In means-ends analysis, problem specific knowledge is used to determine both how to decompose a problem and how to solve each identified sub-problem, possibly by further decomposition. The General Problem Solver (GPS) [Newell and Simon 1963] uses this weak method as its core computational mechanism. As in all weak methods, GPS uses task specific knowledge, in this case to determine the decomposition of problems and their solutions. SOAR [Newell 1992] is in a sense the ultimate version of GPS. It uses task specific knowledge together with a technique called *chunking* to determine decompositions of the task [Newell 1992]. Whenever no production rule is available in SOAR to fulfill any of the current outstanding goals, a default production rule creates appropriate subgoals. Whenever SOAR accomplishes a subgoal, the conditions that lead to the subgoal's creation are "chunked" together with the resulting state changes into a new production rule. When a similar set of conditions occur, SOAR uses the new "chunk" rather than recreating the subgoal. Given a particular problem space, SOAR always creates the same decompositions because of its static, task specific knowledge for when to create subgoals and hence new chunks.

4.4.1.1 The Genetic Library Builder

The Genetic Library Builder (GLiB) [Angeline and Pollack 1993a; Angeline and Pollack 1992] is an extended genetic program that uses a form of emergent intelligence to construct solutions with task specific decompositions without relying on explicit knowledge. In order to allow modular genetic programs to emerge from the dynamics of GP, only two mutation operators need to be added. The first mutation is *compression* which defines a new module. Compression begins by randomly selecting a subtree from the parent. This subtree becomes the body of a newly defined module that is added to the *genetic library,* a logical storehouse for the definitions of all modules in the system. This library associates the chosen subtree with a unique identifier that becomes the module's name. The subtree is then extracted from the parent and replaced by the newly defined module's name. When a genetic program with a compressed module is evaluated, the definition of the module is retrieved from the genetic library and executed as though the original subtree were in the genetic program at the point the module's name appears. Thus the compression operator performs only a syntactic modification to the genetic program rather than a semantic one.

Exactly how a subtree is selected is dependent on the compression method used. One technique, dubbed *depth compression*, chooses a maximum depth and a random node in the tree as the root of the module. Any branch of the subtree with depth greater than the maximum depth becomes a parameter to the module, as shown in Figure 4.1. Another modularization method is *leaf compression*, which again selects a random node as the root

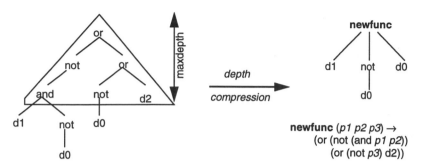

Figure 4.1
GLiB's Depth compression applied to a subtree larger than the maximum depth. Those portions of the selected subtree that fall below the selected depth are used as parameters to the defined function. Expansion is the reverse of this process.

of the subtree. However, in leaf compression, each unique leaf is replaced with a unique parameter name, as shown in Figure 4.2.

The second mutation operator added to GLiB is the inverse operator to compress, called *expand*. When applied, expand searches the parent for all modules at the top level and selects one at random. The chosen module name is then replaced by its definition stored in the genetic library. Because mutations are executed locally in an individual, other population members, which also refer to the module being expanded, are not affected. This operator is exactly the reverse of the process shown in Figures 4.1 and 4.2

The combination of the compression and expansion mutations allow for the non-monotonic development of the solution's decomposition. For instance, a compressed module can be created and the genetic program can be evolved further using the compressed module as an immutable component of its decomposition. Once further refinement is made to the program, the module can be expanded and modified. Such non-monotonic manipulations of genetic material may be indispensable when solving complex problems. Similar non-monotonic decompositions are not possible in SOAR since once a chunk is created it always present as a production rule in the system.

Hierarchically decomposed solutions form in GLiB when a subtree containing an already compressed module is itself compressed by a subsequent application of the compress mutation. Over the course of the evolution of the solution, the hierarchy deepens and widens as more compressions are performed. Because expansion only works on the highest level of modules in an individual, modules further down the hierarchy are protected from being expanded and further manipulated. Thus the problem is decomposed

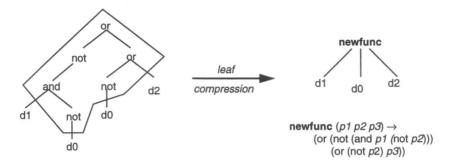

Figure 4.2
Leaf compression applied to a selected subtree. Note that identical leaves are represented by the same parameter name in the defined function. Expansion is the reverse of this process.

from the bottom of the hierarchy towards the top as a natural consequence of the dynamics of the compression and expansion operators.

4.4.1.2 Emergent Evaluation of Modules

The only way that a compressed module can be propagated in GLiB is through reproduction and crossover. A module is not added explicitly to the global primitives of the genetic program since this would increase the number of global primitives greatly and affect performance. Thus any mutation operator that generates a random subtree from the available primitives does not insert a dynamically defined module. Compression and expansion are local manipulations in that they affect only the genetic program being mutated.

As an added benefit, the constrained propagation of compressed modules provides the foundation for their evaluation. Their evaluation comes as a by-product of the dynamics of the evolutionary process and is a testament to the power of empirical credit assignment. Once a subtree is compressed, no further manipulation of the module's contents by crossover or other mutations is permitted until an expansion restores it. If the compression protects a subtree that is a beneficial decomposition, meaning that it solves some subproblem sufficiently, then it will contribute to the overall success of the program. If the expression tree's fitness is high compared to the other population members, it produces children that also contain the module. As long as a compressed module provides an evolutionary advantage to the members that contain references to it, or at least does not inhibit their reproductive success, the module spreads through the population. If, however, a module protects a subtree that is detrimental to solving the problem, it eventually becomes a liability to each offspring that uses it, limiting the fitness of the individual.

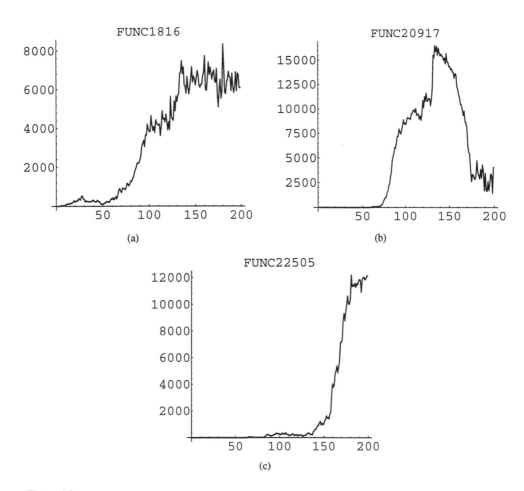

Figure 4.3
Graphs showing number of calls (y-axis) per generation (x-axis) for three evolved modules. Note the scales are different in each graph.

Eventually, all individuals relying on the malformed module disappear from the population through attrition.

Modules that survive over several generations and become entrenched in the population are termed *evolutionarily viable* [Angeline and Pollack 1992]. In this case, a beneficial module is "induced" when more and more of the population relies on its content. Often beneficial modules spread through the population very quickly so that their usage grows dramatically, as shown for three modules in Figure 4.3. Such quick proliferations are indicative of phase changes that Pollack [Pollack 1991] suggests is a form of induction in complex systems including human cognition.

A module is evaluated dynamically by how useful it is to the population members that use it or contain a reference to it. The global worth of a particular compression emerges from the local interactions of population with the fitness function and each other. Such a usage centered method for evaluation of modules is a result of the opportunistic nature of emergent intelligent techniques.

4.4.1.3 Comparison to ADFs

Koza [Koza 1992] also describes a method for constructing modular genetic programs termed Automatically Defined Functions (ADFs). In this method, the hierarchical arrangement of the functions are designed prior to evolving the genetic programs, each ADF having a unique primitive language that is defined by the programmer. Each individual contains its own local definition of the designed functions. Unlike GLiB's modules, when using ADFs the individual's functions are initialized to be random and are evolved along with the individual by an occasional crossover operation with another individual's local definition. Eventually, an individual and its ADFs settle into a complete solution to the problem.

While the definition of each individual ADF is emergent, the hierarchy of modularity, including the number of parameters to each module, is necessarily static. This makes the ADF a special case of applying GP at several levels in parallel. If explicit knowledge of the solution's decomposition into functions is available, then it is better to allow only the definition of the functions to emerge as in ADFs. Systems with more emergent components are more general. Therefore, because GLiB allows both the hierarchy and the definitions of the functions to emerge from the dynamics of the systems, GLiB's method of creating task decompositions is more general than ADF. As is always true, if one system is more general than another then the less general algorithm often has certain other advantages. Kim Kinnear's chapter in this volume describes such an advantage for ADFs.

4.4.1.4 Emergence of High-Level Representations

One of the most important aspects of GLiB and its emergent modules is that each module is an abstraction away from the primitive language of the genetic program and towards a task specific language that is more conducive to solving the problem. Much like the progression of computer languages from machine code to high-level languages, the dynamics in GLiB allow increasingly general abstractions to form and be tested for their ability to solve the problem. Because the dynamics of GLiB propagates only those abstractions that do not inhibit the evolution of better solutions to the problem, the resulting high-level representations in GLiB are necessarily task specific. However, like the solutions themselves, these abstractions are highly opportunistic and consequently difficult to describe.

The emergence of high-level representational abstractions is the chief difference between GLiB and Koza's ADFs. Where GLiB is dynamic in its pursuit of all aspects of the hierarchical structure for a particular task, the pre-specification of hierarchical organization of ADFs forces them to create the same modularity each time, although the semantics of the individual functions may be different. While the definitions of ADFs are emergent, their organization is not. This leads to the conclusion that while ADFs can allow modular organizations to emerge, they cannot allow higher-level abstractions to emerge. In any open-ended evolutionary system, both the hierarchical organization and the semantics of each level of the hierarchy must be manipulated to avoid limiting the organization of the developing individuals.

4.4.1.5 Generality of Emergent Modules

GLiB's form of emergent module definition is general across all evolutionary computations since it relies only on empirical credit assignment which is a natural dynamic in all evolutionary weak methods. For instance, experiments in [Angeline and Pollack 1993c] demonstrate that variants of compression and expansion, called *freeze* and *unfreeze*, speed-up the acquisition of finite state machines (FSMs) in an evolutionary program. Similar methods could be used in genetic algorithms and evolution strategies.

4.4.2 Emergent Goal-Directed Behavior

In problem solving, the goal of the system is always to find a way to accomplish the task. In this respect, all evolutionary computations used for problem solving are teleological, i.e., goal-directed, at least in the sense of moving toward increasingly better solutions. In standard genetic algorithms, the fitness function is generally an absolute objective measure over the range of possible bit strings. Because these fitness functions are objectively defined they are independent of the content of the population. Comparison between population members typically happens *after* each is evaluated by the objective fitness function.

However, absoluteness and objectivity in a fitness function come only at the expense of significant knowledge about the task being solved. Such knowledge typically indicates that the problem has already been solved or that a non-evolutionary approach would be more efficient. This is especially true for difficult problems with no analytical description. For instance, consider the expense of an objective fitness function to evolve an optimal strategy for a particular game. Such a function would need to test members of the population against all possible strategic situations to garner an objectively accurate fitness value. For anything but a trivial game such a computation is immense. An alternative

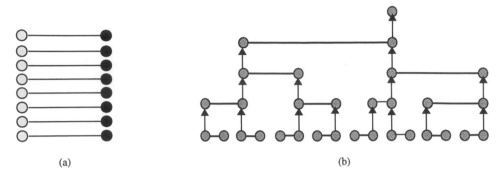

(a)   (b)

Figure 4.4
Two types of competitive fitness functions. (a) Bipartite competition used in Hillis (1992); and (b) Tournament
fitness function used in Angeline and Pollack (1993) with each horizontal line designating a competition and
each upward arrow designating the winner progressing in the tournament.

fitness function for such difficult problems could evaluate evolving solutions by
comparing them against an exemplar solution for the problem (e.g., [Angeline and Pollack
1993a] and [Fogel 1993]). The price for objectivity in this case is that the evolved
solutions are only "optimal" with respect to the exemplar used in the fitness function
rather than the intended task.

An alternative to an objective fitness function is to use a competitive fitness function,
one that is *dependent* on the contents of the population [Angeline and Pollack 1993b]. In a
competitive fitness function, the fitness of an individual is dependent on its performance
against other members of the population. Often such fitness functions can be described
succinctly since only a notion of "better" is required. For instance, to encode a competitive
fitness function for a game, only the rules of the game and means to determine which
player is the "winner" are necessary.

Two examples of evolutionary computations using competitive fitness functions for
problem solving have been investigated. Hillis [Hillis 1992] uses two separate populations
and a variant of a standard genetic algorithm to induce a sorting network for 16 integers,
nearly equivalent to the best known. One population is comprised of candidate sorting
networks while the individuals in the second population each contain several example
sequences to be sorted by the individual sorting networks. The fitness function pairs the
members of the two populations, as shown in Figure 4.4a, and awards the sorting network
on how well it sorts the sequences seen. The group of integers receives points based on
how poorly the sorting network performed on its contents. The two populations are in all

other manners separated, each performing selection and reproduction within their own population.

A second competitive fitness function used to evolve genetic programs to play Tic Tac Toe is described in [Angeline and Pollack 1993b]. This function pits the evolving genetic programs in the same population against each other in a single elimination tournament, as shown in Figure 4.4b. The results of this competitive fitness function are compared to genetic programs evolved for the same task with fitness functions that used several different exemplar strategies to provide an objective measure. Angeline and Pollack show that their competitive fitness function induces more robust genetic programs than fitness functions using a variety of strategic exemplars for objective comparison.

Performing competition *within* a fitness function has distinct properties from evolutionary computations that use tournament selection [Goldberg 1989a] *after* applying an independent fitness function. Dependent fitness measures are self-scaling in that as the complexity and power of the population increases, so does the difficulty of the fitness function. Since the fitness function is dependent on the population, it tracks through the population's non-monotonic development without the need for measuring the average member's ability explicitly.

By placing members of a population in opposition, competitive fitness functions set up interesting evolutionary dynamics. For instance, as the average strategy increases in complexity, simplistic strategies that take advantage of the more complex strategy's idiosyncracies appear. Such "parasites" present specific challenges to their "hosts" and help to promote strategic diversity in the population. Furthermore, as evolution continues, the ecological balance shifts in the population to take advantage of exploitable strategic niches. For complex problems, ecologies of strategies develop similar to those reported in [Ray 1992] but which are specific to the problem being solved [Angeline and Pollack 1993b]. As a result, the competitive populations automatically progress to more and more general solutions.

The question still remains as to what prevents the population from wandering aimlessly through the space of strategies rather than moving towards more complex solutions. Given that the population maintains some level of diversity, this is straightforward. The evolutionary pressure toward general solutions emerges from the progressive diversity of the population. Because there are many differing strategies that a particular genetic program could meet in any generation, only solutions that can perform well against a number of them consistently appear in the upper tiers of the tournament and consistently reproduce. The constant diversity of the competitive environment provides a fertile environment for robust solutions to emerge providing a teleological effect without designing it into the fitness function or previously solving the problem.

4.5 Guidelines for Emergent Intelligence

In the spirit of [Goldberg 1989b], this section provides some guidelines for directed introduction of emergent intelligence into evolutionary computations.

4.5.1 Coaxing Rather than Coercing

Brute force methods, i.e., ones that blindly attack the problem, typically provide little benefit, although they may provide a short term benefit for simple problems. Such strong coercion of the genetic programming dynamics only compromises the natural dynamics of the process. In fact, directly attacking the problem often removes the ability for features to emerge from the dynamics altogether.

What is required is a softer touch, one with some finesse. The desired properties should be coaxed out naturally by the dynamics of the computation. Much like instructing a young child on correct behavior, forcing high level properties of intelligence out with direct effort often leads to stubborn denial. But designing a situation where the correct behavior can emerge on its own, better serves the task and the child. This is because there is usually more than one way to achieve the desired effect and the most appropriate method is often specific to the situation. Thus, the solution should come from within the task. The algorithm's job, much like the young child's parent, is only to reinforce the progression toward improvement rather than restrict the method that accomplishes it.

4.5.2 Hitchhiking as a Technique

One of the easiest methods for allowing intelligence to emerge in evolutionary weak methods is to use *hitchhiking*. In genetic algorithms, a set of genes that do not benefit a genotype are said to hitchhike when they are propagated in the population by virtue of other genes in the genotype [Forrest and Mitchell 1992]. Typically, hitchhiking is detrimental to the evolutionary process since the poorer genes that proliferate with the beneficial genes displace their competitors from the population. This, again, is a result of positional encodings.

In genetic programming, hitchhiking is exploitable as a technique for emergent intelligence. For instance, the emergence of introns in genetic programing happen as a natural consequence of the dynamics. The superfluous components provide no benefit in terms of fitness to the program but "ride along" with the subtrees they surround. The propagation of GLiB's modules also rely on hitchhiking. The graph of Figure 4.3c shows a module that was in the population for an extended period of time but was used only occasionally. When the population determined a use for the module, around generation 137, its presence in the population increased dramatically. Up until that time the module propagated on the strength of other subtrees in the same genotype to remain present in the population.

4.5.3 Opportunism Over Prespecification

For most problems, there is no single solution that is uniformly proficient across all possible instances. In standard AI systems, the static engineered knowledge ends up being either too general or too specific to cover many of the problem instances. Opportunistic methods like emergent intelligence latch on to whatever provides a benefit in the current problem solving situation. The empirical credit assignment in evolutionary weak methods is ideal for opportunistic acquisition of solutions. Each problem solving instance may produce a completely distinct solution since the features that arise during processing are dependent on the points in the problem space visited by the algorithm. The generality of the features that emerge are often minimal, but they are pertinent to the specific problem being attacked.

Because of the opportunistic nature of empirical credit assignment, the evolved genetic programs are often complex, unreadable and incorporate programming practices that are distinct from strategies used by humans. For instance, the problem decompositions induced by GLiB do not reflect the decompositions a programmer would devise for the same problem. While this makes analyzing the results difficult, it identifies the bias humans have as problem solvers and program designers. The opportunism of emergent intelligence and the diversity of the resulting structures may enlighten us as to what our biases are and how they limit our ability to solve problems.

4.5.4 Explicit Knowledge As A Last Resort

The purpose of emergent intelligence is not necessarily to replace all explicit knowledge, but to augment it by removing the need to engineer out knowledge that is a natural consequence of the task. When absolute knowledge about a task is available, an engineer can incorporate it into an emergent intelligent system much as knowledge is incorporated into other evolutionary computations (e.g. [Grefenstette 1989] and [Michalewicz 1993]). However, by including explicit knowledge for a particular problem, the opportunity for contradictory knowledge to emerge, if necessary, is removed. Consequently, when knowledge is available, it should be used judiciously and only in a manner that does not discourage the aspects of the solution that should emerge.

4.6 Conclusion

Artificial intelligence has made great strides in computational problem solving using explicitly represented knowledge extracted from the task. If we continue to use explicitly represented knowledge exclusively for computational problem solving, we may never computationally accomplish a level of problem solving performance equal to humans. Emergent intelligence de-emphasizes the role of explicit knowledge and encourages the development of solutions that incorporate the task description as a component of the problem solver. This allows the constraints of the task to be represented more naturally and permits only pertinent task specific knowledge to emerge in the course of solving the problem.

Acknowledgments

I would like to thank Professor Jordan Pollack, my advisor, who allowed me the mental space and computational resources to explore this area. Thanks also to the GPers that attended ICGA-93, especially Kim Kinnear and Lee Altenberg. The off-line discussions at this conference and the associated GP workshop were significant and invigorating. Thanks also to Greg Saunders, Ed Large, John Kolen, Jan Jannick and my reviewers for comments and proofreading help on various drafts of this chapter.

Bibliography

Angeline, P., G. Saunders and J. Pollack (1994). An evolutionary algorithm that constructs recurrent neural networks. To appear in *IEEE Transactions on Neural Networks*.

Angeline, P. and J. Pollack (1993a). Coevolving high-level representations. To Appear in *The Proceedings of Artificial Life III*.

Angeline, P. and J. Pollack (1993b). Competitive environments evolve better solutions for complex tasks. In *Genetic Algorithms: Proceedings of the Fifth International Conference (GA93)*, S. Forrest (ed.), San Mateo, CA: Morgan Kaufmann Publishers Inc., pp 264-270.

Angeline, P. and J. Pollack (1993c). Evolutionary module acquisition. In *Proceedings of the Second Annual Conference on Evolutionary Programming*, D. Fogel (ed.). Evolutionary Programming Society, pp 154-163.

Angeline, P. and J. Pollack (1992). The evolutionary induction of subroutines. In *The Fourteenth Annual Conference of the Cognitive Science Society*. Hillsdale, New Jersey: Lawrence Erlbaum Associates Inc., pp 236-241.

Bäck, T., F. Hoffmeister and H. P. Schwefel (1991). A survey of evolution strategies. In *Proceedings of the Fourth International Conference on Genetic Algorithms*, R. K. Belew and L. B. Booker (eds.), San Mateo, CA: Morgan Kaufmann Publishers Inc., pp 2-9.

DeJong, K., (1975), *An Analysis of the Behavior of a Class of Genetic Adaptive Systems*. Doctoral dissertation, University of Michigan.

Fogel, D. B (1993). Using evolutionary programming to create neural networks that are capable of playing Tic-Tac-Toe. In *International Conference on Neural Networks*, San Francisco, CA: IEEE Press, pp 875–880.

Fogel, D. B. (1992). *Evolving Artificial Intelligence*. Doctoral Dissertation, University of California at San Diego.

Fogel, L., A. Owens and M. Walsh (1966). *Artificial Intelligence through Simulated Evolution*. New York: John Wiley & Sons.

Forrest, S. (1991) Emergent computation: self-organizing, collective, and cooperative phenomena in natural and artificial computing networks. In *Emergent Computation*, S. Forrest (ed.). Cambridge, MA: MIT Press, pp 1-11.

Forrest, S. and M. Mitchell (1992). Relative building block fitness and the building-block hypothesis. In *Foundations of Genetic Algorithms 2*, D. Whitley (ed.). San Mateo, CA: Morgan Kaufmann Publishers Inc., pp 109-126.

Goldberg, D. (1989a). *Genetic Algorithms in Search, Optimization, and Machine Learning*. Reading, MA: Addison-Wesley Publishing Company, Inc.

Goldberg, D. (1989b). Zen and the art of genetic algorithms. In *Proceedings of the Third International Conference on Genetic Algorithms*. J. Schaffer (ed.) San Mateo, CA: Morgan Kaufmann Publishers Inc., pp 80-85.

Goldberg, D., B. Korb and K. Deb (1989). Messy genetic algorithms: Motivation, analysis, and first results. *Complex Systems*, **3**, pp 493-530.

Grefenstette, J. (1989). Incorporating problem specific knowledge into genetic algorithms. In *Genetic Algorithms and Simulated Annealing*, L. Davis (ed.). San Mateo, CA: Morgan Kaufmann Publishers Inc., pp 42-60.

Hillis, D. (1992). Co-evolving parasites improves simulated evolution as an optimization procedure. In *Artificial Life II*, C. Langton, C. Taylor, J. Farmer and S. Rasmussen (eds.). Reading, MA: Addison-Wesley Publishing Company, Inc., pp 313-324.

Holland, J. (1975). *Adaptation in Natural and Artificial Systems*, Ann Arbor, MI: The University of Michigan Press.

Jefferson, D., R. Collins, C. Cooper, M. Dyer, M. Flowers, R. Korf, C. Taylor, and A. Wang (1992). Evolution as a theme in artificial life: The genesys/tracker system. In *Artificial Life II*, C. Langton, C. Taylor, J. Farmer and S. Rasmussen (eds.). Reading, MA: Addison-Wesley Publishing Company, Inc., pp 549-578.

Koza, J. (1992). *Genetic Programming*, Cambridge, MA: MIT Press.

Levenick, J. (1991). Inserting introns improves genetic algorithm success rate: Taking a cue from biology. In *Proceedings of the Fourth International Conference on Genetic Algorithms*, R. K. Belew and L. B. Booker (eds.) San Mateo, CA: Morgan Kaufmann Publishers Inc., pp 123-127.

Michalewicz, Z. (1993). A hierarchy of evolution programs: an experimental study. *Evolutionary Computation*, **1**, pp 51-76.

Newell, A. (1990). *Unified Theories of Cognition*. Cambridge, MA: Harvard University Press.

Newell, A. (1982). The knowledge level," *Artificial Intelligence*, **18**, pp 87-127.

Newell, A. and H. A. Simon (1963), "GPS: A program that simulates human thought," In *Computers and Thought*, E. A. Feigenbaum and J. Feldman (eds.) New York: McGraw-Hill, pp 279-293.

Pollack, J. (1991). The induction of dynamical recognizers. *Machine Learning*, **7**, pp 227 - 252.

Ray, T. (1992). An approach to the synthesis of life. In *Artificial Life II*, C. Langton, C. Taylor, J. Farmer and S. Rasmussen (eds.), Reading, MA: Addison-Wesley Publishing Company, Inc., pp 371-408.

Rich, E. (1983). *Artificial Intelligence*, New York: McGraw-Hill.

Rumelhart, D., G. Hinton and R. WIlliams (1986). Learning internal representations through error propagation. In *Parallel Distributed Processing: Explorations in the Microstructure of Cognition, Volume 1: Foundations*, D. Rumelhart and J. McClelland (eds.) Cambridge, MA: MIT Press., pp 25-40.

Winston, P. H. (1984). *Artificial Intelligence*, Reading MA: Addison-Wesley.

5 Scalable Learning in Genetic Programming using Automatic Function Definition

John R. Koza

This chapter uses three differently sized versions of an illustrative problem that has considerable regularity, symmetry, and homogeneity in its problem environment to compare genetic programming with and without the newly developed mechanism of automatic function definition. Genetic programming with automatic function definition can automatically decompose a problem into simpler subproblems, solve the subproblems, and assemble the solutions to the subproblems into a solution to the original overall problem. The solutions to the problem produced by genetic programming with automatic function definition are more parsimonious than those produced without it. Genetic programming requires fewer fitness evaluations to yield a solution to the problem with 99% probability with automatic function definition than without it.

When we consider the three differently sized versions of the problem we find that the size of the solutions produced *without* automatic function definition can be expressed as a direct multiple of problem size. In contrast, the average size of solutions *with* automatic function definition is expressed as a certain minimum size representing the overhead associated with automatic function definition; however, there is only a very slight increase in the average size of the solutions with problem size. Moreover, the number of fitness evaluations required to yield a solution to the problem with a 99% probability grows very rapidly with problem size without automatic function definition, but this same measure grows only linearly with problem size with automatic function definition.

5.1 Introduction

Hierarchical problem-solving ("divide and conquer") may be advantageous in solving large and complex problems because the solution to an overall problem may be found by decomposing it into smaller and more tractable subproblems in such a way that the solutions to the subproblems are reused many times in assembling the solution to the overall problem.

In the top-down way of describing this three-step hierarchical problem-solving process, one first tries to discover a way to decompose a given problem into subproblems (modules). Second, one tries to solve each of the presumably simpler subproblems. Third, one seeks a way to assemble the solutions to the subproblems into a solution to the original overall problem. Presumably, solving the subproblems will prove to be simpler than solving the original overall problem. In practice, solving some of the subproblems may lead to a recursive reinvocation of this three-step process. In any event, if this three-step process is successful, one ends up with a hierarchical (modular) solution to the problem.

Hierarchical solutions to problems are potentially very favorable because they avoid tediously re-solving what are essentially identical subproblems. Consequently, such solutions may be more parsimonious. They may also require less total effort to solve the overall problem. Leverage gained from the successful reuse of solutions to subproblems by means of some kind of hierarchical approach appears to be necessary if machine learning methods are ever to be scaled up from small "proof of principle" problems to large problems.

In the terminology of computer programming, the three-step process of solving problems hierarchically starts by analyzing the overall problem and dividing it into parts. Second, one writes subprograms (subroutines, procedures, defined functions) to solve each part of the problem. Third, one writes a main program or other calling program that invokes the subprograms and assembles the results produced by the functions into a solution to the overall problem. In practice, the result produced by a computation can be a single value, a set of values, the side-effects performed on a system, or a combination thereof.

This same three-step process of solving problems hierarchically can also be described in a bottom-up way. First, one seeks to discover regularities and patterns at the lowest level of the problem environment. Second, one restates or recodes the problem in terms of these regularities so as to create a new problem stated in new terms. Third, one tries to discover a solution to the presumably simpler recoded problem. In practice, the process of discovering a solution to the recoded problem may recursively involve further discovery of regularities and patterns and further recoding. If this process of regularity finding and assembly is successful, one ends up with a hierarchical solution to the problem.

The bottom-up way of solving problems hierarchically is often viewed as a change of representation. The recoding of the original problem is a change of representation from the original representation of the problem to a higher level. New regularities often become apparent when the representation is changed in this way.

The goal of automatically solving problems hierarchically (whether top-down or bottom-up) has been a central issue in machine learning and artificial intelligence since the beginning of these fields.

The one obvious question concerning this alluring three-step process of solving problems hierarchically is how does one go about implementing this process in an automated and domain independent way? The discovery of a solution to a subproblem (i.e., the second step of the top-down approach) can potentially be accomplished by means of genetic programming. Indeed, The book *Genetic Programming: On the Programming of Computers by Means of Natural Selection* [Koza 1992a] demonstrated that a variety of problems can be solved, or approximately solved, by genetically breeding a population of computer programs over a period of many generations. See also [Koza and Rice 1992].

But what about the other steps of this three-step problem-solving process? This chapter illustrates, for a particular problem, that when the recently developed mechanism of automatic function definition [Koza 1992a, 1992b, 1993; Koza, Keane, and Rice 1993] is added to genetic programming, *all three steps* in the hierarchical problem-solving process described above can be simultaneously performed within a single run of genetic programming. An *automatically defined function* (an *ADF*) is a function (i.e., subroutine, procedure, module) that is evolved during a run of genetic programming and which may be called by the main program (or other calling program) that is being simultaneously evolved during the same run.

5.2   The Lawn Mower Problem

In the lawn mower problem, the goal is to find a program for controlling the movement of a lawn mower so that the lawn mower cuts all the grass in the yard.

We first consider a version of this problem in which the lawn mower operates in an 8 by 8 toroidal square area of lawn that initially has grass in all 64 squares.

Each square of the lawn is uniquely identified by a vector of integers modulo 8 of the form (i,j), where $0 \leq i, j \leq 7$. The lawn mower starts at location (4,4) facing north. The state of the lawn mower consists of its location on one of the 64 squares of the lawn and the direction in which it is facing. The lawn is toroidal in all four directions, so that whenever the lawn mower moves off the edge of the lawn, it reappears on the opposite side. There are no obstacles in the yard.

The lawn mower is capable of turning left, of moving forward one square in the direction in which it is currently facing, and of jumping by a specified displacement in the vertical and horizontal directions. Whenever the lawn mower moves onto a new square (either by means of a single move or a jump), it mows all the grass, if any, in the square onto which it arrives. The lawn mower has no sensors.

A human programmer writing a program to solve this problem would almost certainly not solve it by tediously writing a sequence of 64 separate mowing operations (and appropriate turning actions). Instead, a human programmer would exploit the considerable regularity, symmetry and homogeneity inherent in this problem environment by writing a program that mows a certain small area of the lawn in a particular way, then repositions the lawn mower in some regular way, and then repeats the particular mowing action on the new area of the lawn. That is, the human programmer would decompose the overall problem into a set of subproblems (i.e., mowing a small area), solve the subproblem, and then repeatedly reuse the subproblem solution at different places on the lawn in order to solve the overall problem.

5.3   Preparatory Steps Without Automatic Function Definition

The operations of turning left, moving one square and then mowing, and jumping and then mowing each change the state of the lawn mower and are side-effecting operators that take no arguments. They can be treated as terminals.

Since it may be desirable to be able to manipulate the numerical location of the lawn mower using arithmetic operations, both random constants and arithmetic operations should be available as ingredients of programs for solving this problem. The terminal set should thus include random constants. The random constants, $\mathfrak{R}$, appropriate for this problem are vectors (i,j) of integers modulo 8.

Thus, the terminal set for this problem consists of two zero-argument side-effecting operators and random vector constants. That is,

T = { (LEFT), (MOW), $\mathfrak{R}$}.

The operator LEFT takes no arguments and turns the orientation of the lawn mower counter-clockwise by 90° (without moving the lawn mower). Since the programs will be performing arithmetic, it is necessary that all terminals and functions return a value that can serve as a legitimate argument to the arithmetic operations. Thus, to assure closure, LEFT returns the vector value (0,0).

The operator MOW takes no arguments and moves the lawn mower in the direction it is currently facing and mows the grass, if any, in the square to which it is moving (thereby removing all the grass, if any, from that square). MOW does not change the facing direction of the lawn mower. For example, if the lawn mower is at location (1,3) and facing east, MOW increments the first component (i.e., the x location) of the state vector of the lawn mower thus moving the lawn mower to location (2,3) with the lawn mower still facing east. As a further example, if the lawn mower is at location (7,3) and facing east, MOW moves the lawn mower to location (0,3) because of the toroidal geometry. To assure closure, MOW also returns the vector value (0,0).

The function set consists of

F = {V+, FROG, PROGN},

with these functions taking 2, 1, and 2 arguments, respectively.

V+ is two-argument vector addition function modulo 8. For example, (V+ (1,2) (3,7)) returns the value (4,1).

FROG is a one-argument operator that causes the lawn mower to move relative to the direction it is currently facing by an amount specified by its vector argument and mows the grass, if any, in the square on which the lawn mower arrives (thereby removing all the grass, if any, from that square). FROG does not change the facing direction of the lawn mower. For example, if the lawn mower is at location (1,2) and is facing east, (FROG (3,5)) causes the lawn mower to end up at location (6,5) with the lawn mower still facing east. FROG acts as the identity operator on its argument. Thus, (FROG (3,5)) returns the value (3,5).

The problem can be solved with either the MOW or FROG operator; however, we include both operators to enrich the function set by allowing alternatives approaches for solving the problem.

The goal is to mow all 64 squares of grass. The movement of the lawn mower is terminated when either the lawn mower has executed a total of 100 LEFT turns or a total of 100 movement-causing operations (i.e., MOWs or FROGs). The raw fitness of a particular program is the amount of grass (from 0 to 64) mowed within this allowed amount of time. Since the yard contains no obstacles and the toroidal topology of the yard is perfectly symmetrical, it is only necessary to measure fitness over one fitness case for this problem.

Table 5.1 summarizes the key features of the lawn mower problem with 64 squares. The last seven rows of this table apply to automatic function definition (described below).

5.4 Lawn Size of 64 Without Automatic Function Definition

The only way to write a computer program to mow all 64 squares of the lawn with the available movement-causing and turning operators involves tediously writing a program consisting of at least 64 MOWs or FROGs so that all 64 squares of the lawn are mowed. One possible orderly way of writing this tedious program involves mowing all eight squares of lawn in the vertical column beginning at the starting location (4,4), turning left upon returning to (4,4), moving and mowing one column to the west, turning left three times so as to face north again, and mowing all eight squares of lawn in the new vertical column.

Subtrees of somewhat effective programs in this problem typically mow small portions of the lawn. If two programs are selected from the population based on their fitness, both of the selected programs will usually mow an above-average amount of lawn for their generation. Moreover, a random subtree from either of these selected individuals will, on average, mow an above-average amount of lawn for its generation. Thus, the effect of the crossover operation is to create new programs which will, on average, mow an increasing and above-average amount of lawn.

Table 5.1
Tableau for the lawn mower problem with 64 squares.

Objective:	Find a program to control a lawn mower so that it mows the grass on all 64 squares of lawn in an unobstructed yard.
Terminal set without automatic function definition:	(LEFT), (MOW), ℜ.
Function set without automatic function definition:	V+, FROG, PROGN.
Fitness cases:	One fitness case consisting of a toroidal lawn with 64 squares, each initially containing grass.
Raw fitness:	Raw fitness is the amount of grass (from 0 to 64) mowed within the maximum allowed number of state-changing operations.
Standardized fitness:	Standardized fitness is the total number of squares (i.e., 64) minus raw fitness.
Hits:	Same as raw fitness.
Wrapper:	None.
Parameters:	$M = 1,000$. $G = 51$.
Success predicate:	A program scores the maximum number of hits.
Overall program structure with automatic function definition:	One result-producing branch and two function definitions with ADF0 taking no arguments and ADF1 taking one argument ARG0.
Terminal set for the result-producing branch:	(LEFT), (MOW), ℜ.
Function set for the result-producing branch:	ADF0, ADF1, V+, FROG, PROGN.
Terminal set for the function definition ADF0	(LEFT), (MOW), ℜ.
Function set for the function definition ADF0	V+, PROGN.
Terminal set for the function definition ADF1	ARG0, (LEFT), (MOW), ℜ.
Function set for the function definition ADF1	ADF0, V+, FROG, PROGN.

The following 296-point individual achieving a raw fitness of 64 emerged on generation 34 of this run without automatic function definition:

```
(V+ (V+ (V+ (FROG (PROGN (PROGN (V+ (MOW) (MOW)) (FROG (3,2))) (PROGN
(V+ (PROGN (V+ (PROGN (PROGN (MOW) (2,4)) (FROG (5,6))) (PROGN (V+ (MOW)
(6,0)) (FROG (2,2)))) (V+ (MOW) (MOW))) (PROGN (V+ (PROGN (PROGN (0,3)
(7,2)) (FROG (5,6))) (PROGN (V+ (MOW) (6,0)) (FROG (2,2)))) (V+ (MOW)
(MOW)))) (PROGN (FROG (MOW)) (PROGN (PROGN (PROGN (V+ (MOW) (MOW)) (FROG
(LEFT))) (PROGN (MOW) (V+ (MOW) (MOW)))) (PROGN (V+ (PROGN (0,3) (7,2))
(V+ (MOW) (MOW))) (PROGN (V+ (MOW) (MOW)) (PROGN (LEFT) (MOW)))))))))))
(V+ (PROGN (V+ (PROGN (PROGN (MOW) (2,4)) (FROG (5,6))) (PROGN (V+ (MOW)
(6,0)) (FROG (2,2)))) (V+ (MOW) (MOW))) (V+ (FROG (LEFT)) (FROG
```

```
(MOW))))) (V+ (FROG (V+ (PROGN (V+ (PROGN (V+ (MOW) (MOW)) (FROG (3,7)))
(V+ (PROGN (MOW) (LEFT)) (V+ (MOW) (5,3)))) (PROGN (PROGN (V+ (PROGN
(LEFT) (MOW)) (V+ (1,4) (LEFT))) (PROGN (FROG (MOW)) (V+ (MOW) (3,7))))
(V+ (PROGN (FROG (MOW)) (V+ (LEFT) (MOW))) (V+ (FROG (1,2)) (V+ (MOW)
(LEFT)))))) (PROGN (V+ (FROG (3,1)) (V+ (FROG (PROGN (PROGN (V+ (MOW)
(MOW)) (FROG (3,2))) (FROG (FROG (5,0))))) (V+ (PROGN (FROG (MOW)) (V+
(MOW) (MOW))) (V+ (FROG (LEFT)) (FROG (MOW)))))) (PROGN (PROGN (PROGN
(PROGN (LEFT) (MOW)) (V+ (MOW) (3,7))) (V+ (V+ (MOW) (MOW)) (PROGN
(LEFT) (LEFT)))) (V+ (FROG (PROGN (3,0) (LEFT))) (V+ (PROGN (MOW)
(LEFT)) (FROG (5,4)))))))) (PROGN (FROG (V+ (PROGN (V+ (PROGN (PROGN (V+
(PROGN (PROGN (MOW) (2,4)) (FROG (5,6))) (PROGN (V+ (MOW) (1,2)) (FROG
(2,2)))) (V+ (MOW) (MOW))) (FROG (3,7))) (V+ (PROGN (PROGN (MOW) (2,4))
(FROG (5,6))) (PROGN (V+ (MOW) (6,0)) (FROG (2,2))))) (PROGN (PROGN (V+
(FROG (MOW)) (V+ (1,4) (LEFT))) (PROGN (FROG (MOW)) (V+ (MOW) (3,7))))
(V+ (PROGN (FROG (MOW)) (V+ (LEFT) (MOW))) (V+ (FROG (1,2)) (V+ (MOW)
(LEFT)))))) (PROGN (V+ (PROGN (FROG (2,4)) (V+ (MOW) (MOW))) (V+ (FROG
(MOW)) (LEFT))) (PROGN (3,0) (LEFT))))) (FROG (V+ (7,4) (MOW)))))) (V+
(V+ (PROGN (MOW) (4,3)) (V+ (LEFT) (6,1))) (MOW)))
```

This 296-point program solves the problem by agglomerating enough erratic movements so as to cover the entire area of the lawn within the allowed maximum number of operations. In fact, the way that this program solves the problem is so tedious and convoluted that it can be easily visualized only after dividing the trajectory of the lawn mower into three epochs.

Figure 5.1 shows a partial trajectory of this best-of run 296-point individual for a first epoch consisting of mowing operations 0 through 30 for the lawn mower problem; figure 5.2 shows a partial trajectory for a second epoch consisting of mowing operations 31 through 60; and figure 5.3 shows a partial trajectory for a third epoch consisting of mowing operations 61 through 85. As can be seen, even though the problem environment contains considerable regularity, symmetry, and homogeneity in that it requires mowing all 64 squares in an unobstructed toroidal yard, this solution operates in an entirely *ad hoc* fashion. For example, between mowing operations 2 and 3, the lawn mower FROGs up two rows and three columns to the right; between operations 4 and 5, the mower FROGs up six rows and three columns to the left; and between operations 6 and 7, the mower FROGs up two (i.e., down six) and two columns to the right.

Over 38 runs, the average structural complexity (i.e., total number of functions and terminals in the program) of the 35 successful solutions to the lawn mower problem without automatic function definition was 280.82 points. The structural complexity of the successful solutions is about 4.4 times the size of the lawn. The successful programs are so large because they make no use of the inherent regularity of the problem environment.

Using methods described in detail in [Koza 1992a], we find that the total number of individuals $I(M,i,z)$ that need to be processed in order to solve this problem with a 99% probability is 100,000.

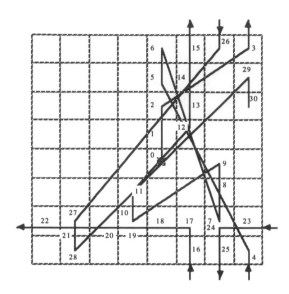

Figure 5.1
First partial trajectory of 296-point program for mowing operations 0 through 30 without automatic function definition.

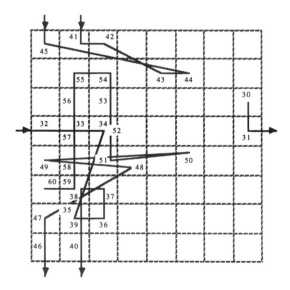

Figure 5.2
Second partial trajectory of 296-point program for mowing operations 31 through 60 without automatic function definition.

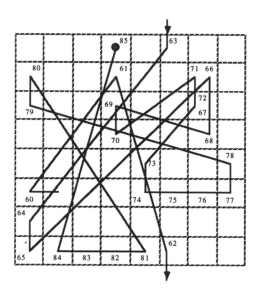

Figure 5.3
Third partial trajectory of 296-point program for mowing operations 61 through 85 without automatic function definition.

5.4.1 Lawn Size of 96 Without Automatic Function Definition

When the size of the problem is scaled up by 50% from 64 to 96 squares of lawn, the average size of the successful solutions among 197 runs increases to 426.9 (i.e., about 4.4 times the lawn size). The number of fitness evaluations required to yield a solution of the problem with 99% probability increases substantially to 4,539,000.

5.4.2 Lawn Size of 32 Without Automatic Function Definition

When the size of the problem is scaled down by 50% from 64 to 32 squares of lawn, the average size of the successful solutions among 64 runs decreases to 145 (i.e., about 4.5 times the lawn size).   The number of fitness evaluations required to yield a solution of the problem with 99% probability decreases substantially to 19,000.

5.5 Preparatory Steps With Automatic Function Definition

Each of the programs presented in the previous section for solving the lawn mower problem without automatic function definition contained at least 64 MOWs or FROGs when the lawn size is 64. However, a human programmer would never consider solving this problem in this tedious way. Instead, a human programmer would write a program that first mows a certain small sub-area of the lawn in some orderly way, then repositions the lawn mower to a new sub-area of the lawn in some orderly (probably tessellating) way, and then repeats the mowing action on the new sub-area of the lawn. The program would contain a sufficient number of invocations of the orderly method for mowing sub-areas of the lawn so as to mow the entire lawn. That is, a human programmer would exploit the considerable regularity and symmetry inherent in the problem environment by decomposing the problem into subproblems and then would repeatedly reuse the solution to the subproblem in order to solve the overall problem.

In applying genetic programming with automatic function definition to the lawn mower problem, we decided that each individual overall program in the population will consist of two function-defining branches (defining a zero-argument function called ADF0 and a one-argument function ADF1) and a final (rightmost) result-producing branch. The second defined function ADF1 can hierarchically refer to the first defined function ADF0.

We first consider the two function-defining branches.

The terminal set $\mathcal{T}_{fd0}$ for the zero-argument defined function ADF0 consists of

$$\mathcal{T}_{fd0} = \{ \text{(LEFT)}, \text{(MOW)}, \; \mathfrak{R} \}.$$

The function set $\mathcal{F}_{fd0}$ for the zero-argument defined function ADF0 is

$$\mathcal{F}_{fd0} = \{\text{V+}, \text{PROGN}\},$$

each taking 2 arguments.

The body of ADF0 is a composition of primitive functions from the function set $\mathcal{F}_{fd0}$ and terminals from the terminal set $\mathcal{T}_{fd0}$.

The terminal set $\mathcal{T}_{fd1}$ for the one-argument defined function ADF1 taking dummy variable ARG0 consists of

$$\mathcal{T}_{fd1} = \{\text{ARG0}, \text{(LEFT)}, \text{(MOW)}, \mathfrak{R}\}.$$

The function set $\mathcal{F}_{fd1}$ for the one-argument defined function ADF1 is

$$\mathcal{F}_{fd1} = \{\text{ADF0}, \text{V+}, \text{FROG}, \text{PROGN}\},$$

taking 0, 2, 1, and 2 arguments, respectively,

The body of ADF1 is a composition of primitive functions from the function set $\mathcal{F}_{fd1}$ and terminals from the terminal set $\mathcal{T}_{fd1}$.

Since LEFT and MOW each evaluate to (0,0) and since FROG acts as an identity function, the value returned by ADF0 and ADF1 is either (0,0) or the result of the vector addition V+ operating on random constants, or on ARG0, the value of calls to ADF0 and random constants in the case of ADF1.

We now consider the result-producing branch.

The terminal set $\mathcal{T}_{rp}$ for the result-producing branch is

$$\mathcal{T}_{rp} = \{ (\text{LEFT}), (\text{MOW}), \Re \}.$$

The function set $\mathcal{F}_{rp}$ for the result-producing branch is

$$\mathcal{F}_{rp} = \{\text{ADF0}, \text{ADF1}, \text{V+}, \text{FROG}, \text{PROGN}\},$$

with the functions taking 0, 1, 2, 1, and 2 arguments, respectively.

The result-producing branch is a composition of the functions from the function set $\mathcal{F}_{rp}$ and terminals from the terminal set $\mathcal{T}_{rp}$.

The last seven rows of table 5.1 summarize the key features of this problem with automatic function definition.

Additional details about automatic function definition and the structure-preserving crossover required with automatic function definition can be found in chapter 2 of this book, the forthcoming book *Genetic Programming II* [Koza 1994], and the forthcoming videotape *Genetic Programming II Videotape: The Next Generation* [Koza and Rice 1994].

5.6 Lawn Size of 64 With Automatic Function Definition

When genetic programming with automatic function definition is applied to this problem, the results are very different from the haphazard solution obtained without automatic function definition.

In one run of this problem with automatic function definition, the following 100% correct 42-point program scoring 64 (out of 64) emerged in generation 5:

```
(PROGN   (DEFUN ADF0 ()
               (VALUES (PROGN (V+ (0,1) (2,0)) (V+ (V+ (PROGN
               (MOW) (LEFT)) (V+ (MOW) (LEFT))) (PROGN (V+
               (LEFT) (LEFT)) (PROGN (MOW) (MOW)))))))
```

```
(DEFUN ADF1 (ARG0)
    (VALUES (V+ (FROG (FROG (ADF0))) (PROGN (PROGN
        (V+ (MOW) (ADF0)) (V+ (ADF0) (MOW))) (V+ (FROG
        (ADF0)) (V+ ARG0 ARG0))))))
    (VALUES (ADF1 (ADF1 (ADF1 (ADF1 (ADF0)))))))
```

Note that this 42-point solution is a hierarchical decomposition of the problem. Genetic programming discovered the decomposition of the overall problem, discovered the content of each subroutine, and assembled the results of the multiple calls to the subroutines into a solution of the overall problem. Specifically, genetic programming discovered a decomposition of the overall problem into five subproblems (four ADF1's and one ADF0) in the result-producing branch at the top level. The result-producing branch does not contain any LEFT, MOW, or FROG operations at all. ADF1 contains four invocations of ADF0, two MOWs, and no LEFT or FROG operations. ADF0 contains four MOWs, and four LEFTs.

Figure 5.4 shows the trajectory of the lawn mower for this 42-point solution. Note the difference between this regular trajectory and the haphazard character of the three partial

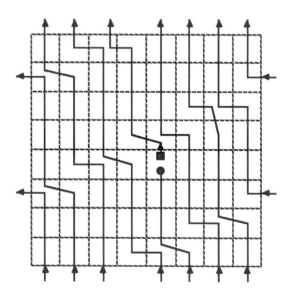

Figure 5.4
Trajectory of lawn mower using 42-point program with automatic function definition.

trajectories shown in figures 5.1, 5.2, and 5.3. The lawn mower here takes advantage of the regularity, symmetry, and homogeneity of the problem environment. It performs a tessellating activity that covers the entire lawn. Specifically, it mows four consecutive squares in a column in a northerly direction, shifts one column to the west, and then does the same thing in the next column. The fact that the entire trajectory can be conveniently presented in only one figure testifies to this solution's regular behavior.

When this 42-point program is evaluated, ADF0 is executed first by the result-producing branch. ADF0 begins with a PROGN whose first argument is (V+ (0,1) (2,0)). Since vector addition V+ has no side effects and since the return value of PROGN is the value returned by its second argument, this first argument to the PROGN can be totally ignored. Since the remainder of ADF0 contains only MOW and LEFT operations, ADF0 returns (0,0). As it turns out, ADF1 never uses its dummy variable.

The basic activity of ADF0 is to mow four squares of lawn in a northwesterly zigzag pattern. This zigzag action is illustrated at the starting point (4,4) in the middle of the figure. ADF0 moves forward (i.e., north) one square and mows that square; it then turns left (i.e., west) and moves forward and mows that square; it then turns left three times (so that it is again oriented north); and it then moves and mows two squares.

The northwesterly zigzag mowing activity of ADF0 is then repeatedly invoked. The result-producing branch invokes ADF1 a total of four times. Each time ADF1 is invoked, ADF0 is invoked four times. This hierarchy of invocations produces a total of 16 calls for the zigzag activity of ADF0. Because of the initial direct call of ADF0 at the beginning of the evaluation of the result-producing branch, the last of the 16 hierarchical invocations of ADF0 is not needed since the program is terminated by virtue of the completion of the overall task.

This zigzagging solution is an hierarchical decomposition and solution of the problem involving three simultaneous, automatic discoveries. First, genetic programming discovered a decomposition of the overall problem into 16 subproblems each consisting of the northwesterly zigzag mowing pattern. Second, genetic programming discovered the sequence of turns and moves to implement the northwesterly zigzag mowing activity. Third, genetic programming assembled the results of this mowing motion into a solution of the overall problem by appropriately repositioning the lawn mower.

In other runs, some 100% correct solutions mowed entire rows and columns while others zigzagged, swirled, crisscrossed, and leapfrogged around the lawn in a tessellating manner.

The average structural complexity of the 76 successful solutions to the lawn mower problem with automatic function definition was 76.95 points.

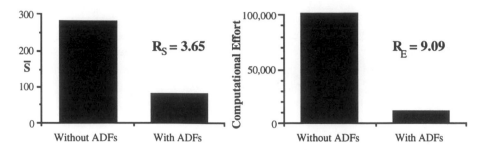

Figure 5.5
Summary graphs for the lawn mower problem with 64 squares.

11,000 fitness evaluations are required to yield a solution to the lawn mower problem with 64 squares of lawn with automatic function definition with 99% probability.

The beneficial effect of automatic function definition can be seen by taking the ratio of the average structural complexities (280.82 and 76.95) for the successful solutions to this problem without and with automatic function definition. This ratio (3.65 here) is called the structural complexity ratio and provides a way of measuring the parsimony (or lack of parsimony) produced by automatic function definition. Similarly, the ratio of the minimal $I(M,i,z)$ numbers (100,000 and 11,000), called the efficiency ratio, provides a way of measuring the improvement (or degradation) in computation effect produced by automatic function definition.

Figure 5.5 shows that the structural complexity ratio for the lawn mower problem with 64 squares is 3.65 and the efficiency ratio is 9.09. Since both ratios are above 1.00, automatic function definition improves both parsimony and efficiency.

5.6.1 Lawn Size of 96 With Automatic Function Definition

When the size of the problem is scaled up from 64 to 96 squares, the average structural complexity of the successful solutions to the lawn mower problem over 52 runs with automatic function definition was 84.3 points.

The number of fitness evaluations required to yield a solution to the problem with 99% probability with automatic function definition does not increase in the same dramatic way as is the case without automatic function definition. In fact, only 16,000 fitness evaluations are required.

Figure 5.6 shows that the structural complexity ratio for the lawn mower problem with 96 squares is 5.06 and the efficiency ratio is 283.7.

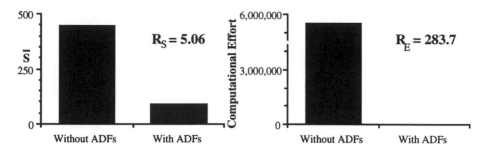

Figure -5.6
Summary graphs for the lawn mower problem with 96 squares.

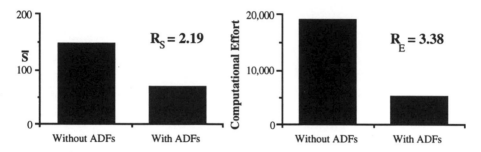

Figure 5.7
Summary graphs for the lawn mower problem with 32 squares.

5.6.2 Lawn Size of 32 With Automatic Function Definition

When the size of the problem is scaled down from 64 to 32 squares, the average structural complexity of the successful solutions to the lawn mower problem over 52 runs with automatic function definition was 66.3 points.

The number of fitness evaluations required to yield a solution to the problem with 99% probability with automatic function definition is 5,000.

Figure 5.7 shows that the structural complexity ratio for the lawn mower problem with 32 squares is 2.19 and the efficiency ratio is 3.8.

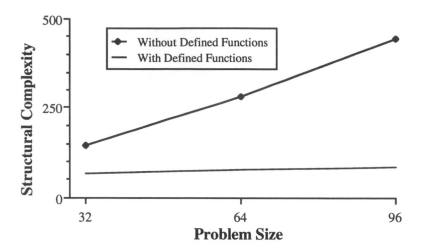

Figure 5.8
Comparison of average structural complexity of the successful programs for lawn sizes
of 32, 64, and 96 both with and without automatic function definition.

5.7 Relationship of Parsimony to Problem Size

Figure 5.8 shows the relationship between the average structural complexity of the
successful programs for lawn sizes of 32, 64, and 96 both with and without automatic
function definition.

As can be seen, the relationship is approximately linear for the curves with and without
automatic function definition; however, the relationships are very different.

When we perform a least squares linear regression on the three-point curve without
automatic function definition, we find that structural complexity S can be stated in terms
of lawn size L_s by

$$S = 2.4 + 4.4L$$

The vertical intercept is small and the slope is 4.4. In other words, a small and
essentially zero size is associated with a solution for a lawn of zero size and there is the
rather steep 4.4 growth in the program size with the lawn size. That is, there was no
economy of scale without automatic function definition. Increments in lawn size are
reflected in substantial linear growth in program size.

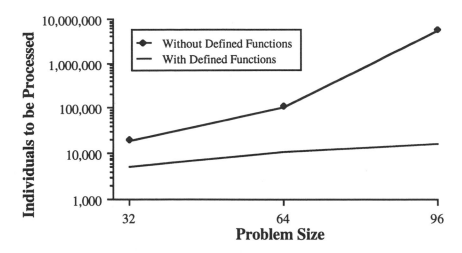

Figure 5.9
Comparison of $I(M,i,z)$, for lawn sizes of 32, 64, and 96 both with and without automatic function definition.

In contrast, when we perform the least squares linear regression on the three-point curve with automatic function definition, we find that structural complexity S can be stated in terms of lawn size L_s by

$$S = 57.85 + 0.28L_s$$

The vertical intercept has the substantial non-zero value of 57.85, but the slope is the very gentle value of 0.28. That is, there seems be a significant fixed minimum overhead associated with automatic function definition and relatively little additional cost associated with growth in the size of the problem. That is, there is a considerable economy of scale associated with automatic function definition.

5.8   Relationship of Computational Effort to Problem Size

Figure 5.9 shows the relationship between the number of fitness evaluations, $I(M,i,z)$, for lawn sizes of 32, 64, and 96 both with and without automatic function definition. Note that the vertical axis for $I(M,i,z)$ uses a logarithmic scale.

As can be seen, the progression of values of $I(M,i,z)$ with lawn size without automatic function definition from 19,000 to 100,000 to 4,539,000 is a steeply nonlinear pattern of growth. This nonlinearity is even greater than it may first appear by inspection of the graph because of the logarithmic scale.

In contrast, the progression of values of $I(M,i,z)$ with automatic function definition from 5,000 to 11,000 to 16,000 is a nearly linear relationship based on the problem sizes of 32, 64, and 96. In fact, when we perform the least squares linear regression on the three-point curve with automatic function definition, we find that the number of individuals I can be stated in terms of lawn size L_S by

$$I = -333 + 172L_S$$

with a correlation R of 1.00.

5.9 Conclusions

This chapter considered a problem with substantial symmetry and regularity in its problem environment. Three differently sized versions of the problem were solved both with and without automatic function definition.

For a fixed lawn size of 64, substantially fewer fitness evaluations are required to yield a solution of the problem with 99% probability with automatic function definition than without it. Moreover, the average size of the programs that successfully solved the problem is considerably smaller with automatic function definition than without it.

When the problem size was varied upwards and downwards by 50% from 64 without automatic function definition, the average size of the programs that successfully solved the problem was almost a direct linear multiple of the problem size and there is no economy of scale without automatic function definition. However, with automatic function definition, the average size of the programs that successfully solved the problem started with a certain fixed overhead and increased only gently with problem size.

There is a steeply nonlinear pattern to the number of fitness evaluations required to yield a solution to the problem with 99% probability without automatic function definition when the problem size is varied upwards and downwards by 50% from 64. However, with automatic function definition, there is only a gentle linear growth in the number of fitness evaluations.

Acknowledgements

James P. Rice of the Knowledge Systems Laboratory at Stanford University did the computer programming of the above on a Texas Instruments Explorer II[+] computer.

Bibliography

Koza, J. R. (1992a) Genetic Programming: On the Programming of Computers by Means of Natural Selection. Cambridge, MA: The MIT Press.

Koza, J. R. (1992b) Hierarchical automatic function definition in genetic programming. In Whitley, Darrell (editor). Proceedings of Workshop on the Foundations of Genetic Algorithms and Classifier Systems, Vail, Colorado 1992. San Mateo, CA: Morgan Kaufmann Publishers Inc.

Koza, J. R. (1993) Simultaneous discovery of detectors and a way of using the detectors via genetic programming. 1993 IEEE International Conference on Neural Networks, San Francisco. Piscataway, NJ: IEEE 1993. Volume III. Pages 1794-1801.

Koza, J. R. (1994) Genetic Programming II. Cambridge, MA: The MIT Press. In Press.

Koza, J. R., M. A. Keane and J. P. Rice (1993), Performance improvement of machine learning via automatic discovery of facilitating functions as applied to a problem of symbolic system identification. 1993 IEEE International Conference on Neural Networks, San Francisco. Piscataway, NJ: IEEE 1993. Volume I. Pages 191-198.

Koza, J. R., and J. P. Rice (1992), Genetic Programming: The Movie. Cambridge, MA: The MIT Press.

Koza, J. R., and J. P. Rice (1994), Genetic Programming II Videotape: The Next Generation. Cambridge, MA: The MIT Press. In Press.

6 Alternatives in Automatic Function Definition: A Comparison of Performance

Kenneth E. Kinnear, Jr.

Two approaches to the automatic definition of functions are compared, Koza's Automatically Defined Functions (ADF) and Angeline and Pollack's Module Acquisition (MA). Their effect on the likelihood of evolving a correct solution to the even-4-parity problem is contrasted, with the use of ADFs causing a significant improvement and MA having no apparent effect. Through a variety of experiments the differences in these approaches are explored. Ultimately it is concluded that the ADF approach creates a particular form of *structural regularity* that strongly increases the likelihood of evolving a correct solution to the even-4-parity problem – a form of structural regularity not present in the MA approach. A similar type of structural regularity can be created by a new genetic operator called modular crossover, created from the primitives used in the MA approach.

6.1 Introduction

Certainly one of the most challenging and potentially rewarding areas for research in genetic programming consists of finding ways to successfully apply it to larger and more complex problems. Progress in this general direction is being made on a number of fronts.

When using genetic programming, it quickly becomes clear that for most problems the primitives chosen for the problem representation have an enormous influence on how difficult the problem is to solve using GP. While it is frequently possible to choose a problem representation which makes an intractable problem tractable, doing so rarely generalizes into a technique which can be used on a different problem.

What is needed is automatic discovery of problem representations which make a given problem more tractable. This is analogous to the creation of subroutines or sub-functions when done by a human programmer. The same process can also be viewed as finding a way to decompose a problem into smaller, potentially more easily solvable subproblems.

Two approaches have been proposed to automatically perform these operations, generically known as *automatic function definition*, and the work described here compares them: Automatically Defined Functions (ADF) [Koza 1992, 1993, 1994a, 1994b, 1994c] and Module Acquisition (MA) [Angeline and Pollack 1992, 1993a, Angeline 1994].

At first glance, both approaches seem fundamentally similar though different in many details, and one might expect them to show similar increases in performance, yet they do not. *Performance*, for the purposes of the work reported here, refers to the probability that a given run will evolve a complete solution to the problem. Comparison of the ADF approach with the MA approach shows that for the even-4-parity function the ADF approach yields a significant increase in performance, while the MA approach does not show a similar improvement compared to a baseline that uses neither technique. The work described here examines the reason for this disparity.

Section 6.2 describes the problem used for the experiments in this chapter. Section 6.3 describes the two approaches to automatic function definition, Section 6.4 discusses the experimental environment used, and Section 6.5 compares and contrasts the processes used by each as well as the effect of each on the performance of the even-4-parity problem. Further, it explores various possibilities for the differences in performance of the two approaches.

6.2 The Even-4-Parity Problem

The test problem used for these experiments is the even-4-parity problem, which is discussed at some length in [Koza 1992, Chapter 20] as well as [Koza 1993]. It takes 4 boolean input variables: **d0**, **d1**, **d2**, **d3**, and the desired output is **t** if an even number of the input variables are **t**, and **nil** otherwise. The functions allowed are: **and**, **or**, **nand** and **nor**. Table 6.1 describes the problem in more detail.

Table 6.1
Tableau for the Even-4-Parity Function, With and Without ADF

Objective:	Evolve an expression which will return **t** when an even number of the arguments are **t** and **nil** otherwise.
Terminal set:	Generally, **d0**, **d1**, **d2**, **d3**, corresponding to the 4 boolean input variables. May include **arg0**, **arg1**, and **arg2** when using ADF.
Function set:	**and**, **or**, **nand**, **nor** throughout. May include **adf0** and/or **adf1** in ADFs, and various modules in MA.
Fitness cases:	All 16 combinations of the 4 boolean input variables.
Raw fitness:	The number of fitness cases for which the expression fails to generate the correct value of the even-4-parity function. Equivalent to the Hamming distance to the correct solution.
Standardized fitness:	Same as the raw fitness.
Hits:	Not used.
Wrapper:	Not used.
Parameters:	Population size (M) = 1000. Maximum generations (G) = 50.
Success predicate:	When one individual has a raw fitness = 0.

Rules of Construction:	None in the Basic or MA tests. In the ADF tests, the template is:
	`(prog3 (defun adf0 (arg0 arg1) (area0))`
	`       (defun adf1 (arg0 arg1 arg2) (area1))`
	`       (values (area2)))`
	area0: functions: **and**, **or**, **nand**, **nor**, terminals: **arg0**, **arg1**
	area1: functions: **and**, **or**, **nand**, **nor**, **adf0**, terminals: **arg0**, **arg1**, **arg2**
	area2: functions: **and**, **or**, **nand**, **nor**, **adf0**, **adf1**, terminals: **d0**, **d1**, **d2**, **d3**

Types of points:	Three evolvable areas. See Section 6.3.1.

6.3   Automatic Function Definition: Two Approaches

At present, there are two fundamental approaches for the generic process of *automatic function definition* which can then be used as part of the solution to a particular problem. This section describes each of them in detail.

6.3.1   Automatically Defined Functions: ADF

Automatically Defined Functions (ADFs) were first discussed by John Koza in [Koza 1992, Chapters 20, 21]. For an excellent introduction to ADFs, see Chapter 2 of this volume [Koza 1994a]. For examples of ADFs in use, see Chapter 5 [Koza 1994b], Chapter 9 [Teller 1994] and Chapter 18 [Handley 1994]. *Genetic Programming II* [Koza 1994c] has ADFs as its central theme.

Essentially, in the ADF approach, each individual in the population contains an expression, called the *result producing branch*, that is evaluated to determine the fitness of the individual, as well as definitions of one or more functions (ADFs) which may be called by the result producing branch. The expression for the result producing branch evolves, as do the expressions which define the ADFs. The result producing branch can call the ADFs, and some of the ADFs can be defined to call others. It is important to avoid the potential for recursive call loops among the ADFs.

A constrained syntactic structure is required in order to allow evolution of the individuals, since distinct function and terminal sets must be created for the bodies of the ADFs as well as the result producing branch. The constrained syntactic structure is expressed as a template.

A template for a two function, hierarchical ADF individual is shown in Figure 6.1. Only the parts of the template shown in Figure 6.1 in *italics* are evolvable, and these are the parts generated randomly in the initial population. The parts not in italics are the same in each individual in the population, both in the initial random generation and throughout the run. During execution, the **prog3** function acts as the connective between the result producing branch headed by the **values** function and the two function defining branches, each headed by a **defun** macro. The two functions **ADF0** and **ADF1** are defined, and then the result producing branch is executed, which may or may not call either **ADF0** or **ADF1**.

The individuals in the population thus have a particular structure, and cannot be evolved in the straightforward way that is usual in genetic programming. Evolution of such constrained syntactic structures requires modification to several genetic operators. The template above defines distinct function and terminal sets for the bodies of the ADFs as well as the result producing branch. The crossover operator usually used in GP only works between expression trees having the same function and terminal set. This is not an issue in a standard GP run since there is only one function and terminal set.

```
(PROG3
 (DEFUN ADF0  (ARG0 ARG1)  (evolvable area 0))
 (DEFUN ADF1  (ARG0 ARG1 ARG2)  (evolvable area 1))
 (VALUES  (evolvable area 2)))
```

evolvable area	allowable functions	allowable terminals
0	and or nand nor	arg0 arg1
1	and or nand nor adf0	arg0 arg1 arg2
2	and or nand nor adf0 adf1	d0 d1 d2 d3

Figure 6.1
ADF Template

In order to allow creation of a crossover operator which operates correctly in the ADF environment, the number of evolvable *areas* is specified when the template is defined, and a function is created which, when given an area number, will access that area. Then, at crossover time, an area is chosen with uniform random selection, that area is accessed, and then the crossover point for each parent is chosen from within that area.

Note that **ADF1** in the template above can call **ADF0**, and that the result producing branch can call either **ADF0** or **ADF1**. This is specified by the function sets for areas 1 and 2, respectively.

An evolved solution for the even-4-parity problem using ADFs is shown below:

```
(PROG3
 (DEFUN ADF0  (ARG0 ARG1)  (NOR ARG0 ARG1))
 (DEFUN ADF1  (ARG0 ARG1 ARG2)  (NAND (OR ARG0 ARG1)
                                      (NAND ARG1 ARG0)))
 (VALUES (ADF1 (ADF1 D0 D3 D2)
               (ADF1 D1 D2 D3)
                D1)))
```

Note that **ADF0** is not used by the result producing branch, and **ARG2**, the third argument of **ADF1**, is likewise not used.

Key points to note about ADFs are:

• The number of available automatically defined functions and the number of arguments of each must be defined by the experimenter.

• The allowable function and terminal sets for each function must also be specified in advance. The function set governs not only the functions used by any particular area (as usual), but implicitly defines the allowable call hierarchy.

• The function definitions are not global, but rather local to each individual. When a call to **ADF0** is moved by crossover from one individual to another, it refers to a **different** **ADF0** in the new individual.

6.3.2   Module Acquisition: MA

Module acquisition (MA) was developed by Peter Angeline and Jordan Pollack [Angeline and Pollack 1992, 1993a], and is discussed in Chapter 4 [Angeline 1994]. Module acquisition creates a library of unique modules, which are globally defined, and which essentially extend the function set for all individuals.

Module acquisition operates like a new genetic operator on a traditional GP population which has no constrained syntactic structure. Every so often, a module is *acquired* by selecting a sub-tree within an existing individual and making it into a globally defined module. This is similar to the *encapsulation* operation discussed by Koza in [Koza 1992, Pages 110 & 601]. The difference appears when the subtree defining the module is trimmed off at a random depth, thus defining a module (function) with parameters, with the parts of the sub-tree that are trimmed off becoming the actual parameters. Figure 6.2 shows this graphically.

In this example, the shaded sub-tree has been selected as a module. The trim depth determines the point below which the sub-tree is considered parameters to the module,

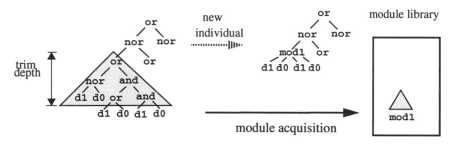

Figure 6.2
Module Acquistion: Schematic Diagram.

instead of part of the definition of the module. In Figure 6.2, the trim depth is set to 3, although usually the trim depth is randomly chosen between two limits. The sub-tree is moved into the module library with four formal parameters and the name **mod1**, and a new individual is created with a heretofore unknown four parameter function, **mod1**. This same sequence of operations is shown in Figure 6.3 from a different perspective.

Modules in the module library do not evolve, and are retained as long as any individual references them. While the initial definition of a module creates only one individual which references it, if the module confers increased fitness on offspring that contain it, the number of references will expand dramatically. This is an example of *constructional fitness* as discussed in Chapter 3 [Altenberg 1994]. Care must be taken with the module acquisition rate in order to keep the size of the module library within bounds.

Another operation called *module expansion* takes an individual which has references to modules within it, and expands all of the modules that it references to create a new individual with no module references. Module expansion allows the genetic material in a module to participate again in crossover and other genetic operations. (Angeline and Pollack [1993a] only expand a single module reference.) Neither acquisition nor expansion of modules has any direct effect on the fitness of the individuals involved.

Some points to note about module acquisition:

- Modules are acquired automatically. The trim depth limits must be specified, although there is random choice between those limits.

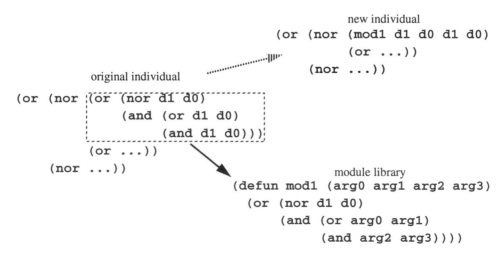

Figure 6.3
Module Acquistion: The same example as Figure 6.2 shown in code.

- Modules, once defined, don't evolve. The genetic material in modules does come back into play through module expansion, but its identity as a module is lost at that point, and it is highly unlikely that a similar module would be defined from that material at a later time.

- Modules can never be recursive, although they may call other modules to any arbitrary depth.

- As defined, modules use each argument exactly once (which is why **d1** and **d0** are repeated in the call to **mod1** above). It would be possible to compare the parameter subtrees for equality and reduce the number of formal and actual parameters, but it is unlikely to yield a significant increase in the multiple use of parameters. Since module definitions don't evolve, there is no possibility of generating multiple uses of a parameter through evolution.

- Modules, as defined above, potentially leave many terminals in the module. This results in modules which use global variables.

- Module definitions contain the same function and terminal set as the original population.

6.4   The Genetic Programming Environment

This section describes the GP environment used to produce the reported results. The experiments reported in Section 6.5 use three distinct approaches, with several variations on each:

- **Basic**: where the individuals are generated with the functions **and, or, nand, nor** and terminals **d0, d1, d2**, and **d3**. This is the basic even-4-parity problem with no embellishments.

- **ADF**: where the template described in Section 6.3.1 is used, and there are two subfunctions designed to aid in the solution to the even-4-parity problem.

- **MA**: where the even-4-parity problem is solved in an environment which uses module acquisition.

The GP environment, implemented by the author in Common LISP, differs from other implementations by using *steady state* GP, as well as employing some unusual genetic operators. The results presented are believed to be independent of the specific GP environment used.

Lisp was used as the implementation language, and the syntax of the expressions presented in this chapter is that of Lisp, however these results could have been obtained with a GP system implemented in any language.

6.4.1   Steady State GP

Steady state GP is used, as developed by Craig Reynolds [Reynolds 1992] from work by Gilbert Syswerda on steady state genetic algorithms [Syswerda 1991]. John Holland originally discussed a variety of genetic plans in [Holland 1975, 1992], including one that was essentially steady state. Since that time, several researchers have pursued steady state genetic algorithms [De Jong 1975, 1993], [Davis 1991], [Whitley 1989].

Briefly, steady state GP differs from generational GP in that individuals are created one at a time, evaluated immediately for fitness, and then merged into the population in place of an existing low fitness individual. In contrast, in generational GP an entirely new population of individuals is created each generation, and then they are all evaluated for fitness. In both cases the population size remains constant. Steady state GP as defined by Reynolds and implemented here also guarantees the uniqueness of each individual in the population, thus maximizing diversity. A performance comparison of steady state GP to generational GP can be found in [Kinnear 1993a].

Since strictly speaking there are no generations in steady state GP, a *generation equivalent* has passed when the number of new individuals that have been generated is equal to the population size – 1000 for the work reported here. Throughout the work reported here a generation should be taken to mean a generation equivalent.

6.4.2   Selection Procedures

Tournament selection with a tournament size of seven is used. Uniform random selection over the population is used to select the seven individuals for the tournament, and the one with the best fitness is used. In the explanations of the genetic operators below, the phrase *based on fitness* is used to mean this form of tournament selection. Since steady state GP requires replacement of a low fitness individual with each newly created individual, a separate inverse tournament (where the worst instead of the best individual is selected from the group), again with a size of seven, is used to select the low fitness individual to be replaced.

6.4.3   Genetic Operators

The following genetic operators are used to generate the new individuals. In all cases where points are chosen inside of existing functions, they are chosen with a 90% probability of being a function point and a 10% probability of being a terminal point, after [Koza 1992].

- **Single Crossover**: where two individuals are selected based on fitness, and a point is chosen inside a copy of each. A single new individual is created by replacing the subtree selected in one individual with the subtree from the other individual. The resulting new individual (or current area) is restricted from growing deeper than a depth of 10, and the crossover points are rechosen until this restriction is satisfied.

- **Non-fitness Single Crossover**: just like single crossover, except that the two individuals are selected using uniform random selection over the entire population instead of based on fitness.

- **Mutation**: where an individual is selected based on fitness and then an internal point in a copy of the individual is replaced by a randomly generated tree. The resulting individual (or current area) is restricted from growing in depth by more than 15%.

- **Hoist**: (a special case of single crossover) where a new individual is created by selecting a point inside a copy of an existing individual chosen based on fitness, and elevating ("hoisting") it up to be the entire new individual (or current area).

- **Create**: (a special case of mutation) where an entirely new individual is created in the same way as the initial random generation. This locates a random new point of departure in the search space.

- **Acquire Module**: where a new individual and a new module are both created from an existing individual chosen based on fitness, as described in Section 6.3.2.

- **Expand Module**: where a new individual is created by first, choosing an individual based on fitness which contains at least one call to a module, and then expanding all of the modules in that individual until there are no module calls remaining.

The table at the end of this section shows the default genetic operator probabilities for all tests.

What motivates these operators? The combination of single crossover (essentially a standard genetic operator) and non-fitness single crossover (decidedly non-standard) was shown in [Kinnear 1993a] to improve the performance on the sorting problem described there, and likewise improves performance for the even-4-parity problem using ADFs. When using non-fitness single crossover the cumulative probability of success rises more slowly than without, but tends to plateau out at a higher level. Presumably non-fitness single crossover allows potentially valuable fragments of genetic material held within otherwise undistinguished individuals to come together over several crossover operations, thus more thoroughly exploring the space of possible individuals.

Hoist and create, both rather non-standard operators, were also introduced in [Kinnear 1993a] as special cases of crossover and mutation, respectively, and are designed to increase the likelihood that the resulting individual will be smaller than its parent.

Operator	Probability of Selection	
	Basic and ADF	MA
Single Crossover	.35	.30
Non-fitness Single Crossover	.35	.30
Mutation	.10	.08
Hoist	.10	.08
Create	.10	.08
Acquire Module		.08
Expand Module		.08

6.5   Comparisons of ADF and MA strategies

The following experiments compare the ADF strategy and the MA strategy against the Basic strategy, where neither approach is used. The comparisons are made on the basis of *performance*, as expressed by the *cumulative probability of success* which is the probability of evolving a correct solution by a particular generation on a given run.

6.5.1   Basic Performance Comparison

Figure 6.4 compares the Basic strategy with both the ADF and the MA strategy for the even-4-parity problem. Two things are worth noting; first, that the difference between the ADF and MA is so great and second, that the MA causes essentially *no change* from the basic strategy. It doesn't help, but it doesn't degrade performance either.

Both ADF and Modules were also run against the sorting problem described in [Kinnear 1993a, 1993b], where the goal was to evolve a generalized sorting algorithm which would sort any sequence of numbers presented to it. The MA approach again caused no change in performance. The several ADF approaches tried all strongly degraded the performance over that without ADFs. Part of the reason for this could be because implementing ADFs for this problem was quite complicated, and no particularly natural solution was found. Further work remains to confirm these results and explore the reasons for this behavior.

Throughout the remainder of this chapter, all experiments use the even-4-parity problem.

The following sections explore in greater detail the differences between the ADF approach and the MA approach with experiments to clarify the key distinctions which affect performance.

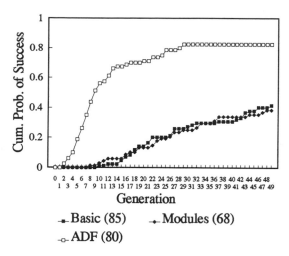

Figure 6.4
Comparison of the three approaches: Basic, using just normal functions and terminals; Modules, like Basic but with module operators employed; and ADF, using automatic function definition. Only ADF shows an increase in performance. Each curve represents an average over the number of runs in parentheses.

6.5.2   Use of *a priori* Knowledge

A fundamental difference between ADFs and MA is that with ADFs a variety of items must be specified prior to the run, while in MA these items evolve during the run. In the ADF approach, the number of the functions, how many arguments they each have, and the allowable calling hierarchy are all decided by the user in advance of the GP run. In the MA approach only the trim depth limits (which affect but do not closely determine the number of parameters) and the rate at which modules are acquired are pre-determined. All else is determined by the natural action of the genetic operators. This distinction, while significant, seems inadequate to explain the difference in performance. Inspection of the modules evolved by MA shows modules with reasonable numbers of parameters, comparable to the human specified functions in ADFs, with no similar increase in performance.

6.5.3   Frequency of Function Calls

Given the way that ADFs are generated, the frequency of function calls to automatically defined functions is considerably higher than with MA. This is because the function names and parameter sizes are known and included in the function sets used to generate the initial random generation. On the other hand, in MA, there are no modules in the initial random generation. All module acquisition happens during the progress of a run, and the module acquisition rate must be limited to keep the module library size within bounds.

Because of the difference in density of ADF calls compared to module calls in the initial population, an experiment was designed to determine whether this could explain the difference in performance between the two approaches. This experiment lowers the frequency of ADF calls in the initial random generation (and during operation of the mutation and create operators).

Figure 6.5 shows the results. The probability of selecting each function is altered from the normal, uniform probability, in two different ways, called *low* and *moderate* (Mod) described below.

Area	Function	ADF	Mod	Low
1	**ADF0**	.20	.12	.12
1	**and, or, nand, nor**	.20	.22	.22
2	**ADF0, ADF1**	.17	.10	.06
2	**and, or, nand, nor**	.17	.20	.22

The ADF approach shows remarkable sensitivity to the frequency with which ADF function calls are generated in the initial random generation (and during operation of the mutation and create operators). The low and moderate runs differ by only 4% in the likelihood that either **ADF0** or **ADF1** will appear as functions at a particular node in the result producing branch, yet the effect is dramatic.

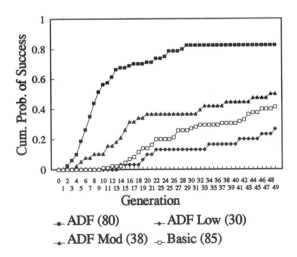

Figure 6.5
Low and moderate ADF call frequencies show a significant difference from the standard ADF approach, which represents a high ADF call frequency. Each curve represents an average over the number of runs in parentheses.

Notice that the graphs are still climbing as they reach generation 49. This probably indicates that the genetic operators (principally crossover) are creating the multiple copies of calls needed, since they are not present in the initial generation. The runs with a very low frequency of function calls (ADF Low) are worse than the basic, non-ADF approach, possibly because the genetic operations are being wasted on the automatically defined functions – functions which are called only infrequently in these runs.

6.5.4 Functions Including Global Variables

The automatically defined functions in ADFs do not, in this problem at least, include any terminals other than the formal parameters of the function. The modules in MA, on the other hand, frequently include terminals, though they can be modified to prevent this.

In order to determine whether the performance of ADF approach is due to the exclusion of terminals from ADF, an experiment was performed where terminals were allowed to appear in the ADFs, but only through mutation during the run. Figure 6.6 shows the degradation in performance when no terminals are allowed in the initial random generation, but they are allowed to leak in **only during mutation**.

The normal case of MA is to allow terminals in the modules, but an experiment to test the effect on performance of excluding terminals from modules by forcing all terminal sites to become parameters is shown in Figure 6.7. No change in performance is observed.

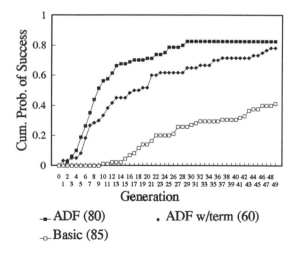

Figure 6.6
When terminals are allowed to appear in small numbers in automatically defined functions, a degradation of performance is noted. Each curve represents an average over the number of runs shown in parentheses.

6.5.5   Multiple Use of Parameters

In the ADF approach, parameters are used and reused many times, especially in the case where the only terminals allowed in the automatically defined functions are the formal parameters of the function. This stands in contrast to the MA approach, where the parameters are in general used exactly once. While MA without terminals allowed in the modules does involve some reuse of parameters, it is unlikely to be significant.

The degradation in performance in ADFs when terminals other than the formal parameters are introduced through mutation into the automatically defined functions (see Section 6.5.4) can be explained as the result of fewer uses of each formal parameter. To be conclusively demonstrated, this would require further testing.

6.5.6   Local vs. Global Function Definitions

In the MA case, once a module is defined, its definition is fixed for all calls. Thus, when a call to **mod2**, say, moves via crossover into a different individual, the definition and operation of **mod2** is the same as it is in the individual where it was originally acquired. Of course its ultimate value is based on the parameters with which it is called, which in general will vary. In the ADF case, on the other hand, when a call to **ADF1** moves from the result producing branch in one individual to another, in general it will be referring to a different **ADF1** – one local to the individual in which the call to **ADF1** appears.

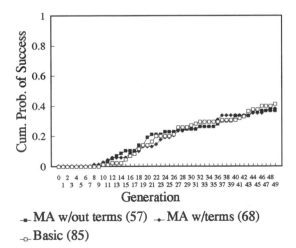

Figure 6.7
When terminals are excluded from modules, no effect on the performance is seen.

An experiment was designed to give some indication of the sensitivity of the even-4-parity problem to the issue of global vs. local function definitions. This experiment uses an approach that is something of a hybrid, and uses some convenient features of the even-4-parity problem. The basic individuals are generated without ADFs or MA enabled. One of the functions that may appear in the initial random generation (with low probability) is a call to **another** individual chosen using uniform random selection. When the call to the other individual is executed, it considers that individual a function of four arguments, evaluates the arguments, binds them to the terminals $d0$, $d1$, $d2$, and $d3$, and then calls the individual.

Of course, at some point the called individual is likely to be replaced by a new individual (with a probability inversely proportional to its fitness). At that time, the definition of the individual will change. This provides a middle ground between the totally global definitions of modules and the totally local definitions of ADFs. Each individual may contribute to the solution to the problem in two ways. One is measured by its fitness as a direct solution to the overall problem. The other is measured by its contribution as a solution to a sub-problem in combination with another individual's attempt at a solution to the overall problem. For the purposes of this test, the only recognition that a particular individual is used by other individuals as a solution to a sub-problem is that the number of times it is used in that manner is counted. That information is referenced only when the tournament is held to choose a low fitness individual to replace with a new individual in the steady state flow of individuals into the population. At that time, an individual's raw fitness (only for the purposes of this tournament) is reduced by a factor multiplied by the number of times it is used.

This approach performs better than the Basic approach in the early generations, but after 15 generations or so it is equivalent to the Basic approach.

One way to think about this test is to view the individuals which had low fitness as solutions to the overall problem but whose fitness is altered during the tournament for replacement because they are good solutions to a sub-problem, as the equivalent of sterile workers in some biological colony. They tend to live on because they are assisting other members of the colony in solving the problem, but they do not, in general, reproduce.

6.5.7 Evolution of Function Definitions

Another significant difference between ADFs and MA is that, in ADFs, the function definitions of the automatically defined functions evolve throughout the run, while in MA the modules that have been defined do not. Certainly, in MA they are brought back into the evolutionary process by module expansion, but this is not the same as undergoing evolution while maintaining their identity as separate modules. The significance of this remains to be tested with an experiment.

6.5.8   Structural Regularity

ADFs contain a large number of calls to the automatically defined functions (and frequently the same function) as well as frequent multiple use of parameters. This creates a situation where the result producing branch exhibits a large degree of *structural regularity*. One way to view this is to imagine the result producing branch of an ADF with all of the calls to all of the automatically defined functions completely expanded. Generally, there would be a large number of partially similar sub-trees in the result. They would be **similar**, because the tree that defines the automatically defined function is repeated wherever that function is called. They would be **partially** similar because the replacement of the formal parameters by the actual parameters would tend to distinguish them, since the actual parameters would rarely be the same.

This form of structural regularity is not directly created by modules, although conceivably it could evolve.

The existence of this structural regularity in the ADF approach is clear, but does it contribute to the performance gains displayed by ADFs? This section contains a number of experiments designed to create similar forms of structural regularity in the even-4-parity problem without using ADFs and to measure the effect on performance.

6.5.8.1   Self-Crossover

A simple experiment in structural regularity is possible. An additional genetic operator called *self-crossover* is defined. This operator chooses a parent based on fitness in the normal way, and then **always** uses the same parent for the crossover operation. This tends to generate a certain amount of sub-tree similarity within the new individual, similar to that found in ADFs.

Self-crossover was first suggested as a genetic operator by Peter Angeline. A precursor to self-crossover called *split mutation* was documented by Angeline and Pollack [1993b].

If the performance improvement of the ADF approach is based on the kind of structural regularity created by self-crossover, then the Basic approach with self-crossover should show some performance improvement.

Experiments using self-crossover as an adjunct to regular crossover were performed with all three approaches (though strictly only the one using the Basic approach was necessary). In each case, self-crossover had no effect on performance — the performance of each approach with and without self-crossover was identical. The crossover probabilities used are shown in Table 6.2, located at the end of Section 6.5.8.

6.5.8.2 Modular Crossover

In an attempt to generate structural regularity more like that of ADFs, *modular crossover* is defined. This genetic operator grew out of a discussion with Lee Altenberg where we jointly defined the operator. He calls this operator *cassette recombination* in Chapter 3 of this volume [Altenberg 1994].

Modular crossover can be viewed in two different ways. In the first view, modular crossover is similar in overall concept to the normal crossover operator, where a sub-tree is chosen in a copy of the first parent and replaced entirely with a copy of a sub-tree from the second parent. It differs in that only a portion of the sub-tree identified for replacement in the first parent is replaced. The root and some of its adjacent branches in the sub-tree chosen for replacement are replaced with the root and some of its adjacent branches from the sub-tree chosen in the second parent. The branches below a certain point in the first sub-tree remain, and are grafted onto the second sub-tree.

A second view of modular crossover is to conceptualize it through the use of modules. In this view of modular crossover, a module is defined in a copy of the first parent and another module is defined in a copy of the second parent. Then, the definition of the module defined in the first parent is **replaced** by the definition of the module defined in the second parent. Finally, the first parent undergoes module expansion, where the module previously defined is expanded in place **using the definition from the module defined in the second parent**. The actual parameters used in the module call in the first parent are still used in the expansion of the module.

In either of these views, there remains a major unresolved issue: how to match up the fragments of the sub-trees when inserting the fragment of the second sub-tree into the space left by the excision of the first or, alternatively, how to deal with mismatches in the number of formal and actual parameters when the module definition from the second parent is expanded in place of the module defined in the first parent.

This is discussed using the concepts of formal and actual parameters and the use of modules. Using this point of view, the number of formal and actual parameters of the module defined in the first parent will not in general match the number of formal parameters in the module defined in the second parent. When this is true, a number of adjustments are made. If the number of actual parameters in the module **call** in the first parent exceeds the number of formal parameters of the module **definition** from the second parent, then the formal parameters are bound to a randomly chosen subset of the actual parameters. If the number of formal parameters of the module defined in the second parent exceeds the number of actual parameters in the call from the first parent, then a new set of actual parameters is generated by extending the current actual parameter set with randomly chosen copies of the existing actual parameters. If there is no actual parameter set to extend, one is created by randomly choosing terminals.

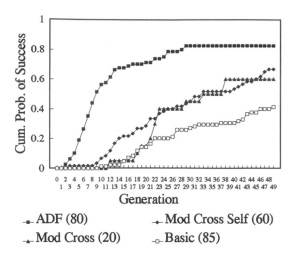

Figure 6.8
Modular crossover (both normal and self crossover) used with regular crossover shows signficant gains over the Basic approach. The curve for the MA approach is omitted because it follows the Basic curve. Each curve represents the average over the number of runs in parentheses.

This basic modular crossover operator can be used in several ways. It can be used as an adjunct to the normal crossover operators – *partial modular crossover*, or it can be used to replace the normal crossover operators – *full modular crossover*. In the latter case modular crossover and non-fitness modular crossover replace both single crossover and non-fitness single crossover. Modular crossover also has a self-crossover mode, yielding *partial modular self-crossover* and *full modular self-crossover*.

If the kind of structural regularity created by modular crossover is even partially responsible for the gains in performance demonstrated by the ADF approach, then the Basic approach using the various modular crossover operators should show performance gains.

Such gains are indeed demonstrated: when any form of modular crossover is used as a genetic operator with the Basic approach to the even-4-parity problem, significant performance gains result. Table 6.2 shows in detail just what operators are used with which probability to produce the results shown in Figures 6.8 and 6.9.

Where modular crossover is used as an adjunct to normal crossover (Figure 6.8) the performance is slightly higher then where modular crossover completely replaces regular crossover (Figure 6.9).

Note that Mod Cross Full Self in Figure 6.9 is better than the Basic approach. This is somewhat surprising, since in these runs **all** crossover is **self**-crossover. There is never any crossover between two parents — all crossover is one parent with itself.

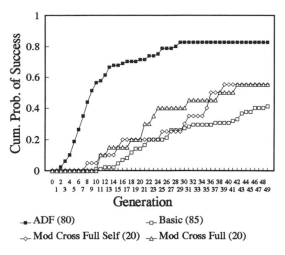

Figure 6.9
Modular crossover as the only crossover operator employed shows some gains over the Basic approach. Mod Cross Full Self is surprising, since *all* crossover in this series of runs was self crossover, and none was between two parents. The curve for MA was omitted since if follows the Basic curve. Each curve represents the average of the number of runs in parentheses.

6.5.8.3   Discussion of Structural Regularity

What is happening here? Why does modular crossover show such significant performance gains?

ADFs have been shown to depend heavily on the frequency of ADF function calls (Section 6.5.3), as well as the multiple use of parameters (Sections 6.5.5). It would seem that at least part of the performance gains shown by the ADF approach comes from creating a particular form of structural regularity — one that is also being created by modular crossover. This structural regularity is that of a sub-tree shorn of its deepest branches transplanted into another (or the same) individual.

Section 6.5.5 discusses the possibility that multiple uses of parameters is one of the factors responsible for the performance improvements of the ADF approach. The results from the modular crossover experiments don't contradict that hypothesis, since there is some likelihood of multiple use of parameters in modular crossover, due to the way that the module parameter lists are rationalized. Further testing is required to determine the extent to which this is a significant factor.

Producing performance gains through the creation of similar sorts of structural regularity first through the use of a particular representation using ADFs, and then through the

Table 6.2
Genetic Operator Probabilities for Experiments in Section 6.5.8

Operator	Probability of Selection			
	Basic and ADF Self Crossover Section 6.5.8.1.	MA Self Crossover Section 6.5.8.1.	Partial Modular Crossover Figure 6.8	Full Modular Crossover Figure 6.9
Single Crossover	.175	.15	.175	
Non-fitness Single Crossover	.175	.15	.175	
Mutation	.10	.08	.10	.10
Hoist	.10	.08	.10	.10
Create	.10	.08	.10	.10
Acquire Module		.08		
Expand Module		.08		
Self Crossover	.35	.30		
Modular Crossover (incl. self)			.35	.35
Non-fitness Modular Crossover (incl. self)				.35

use of a particular genetic operator like modular crossover is an excellent example of the concept of a *transmission function* described in Chapter 3 [Altenberg 1994].

Self-crossover is not really crossover in the sense that two parents' genetic material is being mixed through the process of crossover. Self-crossover is really a sophisticated mutation operator. Thus the curve in Figure 6.9 for Mod Cross Full Self is a curve where the only genetic operators used were various forms of a mutation operator!

The graphs in Figure 6.8 are intriguing, in that the curves for modular crossover appear to still be rising at generation 49. The experiments from Figure 6.8 were rerun to 99 generations to determine whether this was the case. The cumulative probability of success at generation 99 for each curve was unchanged from that shown for 49 generations in Figure 6.8.

6.6 Conclusions

In examining the reasons why automatically defined functions (ADFs) show such a dramatic increase in the likelihood that a given run will correctly solve the even-4-parity problem over both the Basic and Module Acquisition (MA) approach, the following is observed:

- The ADF approach owes some of its performance increase over the Basic approach to a particular form of structural regularity, similar to that created in the Basic approach by the various forms of the modular crossover genetic operator.

- The MA approach doesn't directly create this form of structural regularity.

- The structural regularity created by the ADF approach depends on large numbers of function calls and possibly also depends on multiple uses of the parameters of each ADF. If either of these becomes less frequent, the performance on the even-4-parity problem with ADFs decreases.

- A sophisticated mutation operator in the form of modular self-crossover (with additional mutation operators hoist and create) performs better on the even-4-parity problem than does the more traditional crossover operator.

- Clearly the MA approach protects some parts of the trees from being damaged by crossover. This may well be of real merit in some problems, but any effect it might have in the even-4-parity problem is negligible compared to the increase in performance over the baseline demonstrated by both the ADF approach and the modular crossover operator.

6.7 Further Work

Much remains to be learned about the key factors that make ADFs so powerful. Some of the future areas to explore include the following:

- Define a canonical form of modular crossover. See if it is better to push terminals out of the modules, or leave them in (see Section 6.5.4). See if the trim-depth parameters have a major bearing on the effectiveness of modular crossover. Explore the difference between modular crossover and modular self-crossover more completely.

- See if the results from this work hold for different problems. Is the boolean even-4-parity problem unique or even unusual in how it interacts with ADFs, or do the results shown here generalize to a wide class of other problems? How much (if at all) does modular crossover increase the solution likelihood of other problems? How are ADFs in these problems affected by changing frequencies of function calls on automatically defined functions and introduction of terminals into these functions?

- Are there problems for which other forms of structural regularity are appropriate, different than those introduced by ADFs and modular crossover? How important is structural regularity in general?

- Are there problems for which the sub-tree protection afforded by MA would show a performance increase?

Acknowledgments

A discussion of self-similarity at ICGA-93 with Peter Angeline, Lee Altenberg, and Nicholas McPhee led me directly toward the issues of structural regularity. Peter Angeline suggested self-crossover as a potential operator. Modular crossover was jointly invented in a discussion with Lee Altenberg. There were many fascinating discussions at ICGA-93. It is impossible to thank all of those who participated individually, but I would like to acknowledge them all collectively.

Thanks to John Koza and James Rice for all of the work on GP in general and ADFs in particular.

Special thanks to my wife Karen, who provided constant support and encouragement as well as technical and editorial review.

Bibliography

Altenberg, L. (1994) "The Evolution of Evolvability in Genetic Programming", in *Advances in Genetic Programming*, K. E. Kinnear, Jr., Ed. Cambridge, MA: MIT Press.

Angeline, P. J. (1994) "Genetic Programming and Emergent Intelligence" in *Advances in Genetic Programming*, K. E. Kinnear, Jr., Ed. Cambridge MA: MIT Press.

Angeline, P. J. and Pollack, J. B. (1992) "Coevolving High-Level Representations." LAIR Technical Report 92-PA-COEVOLVE, The Ohio State University, Columbus, OH. To appear in *Artificial Life III*, Santa Fe Institute Studies in the Sciences of Complexity, C. G. Langton, Ed. Reading MA: Addison-Wesley

Angeline, P. J., and Pollack, J. B. (1993a) "Competitive Environments Evolve Better Solutions for Complex Tasks," in *Proceedings of the Fifth International Conference on Genetic Algorithms*, S. Forrest, Ed. San Mateo, CA: Morgan Kaufmann.

Angeline, P. J., and Pollack, J. B. (1993b) "Evolutionary Module Acquisition", in *Proceedings of the Second Annual Conference on Evolutionary Programming*, La Jolla, CA: Evolutionary Programing Society

Davis, L. (1991) *Handbook of Genetic Algorithms*. New York, NY: Van Nostrand Reinhold.

De Jong, K. A. (1975) *An Analysis of the Behavior of a Class of Genetic Adaptive Systems*, Doctoral Thesis, Department of Computer and Communication Sciences, University of Michigan, Ann Arbor, MI.

De Jong, K. A. (1993) "Generation Gaps Revisited," in *Foundations of Genetic Algorithms, 2*, L. D. Whitley, Ed. San Mateo, CA: Morgan Kaufmann.

Handley, S. G., (1994) "The Automatic Generation of Plans for a Mobile Robot via Genetic Programming with Automatically Defined Functions" in *Advances in Genetic Programming*, K. E. Kinnear, Jr., Ed. Cambridge MA: MIT Press.

Holland, J. H. (1975) *Adaptation in Natural and Artificial Systems*. Ann Arbor, MI: The University of Michigan Press.

Holland, J. H. (1992) *Adaptation in Natural and Artificial Systems, (first MIT Press edition)*. Cambridge, MA: MIT Press.

Kinnear, K. E. Jr. (1993a) "Evolving a Sort: Lessons in Genetic Programming," in *Proceedings of the 1993 International Conference on Neural Networks*, New York, NY: IEEE Press.

Kinnear, K. E. Jr. (1993b) "Generality and Difficulty in Genetic Programming: Evolving a Sort," in *Proceedings of the Fifth International Conference on Genetic Algorithms*, S. Forrest, Ed. San Mateo, CA: Morgan Kaufmann.

Koza, J. R. (1992) *Genetic Programming: On the Programming of Computers by Means of Natural Selection.* Cambridge, MA: MIT Press.

Koza, J. R. (1993) "Hierarchical Automatic Function Definition in Genetic Programming," in *Foundations of Genetic Algorithms, 2,* L. D. Whitley, Ed. San Mateo, CA: Morgan Kaufmann.

Koza, J. R. (1994a) "Introduction to Genetic Programming" in *Advances in Genetic Programming,* K. E. Kinnear, Jr., Ed. Cambridge MA: MIT Press.

Koza, J. R. (1994b) "Scalable Learning in Genetic Programming using Automatic Function Definition" in *Advances in Genetic Programming,* K. E. Kinnear, Jr., Ed. Cambridge MA: MIT Press.

Koza, J. R. (1994c) *Genetic Programming II.* Cambridge, MA: MIT Press.

Reynolds, C. W. (1993) "An Evolved, Vision-Based Behavioral Model of Coordinated Group Motion," in *From Animals to Animats 2: Proceedings of the Second International Conference on Simulation of Adaptive Behavior,* J. A. Meyer, H. L. Roitblat, and S. W. Wilson, Eds. Cambridge, MA: MIT Press.

Syswerda, G. (1991) "A Study of Reproduction in Generational and Steady-State Genetic Algorithms.", in *Foundations of Genetic Algorithms,* G. J. E. Rawlins, Ed. San Mateo, CA: Morgan Kaufmann.

Teller, A. (1994) "The Evolution of Mental Models" in *Advances in Genetic Programming,* K. E. Kinnear, Jr., Ed. Cambridge MA: MIT Press.

Whitley, D. (1989) "The GENITOR Algorithm and Selection Pressure: Why Rank-Based allocation of Reproductive Trials is Best." in *Proceedings of the Third International Conference on Genetic Algorithms,* J. D. Schaffer, Ed. San Mateo, CA: Morgan Kaufmann.

7 The Donut Problem: Scalability, Generalization and Breeding Policies in Genetic Programming

Walter Alden Tackett and Aviram Carmi

"Differentiation, due to the cumulative effects of accidents of sampling, may be expected in actual cases to be complicated by the effects of occasional long range dispersal, mutation, and selection but in combination with these it gives the foundation for much more significant evolutionary processes than these factors can provide separately." Sewall Wright

"No aphorism is more frequently repeated in connection with field trials, than that we must ask Nature few questions or, ideally, one question, at a time. This writer is convinced that this view is wholly mistaken. Nature, he suggests, will best respond to a logical and carefully thought out questionnaire; indeed, if we ask her a single question, she will often refuse to answer until some other topic has been discussed." Ronald A. Fisher

The Donut problem requires separating two toroidal distributions (classes) which are interlocked like links in a chain. The cross-section of each distribution is Gaussian distributed with standard deviation σ. This problem possesses a variety of pathological traits: the mean of each distribution, for example, lies in the densest point of the other.

The difficulty of this problem is scalable in two different ways: 1) Overlap: as we increase the standard deviation so that points from the two distributions "intermingle" a classifier which optimally minimizes misclassification will still make mistakes in classifying outlying points; 2) Generalization: by varying the size of the training set, we control the degree to which it is statistically under representative. In testing generalization we examine the effects that occur when training data are uniformly sparse as well as when the training set has spatially nonuniform "bites" taken out of it. Although there is no perfect solution for this problem we formulate an optimal solution that minimizes error as the basis for comparison.

We observe the effects of different breeding policies upon performance under varying conditions, including panmictic versus distributed evolution ("demes") with and without steady-state elitist population update. For distributed breeding, we investigate different sizes and configurations for breeding units. In all, over 2,000 runs are performed, each with a population of 1,024 individuals, over a period of 50 generations.

7.1 Introduction: Depth vs. Breadth

In order for Genetic Programming (GP) to be useful as an engineering tool it must work on difficult problems. How does performance of GP change as we make the problem more difficult, or easier? It is hard to answer these questions when we have no control over the data: indeed, it is important to understand that the notion of difficulty is problem-specific. In the research described here we formulate a problem so that the factors that make it hard are quantifiable and therefore scalable. Koza's work demonstrates the breadth of applicability of genetic programming, (GP) [Koza 1992]; However, many of his problems have known perfect solutions. This is of value in understanding the ability of GP methods to find optimal programs, but invites criticism that knowledge of the solution may bias the choice of function and variable sets towards those which make solutions trivial. This leaves open a question whether GP is of practical use to the scientist and engineer.

Previous work by Tackett concerned a difficult classification problem involving real-world data where the solution is unknown and only an elementary set of functions and variables are used [Tackett 1993]. In particular, features input to a Genetic Programming system were extracted from infrared images using a noisy and error-prone feature extraction process. These features were then classified as belonging two one of two categories: either *targets* (e.g., tanks and trucks) or *non-targets* (e.g., rocks and bushes). There, comparison of performance to existing methods indicates that genetic programming indeed offers a variety of advantages. This is a promising result, but is it representative? Unfortunately, the elements, which make that problem attractive, e.g., significant noise and class overlap, unknown distributions of classes, also make it hard to analyze in order to understand the qualities that make it a good application for GP.

In this work we set out to construct a problem with elements that make it difficult in ways that are representative of problems found in the real world. Difficulty is quantified and parameterized so that we may adjust the scale of difficulty and analyze how the quality of solutions generated by GP changes in response.

In addition, we study breeding policies, particularly the distributed evolution scheme, due to Sewall Wright, as well as the steady state population model. The performance of these variants is studied in the context of difficult and scalable problems.

7.2   The Donut Problem

The Donut problem uses artificially generated data that is purposely formulated to be both difficult and scalable. The Donut problem requires separating two classes, each is toroidal in shape (a donut), and the two are interlocked like links in a chain. For each distribution, or class, sample coordinates are generated by choosing a point p from a uniform distribution lying on a circle of radius 1.0. Next, we construct a plane, which passes through p and through the circle's center, perpendicularly bisecting the circle. The final coordinates are then generated by choosing a point that lies in this plane, whose distance from p is Gaussian distributed with standard deviation σ. That is, each distribution is a torus, with a major radius 1.0, and a cross-section minor radius proportional to σ.

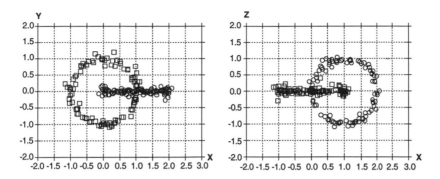

Figure 7.1
Front view and top view of two interlocking toroidal distributions (points from one distribution are shown as circles, the other as squares). For the case shown here, each contains 100 points and has a cross-section that is Gaussian distributed with a standard deviation of 0.1. By increasing the standard deviation we may increase interclass ambiguity to an arbitrary degree.

Here we will arbitrarily designate the two distributions as "class A" and "class B". The circular center line of class A lies in the X-Y plane with its center at the origin, while that of class B lies in the X-Z plane centered a the coordinates $(1, 0, 0)$. A front view and top view of this configuration are provided in Figure 7.1.

This problem possesses a variety of pathological traits: the distributions cannot be linearly separated by a Perceptron rule [Rosenblatt 1962, Minsky & Papert 1969], they cannot be covered by cones or hypercones [Fukushima 1989, Johnson et al 1988], nor can they be enclosed by a pair of radial basis functions [Poggio & Girosi 1989]. Moment statistics cannot adequately describe them: the mean of each distribution, for example, lies in the densest point of the other. Thus the geometric and topological properties of the problem space are an inherent source of difficulty in the Donut problem.

In real-world problems there is often a degree of ambiguity in class membership because features are unreliable, measurements are corrupted by noise, or because some members of one class truly fit in better with another. The Donut problem models this phenomenon in a scalable manner. As we increase the standard deviation σ, some points from the two distributions will intermingle more and more. Thus for sufficiently high values of σ, even a classifier that optimally minimizes misclassification will still make many mistakes in classifying outlying points. This is depicted in Figure 7.2.

Generalization, which is the ability to deal with sparse and nonuniform statistical covering of sample space, plays a particularly important role in everyday engineering applications. As an example of this consider the case of automotive crash tests. These involve features that consist of factors such as speed, angle, and type of object hit. Not only are these tests expensive to perform and hence subject to small sample size, but

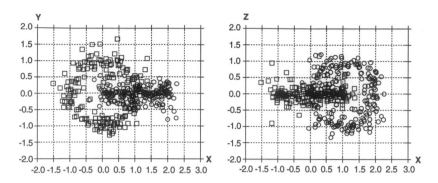

Figure 7.2
By increasing σ to 0.3 a significant degree of overlap develops between the classes.

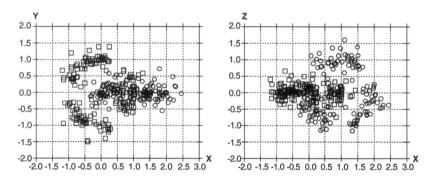

Figure 7.3
The Donut problem training set (σ=0.3) with 3 bites taken out of each class. How well can an evolved function f(x, y, z) which is trained against these data perform against an independent test set containing no gaps?

experiments are biased towards conditions that have features that conform to particular federal standards (e.g., head-on with a brick wall at 15 mph). It is of concern to the auto maker how these results extrapolate to crash features that are more commonly found in everyday traffic.

We can make generalization difficult in two different ways. First, we can provide a sparse set of samples as a training set: the points provided are uniformly representative of the underlying distribution, but are statistically under representative. Second, we can provide data that are not uniformly representative of the underlying distribution (e.g., a training set that looks like the donuts have bites taken out of them). The latter case is shown in Figure 7.3.

There is one other form of generalization that is desirable to test as well: in the real world the data available for training may occur in some idealized form that does not contain noise or distortions. Such factors may manifest themselves as class overlap. Therefore we wish to test the ability of GP to generalize with respect to class overlap: for example, how well does a classifier that was trained against the data of Figure 7.1 perform when tested against the data shown in Figure 7.2, and vice versa? This question may be answered using relatively little extra computation while performing the investigations outlined above.

We require genetic programming to produce solutions (classifying functions) f, where $f(x, y, z) \geq 0.0$ if the point belongs to class A, and $f(x, y, z) < 0.0$ if the point belongs class B. Fitness is based on minimizing the percentage of samples that are incorrectly classified. The function set used throughout our primary experiments consists of the primitive functions {+, -, *, %, IFLTE}, where % is protected division, IFLTE is if-less-then-else. The terminal set consists of {X, Y, Z, RANFLOAT}, where RANFLOAT is a random floating point constant: whenever this terminal is chosen for insertion into a randomly generated (sub)tree, is replaced by a real-valued constant drawn from a uniform random distribution on the interval [-10.0... 10.0]. Table 7.1 summarizes the key features of the Donut problem.

7.2.1 Purposely Introduced Imperfections

We may expect to see some functions that do better against the training set, that is, against the data, upon which they are trained, the error rate will be lower than that of the optimal classifier. This is a clear indication of overfitting, in which the functions

Table 7.1
Tableau for the Donut problem.

Objective:	Evolve an expression that will return ≥ 0.0 if the point belongs to class A, and < 0.0 if the point belongs to class B.
Terminal set:	X, Y, Z, Random constant
Function set:	+, -, *, %, IFLTE (or SQRT for some experiments)
Fitness cases:	A set of points and the class to which they belong, the set size varies with the experiments.
Raw fitness:	Number of correct classifications.
Standardized fitness:	1 - (Raw Fitness / Number of Fitness Cases)
Hits:	Not used.
Wrapper:	Not used.
Parameters:	Population size (M) = 1,024, Maximum generations (G) = 50.
Success predicate:	When one individual has standardized fitness = 0.

produced excel at correctly classifying the data that they have been trained with, at the expense of a poorer fit to the general distribution from which those samples were drawn. Such functions will not do as well against a statistically independent test set not presented during the training phase. Thus there is some imperfection in the fitness function itself, which is based strictly upon misclassification of the training set, since it may assign higher fitness to some functions than it would to the optimal one. This is in line with what happens in the real world, where information about the optimal solution cannot be used. Similarly, we note that the function set we have provided does not include the square-root function, which is an element of the optimal solution. As an experimental control, we run some tests using the square-root function to determine if it yields significant improvement.

7.2.2 There is a Solution (Sort of)

For class A, the expected value of a point lies on the circle $\{(X^2+Y^2 = 1), Z = 0\}$; for class B it lies on the circle $\{((X+1)^2+Z^2 = 1), Y = 0\}$. Given an unlabeled point (X, Y, Z), it is most likely to have come from the class whose expected value lies nearer: thus the best guess is to compute which circle lies nearer. (Actually, this best guess is an approximation to the true likelihood, and is strictly true only for $\sigma = 0$. For values of σ less than or equal to 0.5, the range used for our experiments, it is accurate to small fractions of a percent). The distance from any point to the circle μ_A, $(X^2+Y^2 = 1)$, upon which lies the expected value of class A is given by:

$$\sqrt{X^2 + Y^2 + Z^2 + 1 - 2\sqrt{X^2 + Y^2}} \tag{7.1}$$

and the distance to the circle μ_B, $((X+1)^2+Z^2 = 1)$, of class B is:

$$\sqrt{(X-1)^2 + Y^2 + Z^2 + 1 - 2\sqrt{(X-1)^2 + Z^2}} \tag{7.2}$$

By subtracting the distance to μ_A from the distance to μ_B we obtain a function that produces a value greater than 0 when the sample is most likely to have come from class A and less than 0 when the sample is most likely to have come from class B. A simpler but equivalent function is obtained by first squaring the distance formulae prior to subtraction and canceling terms:

$$2\sqrt{X^2 + Y^2} - 2\sqrt{(X-1)^2 + Z^2} - 2X + 1 \tag{7.3}$$

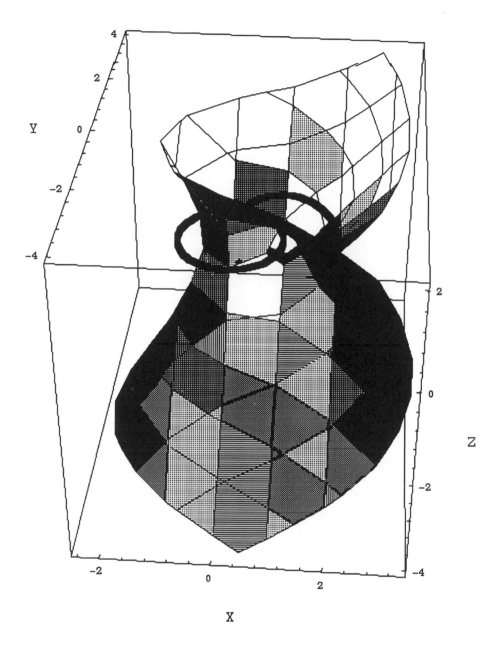

Figure 7.4
The (approximately) optimal solution to the Donut Problem is the locus of points for which Equation (7.3) equals 0: points falling on one side of this surface belong to Class A, while those falling upon the other side belong to Class B.

Although no formula can predict the class of an unlabeled point with certainty, this one will do so with the greatest likelihood of being correct. This is true regardless of the value of the point-spread parameter σ, which governs class overlap. Equation 7.3 is plotted in Figure 7.4.

7.3   Breeding Policies

The method used in Koza [1992] for selection of mates from the population is nominally that originally proposed in Holland [1975, 1992], with the notable exception that the latter only selects one of the two parents according to fitness. Since then, other breeding policies have become popular in the literature of artificial genetics. We introduce these models of population and selection to genetic programming in order to observe their benefits and setbacks.

7.3.1   "Demes" and Spatially Distributed Evolution

Sewall Wright (1889-1988) stands as one of the major figures of evolutionary biology with a career spanning more than 75 years [Provine 1986]. Among his many contributions to the field, one of the most important (and controversial) is his "shifting balance theory" of genetic diversity [Wright 1931, 1978]. In this theory, populations are spatially distributed throughout their environment into local breeding groups called "demes" [Gilmour and Gregor 1939] between which relatively little migration occurs. Within each deme, selection and genetic drift, (a phenomenon occurring in small populations by which traits may become widespread throughout although they are not particularly more fit than others), combine to cause rapid convergence to an evolutionary stable state. Separation by distance allows exploitation at a local level while individual demes explore separate evolutionary paths. Although it is a subject of controversy in the evolutionary biology community [Fisher 1958], such a spatial organization has proven advantageous in the context of artificial genetics, as demonstrated in Collins [1992]: the author shows dramatic improvement of distributed over panmictic ("everybody-breeds-with-everybody") mating policies that are commonly used in traditional GA and GP. The balance between exploration and exploitation can be demonstrated in computer simulations as well. This is graphically illustrated in Ackley [1993] and in Collins [1992], where it is shown that migration of individuals between demes causes particularly fit genotypes to spread slowly through the environment. In time, the environment becomes characterized by patches of successful strategies, genetically similar within each patch, but genetically different from other patches separated by boundary regions composed of less-fit hybrids.

7.3.2   Implementation of Distributed Evolution

Our method for implementing distributed evolution divides the population evenly among a rectangular grid specified by the user. Thus a population of 2,048 individuals could for example be allocated a 16x32-deme grid, creating a total of 512 demes, each populated locally by four individuals. Note that this sizing of deme is not particularly realistic in the biological sense, but rather is copied from distributed schemes of GA implemented on SIMD machines [Collins 1991, Ackley 1993]. Mating is done via the K-tournament selection method similar to that described by Koza: in K-tournament selection, we conduct a tournament with an arbitrary number (K) of individuals rather than 2. Empirically we have found that $K = 6$ works well under almost all circumstances, with a modification that accounts for locality of mating. In the scheme outlined by Koza, the individuals for the K-tournament are drawn from the entire population, whereas in the deme-based scheme they are drawn from spatially local sub-populations via a method we have termed "canonical" selection. To understand this, consider a deme located on element (M, N) of the grid. In canonical selection we fire a cannon sitting in the middle of the deme. Since each deme occupies a unit square, this means that the cannon is at location (M+0.5, N+0.5). The cannon is shot in a random direction (azimuth). The distance, which the shot travels, is zero-mean Gaussian distributed, with a standard deviation **D** provided by the user. A larger **D** will cause more shots to land in adjacent demes, on average, while a smaller **D** will cause most shots to land within the local deme (the grid is treated as a toroid for this purpose). An individual is randomly drawn from the local population of the deme in which the shot landed and is added to the local tournament list associated with deme (M, N). This process is repeated K times to form the entire list, and the winning individual is selected to be a parent. A second parent is chosen by the same canonical-K-tournament process, mating takes place, and two children are inserted into the local new-population list associated with deme (M, N). Thus each deme maintains its own population and its own new-population list, but some of the parents of the new populations may be drawn from neighboring demes. The population of each deme is not replaced by the members of the new-population list, until all demes in the grid have been processed.

7.3.3   Elitism and the Steady State Model

The standard population model used in GA and GP may be considered a batch process in which a new population is created all at once: this is in contrast to most real populations in which births and deaths occur continuously at random intervals. In addition, once the children of the next generation have been bred the previous generation is discarded. Thus it is possible and perhaps even likely that the most fit individuals will not survive into the next generation, although much of their genetic material will be retained. Syswerda first attempted to rectify these properties by proposing steady state genetic algorithms [Syswerda 1989]. Around the same time, Whitley proposed the Genitor algorithm that operates via a similar scheme [Whitley 1989]. The idea behind the steady state model is to perform individual selection and mating as previously described, but with the modification that the two newly generated offspring replace two less successful individuals in the existing population. This conserves the total number of individuals while allowing a single population to continuously evolve as a whole. There are many ways in which one may choose who dies. We employ the K-tournament method previously described, to select the two parents, but in addition, from each tournament we also keep the worst of the K individuals as candidates for replacement.

The steady-state method confers immortality upon the best individual in the population: it can never be selected for replacement until a better individual comes along. More generally this gives genotypes varying lifespans based upon their fitness relative to others in the population. The fact that the best individuals are preserved in the population gives rise to another term for this family of strategies: "elitism".

7.3.4   Implementation of Steady-State Elitism

In our implementation, steady-state elitism requires a tournament selection method as described in Section 7.3.3. It may be used in conjunction with demes, or may be used alone with a panmictic breeding policy. Since individuals are independently drawn to form the two K-tournament lists we check and retry-on-failure to ensure that they do not share the same worst member, since that would lead to an attempt to replace a single worst individual with the two new offspring. Note that when this steady state model is applied to demes, there is no new-population list maintained by each deme: offspring produced via crossover are inserted directly into the local or nearby population. Thus, when we perform mating at deme (M, N) it is possible to draw two parents from adjacent demes, breed them, and place their offspring in some other adjacent demes without any change taking place to the population of deme (M, N), which serves only as a sort of cheap motel.

7.4   Experimental Method

All experiments measure fitness as the total fraction of samples from both classes that are misclassified. Thus we are attempting to minimize the value of the fitness measure. We examine how the fitness varies as the function of parameter and experimental configurations described in detail below.

7.4.1 Performance of GP as Class Overlap Increases

We have illustrated how to increase the probability of misclassification by increasing the parameter σ, thereby making the donuts fatter and increasing overlap between the two distributions. Three separate suites (a variety of training data sets and a single test set) of data are provided: one for donuts with $\sigma=0.1$, a second with $\sigma=0.3$, and a third with $\sigma=0.5$. The optimal minimum misclassifications for these data are 0.0%, 3.45%, and 13.75% respectively.

A major question, which we seek to answer, is whether the solutions generated by GP increase in their deviation from the optimal solution as σ is increased: does GP become confounded by noise in the form of class overlap?

7.4.2 Generalization and Uniform Undersampling of the Training Set

Generalization is tested by varying the number of samples in the training set for a fixed disjoint test set of 1,000 samples. As few as 40 samples and as many as 1,000 samples are used in training. We wish to observe how misclassification varies as a function of the sparseness of the training set, and how this effect varies jointly with point spread.

7.4.3 Generalization and Nonuniform Sampling of the Training Set

Generalization is also tested by training on distributions that have had bites taken out of the donuts: large contiguous regions are unpopulated by sample data points, as was illustrated in Figure 7.3. The nonuniformity is varied parametrically by the number of bites: up to three bites may be removed from each donut, each subtending about 30 degrees, or 8.3% of the sample space.

7.4.4 Assessing the Effects of Demes and Elitism

The Donut problem provides an ideal starting point for benchmarking the effects of breeding policies both alone and in combination. In theory, demes should promote exploration of solution space and prevent global convergence to local minima, while elitism accelerates exploitation and speeds algorithm convergence. Most problems used to study both of these breeding policies have focused upon fitness relative to the original training set [Goldberg & Deb 1991, Collins & Jefferson 1991, Ackley 1993]. In our own studies with GP, the training-set fitness is available as well, but we are primarily concerned with the impact of these methods on generalization and overfitting of data. Again, it is important to understand how effects of breeding policy vary jointly with class overlap and undersampling effects as described in previous sections.

Unlike steady-state elitist policy, distributed evolution has some parameters associated with it: recall that in our "canonical" search for local mates, we hop to an adjoining location whose distance is Gaussian distributed with standard deviation **D**. The value of **D** governs the rate at which genetic material may migrate across the landscape. This is roughly an embodiment of Wright's parameter for size of breeding unit. As **D** is decreased, genetic diversity is greater with change in location. A second parameter is the geometric configuration of the landscape itself: in particular, we concern ourselves with a square toroidal grid of 16x16 demes, and with a coastline configuration, consisting of a 2x128 toroidal grid. The latter case should promote far greater genetic diversity than the square grid for a given value of **D**, since it provides much greater restrictions on travel between demes. Thus the choice between the square versus linear configurations is postulated as a choice between high mobility and low mobility, while the parameter **D** provides a fine adjustment.

7.4.5   Summary of Experimental Configurations

For each of the two deme configurations three values of **D** are used: 0.25, 0.5, and 0.75, where the distance between two adjacent grid elements is 1.0 units. This translates to hops that have a 2.8%, 26%, and 40% chance of landing outside the local deme, respectively. Three values of **D,** times two geometric layouts (square and coastline) makes a total of six distributed evolution types. In addition, we perform panmictic breeding as an experimental control, making a total of seven landscape configurations. Each of these is used with and without steady-state breeding policy, providing a total of 14 breeding policy combinations. As stated in Section 7.4.1, there are three values of the overlap (aka point-spread) parameter σ tested. Each of these is used with four different sizes of uniformly sparse training data sets, and eight sets of nonuniformly sparse training data, totaling 3x(4+8) = 36 different data sets. For each combination of breeding policy and training set, four independent runs are made using different initial conditions (random seeds). Thus, the total number of runs performed is 14x36x4 = 2,016 runs. Each has a population of 1,024 individuals reproducing over fifty generations, resulting in the analysis of a total of 103,219,200 individuals. It is important to note, however, that only four unique random seeds are used. This means that the four initial populations of 1,024 individuals, used for each experiment, are re-used for each configuration, and thus the divergent courses of evolution, are due to the changed factors such as deme configuration, migration rate, steady-state, etc., rather than differing initial conditions.

   In addition, selected experiments are repeated using a function set, which contains the SQRT function, thus in theory providing all mathematical tools necessary to achieve the optimal solution. Also, selected individuals, which are trained against data at a particular value of σ, are additionally evaluated against test data with a different σ value, to test the ability of GP to generalize with respect to class overlap.

7.5   Results

Figures 7.5, 7.6, and 7.7 show the raw data for a sampling of the 2,016 runs performed. Each plot contains a total of 12 curves, representing three figures of merit, plotted against time (in generations) for each of four runs having identical parameters and different initial conditions (induced by different random seeds).   The three figures of merit are: (1) average fitness over the entire population, as measured against the training data set (plotted with circles); (2) fitness of the individual that achieved highest fitness against the training set as measured against the training set, this is referred to by Koza as the "Best-of-Generation Fitness" (plotted with squares); (3) fitness of the individual that achieved highest fitness against the training set as measured against the test set, we will refer to this as the "Generalization Fitness" (plotted with diamonds).   These curves illustrate some general principles that characterize the various configurations and experimental parameters used.

Figure 7.5 depicts a run using donuts with σ=0.3 and a training set containing 680 samples (this can be considered a very non-sparse training set). It employs steady-state population and a 16x16 deme grid with migration parameter **D**=0.75. It possesses some characteristics typical of the steady-state model, notably rapid convergence and average

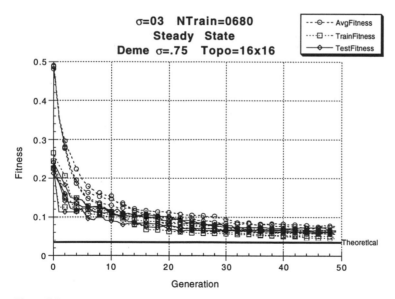

Figure 7.5
Fitness vs. Generation using steady-state population on a 16x16 grid of demes, high migration rate.

population fitness that approaches that of the best individual fitness (compare to 7.6 and 7.7). It also shows good generalization, with fitness against the test set closely approximating that achieved against the training set: this is a typical consequence of the relatively large number of training samples used.

Figure 7.6 shows a run, also using donuts with σ=0.3, but with a much more sparse 100 samples training set. The batch-population update policy is used with a 128x2 coastline deme-grid configuration, with **D**=0.75. A much slower convergence rate can be observed, particularly in the average fitness, as is typical of distributed evolution with low mobility and no steady-state breeding policy. Also note the disparity between the values to which the figures-of-merit converge (compare to Figure 7.5). This is largely due to training set sparsity.

Figure 7.7 depicts another 100 sample sparse-training-set case, this time with σ=0.5. Neither distributed evolution, nor steady-state are used. Note that whereas the σ=0.3 cases depicted in 7.5 and 7.6 may approach a minimum misclassification of about 3.45%, the σ=0.5 case should in theory approach a 13.75% minimum. In practice, we observe *overfitting*, as predicted in Section 7.2.1: the training set fitness and average fitness figures drop well below the theoretical minimum, which indicates that the functions formed by GP have become adapted to the particular samples of the training set, rather

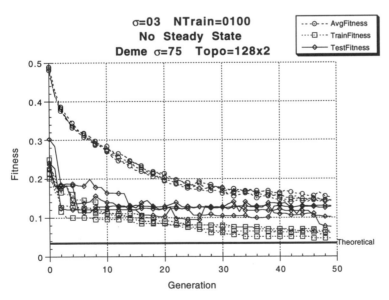

Figure 7.6
Fitness vs. Generations with a relatively small number of training samples.

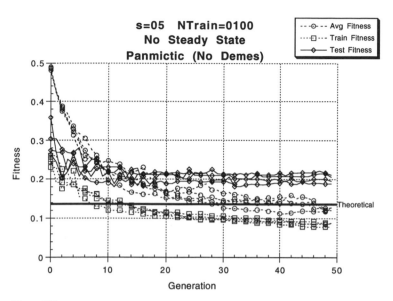

Figure 7.7
Fitness vs. Generation for a small number of training samples and a high degree of class overlap.

than to the true underlying distribution. As expected, the fitness against the independent training set remains far from the theoretical minimum having become fixed at about 20% misclassification at around the 15th generation.

Having examined some of the qualitative properties of the experimental variables we now move on to describe specific measures of performance.

7.5.1 Scalability With Respect to Class Overlap

In order to judge the effectiveness of genetic programming, it is important to understand whether the search procedure is confused by noise in the form of class overlap. Figures 7.8 and 7.9 indicate that, at least for the Donut problem, it is not. Figure 7.8 shows the generalization fitness averaged over all 14 breeding policy configurations and all four initial conditions for the last 10 generations of each run, as a function of the class-overlap parameter σ (plotted with circles). The number of training set elements is fixed at 360.

On the same plot we show the minimum theoretical limit for misclassification as a function of σ (plotted with heavy line). It appears that on average, the fitness of functions generated by GP roughly deviates from the theoretical limit by a constant factor. This is further borne out in Figure 7.9, which depicts the performance of the three single best-test-set-fitness individuals for each value of σ, with training set size fixed at 360 samples.

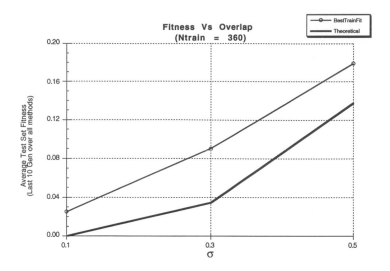

Figure 7.8
Comparison of best average vs. optimal fitness as a function of overlap between classes in training samples.

In plot 7.9 the fitness of the individual at σ=0.1 is not a perfect 0.0, but rather misclassifies only a very few samples from the independent 1,000 samples training set.

7.5.2   Generalization With Respect to Class Overlap

In Section 7.2.2 we showed that our problem has a single underlying solution whose difficulty can be parameterized with respect to ambiguity in the form of class overlap. In order to evaluate generalization with respect to the level of overlap and/or noise, it is not necessary to perform additional training, but rather to take the best individuals discussed in Section 7.5.1 and re-test them using each other's validation test set. This is illustrated in Figure 7.10. Four curves are shown, which depict theoretical minimum misclassification (heavy black circles), and the performance of the best generalization fitness individual trained with   (1) σ=0.1 (triangles) (2) σ=0.3 (x's) and (3) σ=0.5 (circles) as a function of the test-set σ. This yields an interesting result, namely that the individual, which was trained with σ=0.3, not only outperforms the other two individuals in classifying the distributions with which they were trained, but also arrives closer to optimal solution with the cross-tested distributions than it does against its own. The results using σ=0.1 and σ=0.5 are more in line with the author's preconceptions.

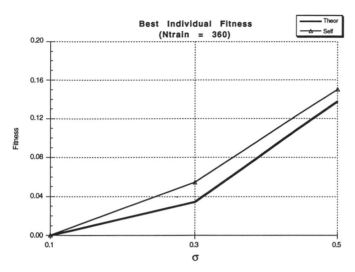

Figure 7.9
Generalization fitness of best individuals for each value of σ vs. optimal fitness.

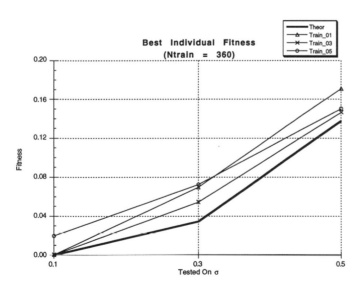

Figure 7.10
Fitness of best individuals trained with three different σ values and tested with three σ values.

The unusual point spread for these results may be due to statistical variations, since they test the performance of only three individuals, but again indicate the same constant error trend previously observed in section 7.5.1.

7.5.3   Generalization and Uniform Undersampling of Training Data

Figure 7.11 summarizes performance as a function of training set size. A total of six curves is presented: two each for $\sigma=0.1$ (circles), $\sigma=0.3$ (diamonds), and $\sigma=0.5$ (triangles). All of them depict the generalization fitness performance averaged over the last 10 generations. In each case, the upper curve of the pair depicts this figure averaged over all 14 breeding-policy configurations, while the lower curve represents the performance of the single best of the 14. Theoretical minimum misclassification for $\sigma=0.3$ (3.45%) and for $\sigma=0.5$ (13.75%) are shown as solid horizontal lines (the minimum for $\sigma=0.1$ is approximately 0.0%). To a first approximation, the rate at which fitness converges towards the optimum, as a function of training set size, is roughly the same for all 3 sets of curves. It is notable that on average, the difference between actual fitness and optimal fitness is significantly greater for the $\sigma=0.3$ case than for the others.

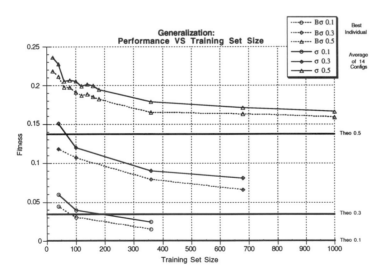

Figure 7.11
Performance vs. training set size for uniformly distributed training samples.

7.5.4 Generalization and Nonuniformly Distributed Training Data

Figure 7.12 shows performance figures in a manner similar to Figure 7.11, except that performance is plotted as a function of the number of bites removed from the training set (see also Figure 7.3). Note that unlike Figure 7.11, the number of points in the training set decreases to the right, since each bite removes about 8.3% of the training set samples. The curves for σ=0.3 and σ=0.5 use 360 samples in the no bites training set, while the σ=0.1 set uses 100. Thus for the σ=0.3 and σ=0.5 cases, there are about 270 samples in the three-bites-removed training set. In both of those cases, the removal of the nonuniformly distributed point sets results in an average increase in misclassification of about 1%. This is comparable to the performance degradation that can be interpolated from the corresponding curves of Figure 7.11 for a training set of size 270. Results for the σ=0.1 case appear inconsistent since they fluctuate with increasing σ, but both average and best results curves display a similar characteristic. We speculate that this behavior is an artifact of the relatively sparse training set used.

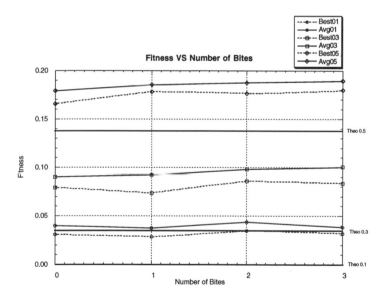

Figure 7.12
Fitness with respect to nonuniformly distributed training samples ("bites" taken from the donuts).

7.5.5   Comparative Performance and the Optimal Function Set

In Section 7.2.2, an approximately optimal solution to the Donut problem was introduced, which involved the use of the square root function. The function set used in experiments up to this point has not included a square root operator. How much performance is being sacrificed in this problem by not including it? Figures 7.13 and 7.14 compare the performance of the standard function set used throughout the problem with one that replaces the IFLTE function with SQRT. The SQRT function that we provide differs from the standard formulation in that it is protected: we take the absolute value of its one argument, prior to the square root computation. This is necessary since we have no direct control over the value of arguments, although Darwinian purists might wish to take the approach of letting negative arguments passed to the square root be lethal to the genotype.

Figure 7.13 shows the average of the generalization performance taken over all breeding policy configurations and all random initial seeds for the last 10 generations. For this experiment, a training set size was used which increases with σ: 100 samples for $\sigma=0.1$, 360 samples for $\sigma=0.3$, and 680 samples for $\sigma=0.5$. Fitness (classification performance) using IFLTE is plotted with circles, while that using SQRT is plotted using squares. We see that there is indeed a consistent improvement achieved, while the overall

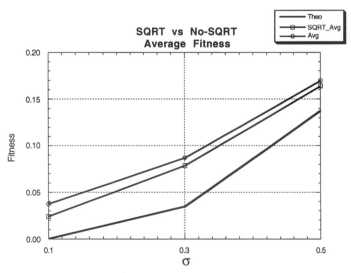

Figure 7.13
Average fitness of individuals with and without square-root included in the function set.

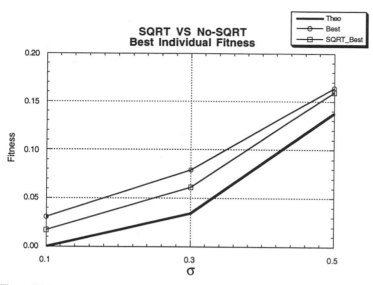

Figure 7.14
Fitness of the best individuals produced with and without the square-root function included in the function set.

error level is still well above the optimal. A similar trend can be seen in Figure 7.14, which depicts the best-test-set performance, averaged over all four random seeds and the last 10 generations, for the single breeding scheme having the best performance at each combination of σ, training set size, and function set. Again, even though overall performance is improved using SQRT, this single best individual does not achieve the optimal solution.

7.5.6 Comparative Performance of Breeding Policies

In order to assess comparative performance between the 14 different breeding policies and configurations used in this study, we must first develop a statistical basis that allows us to draw sound conclusions. This is not as easy as it would seem on the surface: although there were over 2,000 runs performed, many of these had different expected outcomes due to various combinations of training set density, overlap, etc. We cannot simply average the scores together due to their multimodal manner of distribution. Likewise, we need a mean to assess the variability of any composite score, which we arrive upon, so that we may assign a reasonable confidence to it: a degree of uncertainty in our conclusions is acceptable as long as that uncertainty can be accurately bounded. Towards the goal of a sound statistical analysis we apply a methodology, which is derived from the analysis of variance (ANOVA), due in large part to the efforts of Ronald A. Fisher. It is an interesting historical note that R. A. Fisher, the statistician is one and the same as R. A. Fisher the geneticist, with whom Sewall Wright had a lifelong disagreement over the nature of selection and distributed populations, which has outlived them both. It is ironic that they should meet again in this circumstance!

Our methodology divides the experiments into different partitions, which may be treated as independent analyses. Each partition consists of a group of experiments with a single common factor, e.g., all experiments with sparse training data. Within the partition, each of the 14 experimental configurations -- i.e., various deme layouts and parameters in combination with or without steady state -- is kept separate from the others. Perhaps the best analogy for this approach is that we consider each of the 14 configurations as a student in a class: the experiments in a given partition represent a series of tests administered to all students, where each exam may have a different number of points possible, and upon whose sum of scores the grade will be based. (The student analogy breaks down a bit when we consider the lower-is-better scoring system, but it does not affect the nature of the analysis.) Since we wish to be fair in assigning grades, we want to associate a confidence, or standard deviation, with the score of each individual. Because of the fact that different experiments may have wildly different average outcomes, we cannot base estimates of mean and standard deviation upon the mean score of all exams. Instead, we base an estimate of variance upon the four runs consisting of different random seeds with otherwise identical experiments. The mean fitness of these four runs is first computed. The sum of the squared differences between this local mean and each of the four individual fitnesses is summed into the global variance estimate for the individual. The square root of the resulting figure, taken over all experiments in the partition, is the estimate of the standard deviation of the individual's sum of scores from each experiment. It is important to note that the absolute sum of scores is a relatively meaningless measure since it is primarily a function of the number and type of experiments included in the partition. Instead, the true figure of merit is the ratio of the standard deviation for two individuals relative to the difference of their sum of scores. Together these figures tell us the likelihood of making a Type I or Type II error [Larsen & Marx 1981] when stating that one method works better than the other. In the following sections we use the above method to graphically depict the comparative performance for a variety of experiment partitions, with an emphasis on those we consider particularly hard.

7.5.6.1   Performance Across All Experiments

Figure 7.15 depicts a bar chart showing the performance of each configuration taken against all experiments performed. The line running horizontally, through the center of the dark region at the top of each bar, is the sum of best-10-average-test-fitness over all experiments performed for the particular configuration. Best-10-average-test-fitness is the average of the generalization fitness for the 10 best individuals produced during a run. Recall from our definition of generalization fitness that only the individuals, which achieved best-of-generation against the training set, were eligible for testing. The dark region at the top of each bar depicts the fitness sum plus and minus the estimated standard deviation. Table 7.2 summarizes the labels, which are used to describe the experimental configurations in each chart, and the experiments: 27 in all, each with four random seeds. Thus each bar is the sum of fitness measures from 108 runs. In viewing these charts, it is important to remember that lower fitness is better.

The estimated standard deviations are small compared to fitness sums for Figure 7.15, indicating that the differences shown are significant in many cases. When steady state population is used, distributed evolution always does as well as or better than panmictic. Without the steady-state model, however, distributed evolution performs significantly better in three cases, and significantly worse in two, with one being about the same. The parameters and results are summarized in Figure 7.15 and in the following table.

Sigma	Training Set Size	Bites
0.1	40	0, 1, 2, 3
0.1	100	0, 1, 2, 3
0.1	360	0
0.3	40	0
0.3	100	0, 1, 2, 3
0.3	360	0, 1, 2, 3
0.5	100	0
0.5	360	0, 1, 2, 3
0.5	680	0, 1, 2, 3

Figure 7.16 depicts a similar graph that displays fitness relative to the original training data. We observe that fitness appears to be proportional to mobility, being best for panmictic evolution. We have mentioned in Section 7.4.4 that the training-set fitness measure is common to earlier experiments performed with demes and elitism. This would appear, however, to counter results such as Collins [1992]. Why? To see the answer, examine the plots of Figures 7.5, 7.6 and 7.7. The 50-generation cutoff was chosen because it appeared in preliminary experiments that this was sufficient time to allow generalization performance to converge, which it does. Training set performance,

Table 7.2
The meaning of experiment labels in comparative performance diagrams

Configuration	Steady State Elitism	Mobility, D	Deme Shape
nss_25_d128x2	No	0.25	Coastline
nss_25_d16x16	No	0.25	Square
nss_50_d128x2	No	0.50	Coastline
nss_50_d16x16	No	0.50	Square
nss_75_d128x2	No	0.75	Coastline
nss_75_d16x16	No	0.75	Square
nss_00_d00x00	No	N/A	No Demes
ss_25_d128x2	Yes	0.25	Coastline
ss_25_d16x16	Yes	0.25	Square
ss_50_d128x2	Yes	0.50	Coastline
ss_50_d16x16	Yes	0.50	Square
ss_75_d128x2	Yes	0.75	Coastline
ss_75_d16x16	Yes	0.75	Square
ss_00_d00x00	Yes	N/A	No Demes

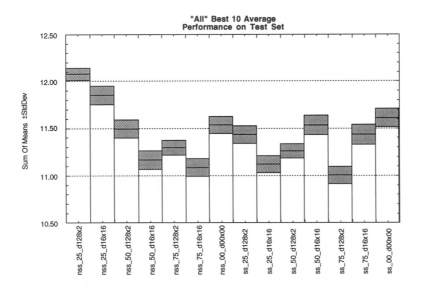

Figure 7.15
Sum of mean fitness, with respect to test set, for all breeding policies across all experiments (std deviation shaded).

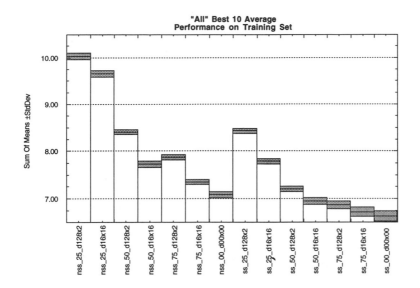

Figure 7.16
Sum of mean fitness for all breeding policies across all experiments (std deviation shaded).

however, is still dropping at generation 50. We postulate, based on the observation of a great number of charts produced in the manner of Figures 7.5, 7.6, and 7.7, that higher mobility (**D**) leads to more rapid convergence, and hence better training fitness at the time training is cut off. This is further demonstrated by the fact that runs with steady state population, well known for speeding up convergence [Goldberg 1989], display even lower sum-of-fitness figures on average than do those without. There is no telling from these data what training fitness would ultimately converge to for each case.

7.5.6.2 Performance Using Uniformly Sparse Training Data

Figure 7.17 depicts best-10-average-test-fitness for a set of experiments using sparsely distributed training data. The experiment parameters are summarized in the following table.

Sigma	Training Set Size	Bites
0.1	40	0
0.3	40	0
0.5	100	0

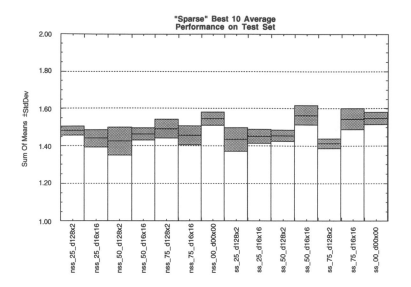

Figure 7.17
Sum of mean fitness for all breeding policies for experiments using small training sets. (std deviation shaded).

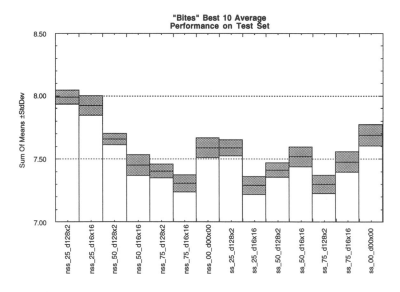

Figure 7.18
Sum of mean fitness for all breeding policies using nonuniform training samples (std deviation shaded).

We consider the sparse training data case to be a particularly hard problem to generalize against. For the steady-state-population case performance of distributed evolution is consistently the same as or significantly better than panmictic. For the batch-population runs, panmictic fitness is consistently at least a standard deviation above (worse than) that offered by distributed evolution.

7.5.6.3 Performance Using Nonuniform Training Data

The best-10-average-test-fitness performance spread achieved with nonuniform training sets (bites), shown in Figure 7.18, appears similar in character to that shown for all data taken together: when using steady-state population, the distributed scheme always produces the same or significantly better performance. With batch learning, the lowest mobility cases ($D=0.25$) perform significantly worse than panmictic, as was seen in the test-set performance of Section 7.5.6.1. The experimental parameters are summarized in the following table.

Sigma	Training Set Size	Bites
0.1	40	1, 2, 3
0.1	100	1, 2, 3
0.3	100	1, 2, 3
0.3	360	1, 2, 3
0.5	360	1, 2, 3
0.5	680	1, 2, 3

7.5.6.4 Performance Using Sparse and Nonuniform Training Data

For both steady-state and batch populations, results using distributed evolution are either significantly better, or similar within much less than one standard deviation. For batch learning, however, the trend of the ($D=0.25$) having less favorable fitness properties appears again, although not nearly as unfavorable as that shown in Figures 7.15 or 7.18.

Figure 7.19 shows test-set performance for sparse training data combined with bites training data, which is also sparse, as summarized in the following table.

Sigma	Training Set Size	Bites
0.1	40	0, 1, 2, 3
0.3	40	0
0.3	40	1, 2, 3
0.5	100	0, 1, 2, 3

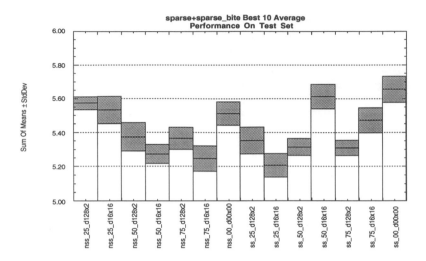

Figure 7.19
Sum of mean fitness, sparse and nonuniform training data (std deviation shaded).

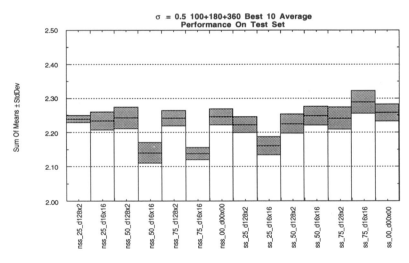

Figure 7.20
Sum of mean fitness for all breeding policies for experiments with large class overlap (std deviation shaded).

7.5.6.5 Performance Using Sparse Data With High Degree of Class Overlap

The test-set performance due to interaction of sparse training data with significant class overlap is summarized in Figure 7.20 and in the following table.

Sigma	Training Set Size	Bites
0.5	100	0
0.5	180	0
0.5	360	0

These runs display a different characteristic previously observed: most methods perform about the same within a reasonable confidence level, but three (distributed populations) display significantly better fitness.

7.5.6.6 Performance Using Non-Sparse Data Sets

Finally, as a control we do a comparative performance analysis using only densely sampled training data on the theory that these data are easy to classify. The parameters and results are summarized in Figure 7.21 and in the following table.

Sigma	Training Set Size	Bites
0.1	100	0
0.1	360	0
0.3	100	0
0.3	360	0
0.5	360	0
0.5	680	0

Here again we see that most configurations, including panmictic, perform about the same, although in the batch-population case, some perform significantly better and worse than the average.

7.6 Conclusions

Several results from this study were rather unexpected. The first and foremost among these is the relationship between training set density and class overlap: in performing up-front experimental design, it was assumed that the spatial density of samples would play a significant role in performance. This is because the volume of a torus is proportional to the square of its cross-section, or minor diameter. For the toroidal distributions of the Donut problem, this diameter should be proportional to the parameter σ. Thus, we might expect that the number of samples required to cover feature space for $\sigma=0.5$ to be about

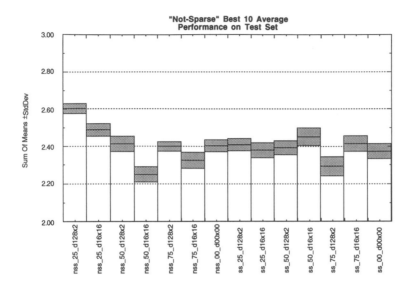

Figure 7.21
Sum of mean fitness for all breeding policies for experiments with large numbers of training samples (std deviation shaded).

$(0.5/0.3)^2 = 2.78$ times as many to achieve equivalent performance as for $\sigma=0.3$. For example, we would expect to see similar quality of results with $\sigma=0.5$ and number of training samples = 1,000, as for $\sigma=0.3$ and number of training samples = 360. We can readily see that this is not so from Figure 7.11. In fact, we observe that the $\sigma=0.5$ case is much closer to its theoretical limit than is the $\sigma=0.3$ case for the 360 sample training set, although their rate of convergence (slope) with respect to this limit is roughly the same at that point.

Another unexpected outcome was the lack of any linear increase in classification error, relative to the theoretical limit, with increase in class overlap (σ). We had expected that the increased presence of noise, in the form of class overlap, would confound the program induction process. Instead, it was a more-or-less constant difference. We explain this in terms of the fact that, considered as an optimization process of searching for a function that defines a surface which minimizes error, increasing class overlap does not change the features of the landscape. It may bias the actual level of the minimum we are searching for, but we know for a fact that the optimal surface itself does not change with σ.

For $\sigma=0.3$ classification error was consistently higher than the optimal error relative to the results achieved for experiments with $\sigma=0.1$ or $\sigma=0.5$. Conversely, when testing generalization with respect to class overlap, it was observed that the classifier trained on data with $\sigma=0.3$ consistently achieved the best performance for all values of σ. This leads us to speculate that there may be some degree of deception to these problems that varies as a function of σ.

7.6.1   Conclusions About Distributed Evolution

Distributed evolution seems to significantly improve performance on the average. Moreover, we have demonstrated this for the case of generalization or test-fitness measurements, whereas previous studies, particularly of GA, have focused only on the training fitness of genetic systems [Collins 1992, Ackley 1993]. In general, no published studies of the effects of demes and elitism in genetic programming have previously appeared. Unfortunately, as discussed in Section 7.5.6.1, our experimental design has not allowed us to explore training set fitness in the sense of prior works. This is perhaps best left to a problem domain other than classification anyway, since for this family of problems the fitness against training data is not a figure of merit and indeed a close fit to training data can be an indicator of trouble (overfitting).

So far as specific recommendations for the problem at hand, we have seen that, with steady-state evolution, the coastline (128x2) distributed breeding configuration with **D**=0.75 consistently performed well. For the batch-population case, 16x16 demes with **D**=0.75 and **D**=0.50 did consistently well, but the parameter **D**=0.25 most frequently provided poor results, not only worse than average, but worse than panmictic in almost all cases. Recall that for the **D**=0.25 case, the actual probability of reaching outside the deme on any given draw is less than 3%. There may very well be a lower limit upon migration rate for GP, below which diversity drops too low to allow the drift-vs-migration tradeoff to be useful.

7.6.2   Conclusions Concerning Elitism

The elitist breeding policy often performed slightly worse on average than did batch population updates. This is most likely due to rapid convergence associated with the method. On the other hand, there was no deme-parameter setting associated with elitism, which routinely performed significantly worse than panmictic, as was the case with batch population. In addition, it is important to note that the elitist method demonstrates significant speedup relative to other methods. Effectively, this allows more search: if we can make our runs 30% shorter while achieving the same degree of convergence, we can perform a larger number of runs in the same period of time. Multiple independent runs with different seeds are indeed an effective method of maintaining a diverse search of the solution space, so this speed factor may make the steady-state option attractive, even if its expected performance on a per-run basis is slightly lower.

7.6.3   Comments on the Procedures Used

For this problem, experimental design was performed up-front, with preconceptions about what problem sizes and parameters should be hard. In general, the GP system dealt much more gracefully with sparseness, nonuniform training data, and class overlap than had been expected, biasing many test runs to explore cases that turned out not to be difficult at all. In doing so, however, this study has set a much tighter bound upon problem parameters most worthy of study.

7.6.4   Need for Benchmark Test Functions

De Jong used a family of five functions for the study of genetic algorithms for optimization [De Jong 1975]. In the context of GP and generalization, it makes a great deal of sense to propose a similar family of distributions with which the analyses here may be repeated for distributions other than donuts.

7.6.5   Big Deme Grids and Other Parameters

Prior studies of distributed evolution have typically used thousands of demes, largely because such a configuration was the only one that efficiently mapped onto the SIMD (Single Instruction, Multiple Data) mesh of the computers used in those studies [Collins 1992, Ackley 1993]. At the same time, those studies, which made a comparison against panmictic evolution, found results that were spectacularly better using the distributed scheme [Collins 1992]. Our own research has implemented virtual demes in a workstation-based system and maintained a relatively small number of them. Would the results we obtain be similarly improved by scaling up the spatial landscape upon which evolution takes place?

Also, the rate of migration makes an order-of-magnitude jump from **D**=0.25 to **D**=0.5. Especially in light of the fact that degraded performance was often seen at the lower value, it would be beneficial to test a more continuous range of values in between.

7.6.6   Gene Frequencies and Distributed Evolution

According to Wright, distributed evolution maintains genetic diversity via separation by distance and "cumulative effects of the accidents of sampling." Can we consider GP trees to have an allelic "building block" structure even though there are no fixed loci? Genetic programming community is confounded by this problem. In Tackett [1993] we have already shown that there are GP building blocks, small (and even relatively large) sub-trees repeated frequently throughout the population. Is it the frequencies of these building blocks, independent of position, which we may consider to be the true gene frequencies? If so, then we make a prediction: in a GP system we should see the frequency distributions of these building blocks shift, and perhaps even contain entirely different sets of building-block subtrees as we move from deme to deme, with disparity being a function of separation by distance. Analytic models of the nature of the floating-loci model of GP inheritance are thus needed, as are the software tools to search populations and verify or refute such hypotheses as that stated above.

Acknowledgments

The authors would like to thank Jerry Burman of GTE Government Systems for valuable statistical advice and counsel, Bret Bryan for helpful discussions of donut geometry, and as always John Koza and James Rice of Stanford University for constant support and encouragement.

Bibliography

Ackley, David H. (1993) A case for distributed Lamarckian evolution. In *Artificial Life III*. In Press.

Adams, Douglas (1982) *The Hitchhikers Guide to the Galaxy*. Bantam.

Collins, R.J., and Jefferson, D. (1991) Selection in massively parallel genetic algorithms. In Belew, R, and Booker, L. (eds.) *Proceedings of the Fourth International Conference on Genetic Algorithms*, pp. 249-256. Morgan-Kaufmann.

Collins, R.J. (1992) *Studies in Artificial Evolution*. Ph.D. Dissertation, University of California at Los Angeles. Available by public ftp to cognet.ucla.edu.

Darwin, Charles (1876) *The Effect of Cross and Self Fertilisation in the Vegetable Kingdom*. Murray, London, UK.

De Jong, K. A. (1975) An analysis of the behavior of a class of genetic adaptive systems. (Doctoral dissertation, University of Michigan). *Dissertation Abstracts International 36(10)*, 5140B. University Microfilms No. 76-9381.

Fisher, R.A. (1958) *The Genetical Theory of Natural Selection*. Dover Press, New York.

Fukushima, K. (1982) Neocognitron: A new algorithm for pattern recognition tolerant of deformations and shifts in position. *Pattern Recognition*, Vol. 15, No. 6, pp. 445-469.

Fukushima, K. (1989) Analysis of the process of visual pattern recognition by the Neocognitron," *Neural Networks*, Vol. 2, pp. 413-420.

Gilmour, J., and Gregor, J. (1939) Demes: A suggested new terminology. *Nature*, 144:333.

Goldberg, David E. (1983) Computer-Aided Gas Pipeline Operation Using Genetic Algorithms and Rule-Learning, Ph.D. Dissertation, University of Michigan. *Dissertation Abstracts International*, 44(10), 3174B (University Microfilms No. 8402282).

Goldberg, David E. (1989)*Genetic Algorithms in Search, Optimization, and Machine Learning*. Reading MA: Addison Wesley.

Goldberg, D.E., and Deb, K. (1991) A comparative analysis of selection schemes used in genetic algorithms. In Rawlins, J.E. (ed) *Foundations of Genetic Algorithms*. Morgan Kaufmann.

Holland, John H. (1975, 1992 2ed) *Adaptation in Natural and Artificial Systems*, Cambridge, MA: MIT Press.

Koza, John R. (1992) *Genetic Programming*. Cambridge, MA: MIT Press.

Larsen & Marx (1981) *An Introduction to Mathematical Statistics*. Prentice-Hall.

Minsky, M., and Papert, S. (1969) *Perceptrons*. MIT Press.

Poggio, T. and Girosi, F. (1989) *A Theory of Networks for Approximation and Learning*. MIT AI Laboratory, A.I. Memo #1140, July.

Provine, W. (1986) *Sewall Wright and Evolutionary Biology*. University of Chicago Press.

Rosenblatt, F. (1962) *Principles of Neurodynamics*. Washington DC: Spartan Books.

Syswerda, G. (1989) Uniform crossover in genetic algorithms. In Schaffer, J. (ed) *Proceedings of the Third International Conference on Genetic Algorithms*. Morgan Kaufmann.

Tackett, Walter A. (1993) "Genetic programming for feature discovery and image discrimination. In Forrest, S. (ed) *Proceedings of the Fifth International Conference on Genetic Algorithms*, Urbana-Champaign IL. Morgan Kaufmann.

Whitley, D. (1989) The GENITOR algorithm and selection pressure: Why rank-based allocation of reproductive trials is best. In Schaffer, J. (ed) *Proceedings of the Third International Conference on Genetic Algorithms*. Morgan Kaufmann.

Whitley, D. (ed) (1993) *Foundations of Genetic Algorithms 2*. Morgan Kaufmann.

Wright, Sewall, (1931) Evolution in Mendelian populations. *Genetics* 16:97-159.

Wright, Sewall, (1932) The roles of mutation, inbreeding, crossbreeding, and selection in evolution. In *Proceedings of the Sixth International Congress of Genetics, v1*, pp356-366.

Wright, Sewall, (1978) *Evolution and the Genetics of Populations. Volume 4: Variability Within and Among Natural Populations*. University of Chicago Press.

8 Effects of Locality in Individual and Population Evolution

Patrik D'haeseleer, Jason Bluming

This chapter describes how introducing locality into the Genetic Programming Paradigm (GP) influences the evolutionary behavior of both the population as a whole and its individual members. We have adopted an Artificial Life (ALife) viewpoint -- focusing more on the population as a whole rather than on individual performance-- for our observations of an illustrative system that uses this approach. We introduce locality into the GP in both the reproductive and evaluation phases. Our implementation of locality uses isolation by distance - on a linear population with wraparound - as opposed to the more commonly used fixed-sized demes.

8.1 Overview

The standard GP uses a uniform fitness function across the entire population, and performs selection for reproduction indiscriminate of positioning of individuals. In nature, it is unrealistic to assume that interactions between individuals are so homogenous. Geographic and temporal displacement plays an important role in determining the pattern of interaction between members of a population. Evolution pushes individuals toward fitness with respect to their immediate environment rather than the population as a whole.

We have adapted the GP to incorporate these kinds of localities into the reproduction and fitness evaluation stages, changing its focus from GA-style optimization applications to ALife population study. ALife studies focus on population dynamics and characteristics rather than convergence of individuals towards an optimum.

The specific problem we study was chosen because it seemed sufficiently complex to ensure continuing dynamic behavior. The search space for this problem is so large that it is unlikely for a globally optimal strategy of survival to be reached. Fitness values result from a contest between individuals, yielding a locally dependent fitness measurement for each element in a population. The relative persistence of positive traits within a local region is indicative of the regional evolutionary pressure to preserve valuable genetic attributes.

It is our intent to illustrate the effects of locality in terms of formations of deme-like structures, increased population diversity, increased population and individual generality, and achievement of local fitness maxima.

Genotypical analysis of the individuals is undertaken by the "tagging" of the genetic material of population individuals to facilitate the study of gene flow between generations. This study encompasses not only the mappings of specific gene frequencies throughout the population, but also generation and recombination of new functional blocks, as well as dynamic effects such as gene diffusion, propagation of favorable traits, and deme migration.

We hope that the compiled results will yield increased understanding in the realms of both ALife applications and biological study.

8.2   Background

Darwin's observations of the effects of natural selection on evolution in the Galapagos Islands [Darwin 1859] was one of the first records calling attention to environmental influence upon individual and species evolution. Whereas the concepts of natural selection and survival of the fittest are commonly referred to in evolutionary programming literature, the role of sub population isolation in species differentiation is cited less frequently. Identifying an individual's environment, including the members of its local population, as a primary influence on the continued development of the local sub population as well as the species as a whole is an important step toward more realistic modeling of evolutionary mechanics.

In the 1930s, Sewall Wright [Wright 1932] developed his "shifting balance" theory of evolution in which he introduced the term *demes* for partially isolated sub populations with relatively little interaction *between* sub populations as compared to *within* the sub populations. This theory postulates that the solution to a problem with multiple local maxima can be more efficiently achieved by partitioning the population into sufficiently small groups such that disparate local regions explore a broader range of possible solutions. The theory combines the findings of evolutionary modeling, population studies, and species husbandry, and was the first to observe and name the concepts of fitness peaks and adaptive landscapes.

Current studies in GP often utilize large populations to effectively sample a very large functional search space. The size of the population is used to ensure a sufficiently large genetic base to reach most areas of the search space in a finite time frame. Modification of the paradigm for multiple parallel sub populations, or colonies, allows the consideration of several possible solutions simultaneously without necessarily converging toward one specific result.

In a recent paper, Collins and Jefferson [Collins and Jefferson 1991] show that, for the problem addressed therein, "local mating is more appropriate for artificial evolution than the panmictic mating schemes that are usually used in genetic algorithms. In addition, local mating appears to be superior to panmictic mating even when considering traditional applications. Local mating (a) finds optimal solutions faster; (b) typically finds multiple optimal solutions in the same run; and (c) is much more robust [than traditional GA]."

8.3 Domain

In addressing the issue of locality, we wanted to use a domain of sufficient complexity to yield multiple viable solutions of locally-optimal performance. Toward this end we selected a simple simulation of robot tanks, pitted against each other in singular combat on a bounded rectangular plane.

8.3.1 Terminal Set

- **(FORW)**: Advance tank one step in direction it is facing
- **(TURNL)**: Turn tank 90° counterclockwise
- **(TURNR)**: Turn tank 90° clockwise
- **(FIRE)**: Fire cannon in the direction tank is facing
- **(SHIELD)**: Activate tank's shields
- **(WAIT)**: No-op, tank waits.

8.3.2 Function Set

- **(IF_WALL** *yes_arg no_arg***)**: Execute first argument if tank is directly facing a wall, otherwise execute second argument.

- **(IF_FB** *front_arg center_arg behind_arg***)**: Execute appropriate argument, depending on relative position of opponent.

- **(IF_LR** *left_arg center_arg right_arg***)**: Execute appropriate argument, depending on relative position of opponent.

- **(IF_RADIUS** *yes_arg no_arg***)**: Execute first argument if opponent is within shield radius (currently a 3x3 square around the tank), otherwise execute second argument.

- **(IF_RANGE** *yes_arg no_arg***)**: Execute first argument if opponent is within firing range (currently a Manhattan distance of 4), otherwise execute second argument.

- **(RPROGN** *sexp1 sexp2***)**: Lisp **PROGN** (multiple execution function) limited to two arguments. Acts as an "accelerator" function to allow the tank to perform multiple operations in a single round.

8.3.3 Fitness Evaluation

The tanks start combat, once per generation, with a fixed amount of energy, currently 1000 units. The energy level of each tank is the only state variable maintained throughout the simulation. When the energy level of either tank drops below zero, that tank is "dead" and the simulation ends. To reduce the advantage that moving first would present, energy accounting is done only at the end of each round, which also allows a combat to end in a draw.

Each non-sensing action (**FORW**, **TURNR**, **TURNL**, **FIRE**, **SHIELD**, and **WAIT**) has an associated energy cost. Damage sustained by a tank is subtracted from its current energy level, which simplifies state accounting. A Tank may be damaged either by cannon fire or by its opponent's shields (if within the opponent's shield radius), and damage from either of these sources is diminished if the tank's own shields are active.

At the end of the combat, fitness is determined from the remaining energy of both combatants. Because fitness is calculated with respect to other members of the population, we establish a system of ongoing coevolution rather than one based on a static fitness measurement. While the advantage of this method in studying population emergence is evident, it has the disadvantage of not providing a absolute measure for the fitness of an individual.

Note that this fitness measure is only meaningful when the tanks actually engage in combat. When we put two defensive tanks in the arena, their fitness will usually be equal, no matter what their defensive strategy is, eliminating evolutionary pressure. We therefore provide an environment that leads to aggressive behavior. We experimented with several parameter sets, varying the cost of each action (such as: high payoff for hitting opponent, cost for waiting is equal to cost for moving, etc.) Based on this experimentation, we also decided against putting an extra cost on using a **RPROGN**.

Following is a table of associated costs and penalties. These values were used for all test simulations to keep the results consistent.

Operation	Cost	Damage to Opponent
TURNL	10	N/A
TURNR	10	N/A
FORW	10	N/A
FIRE	30	400 (normal), 150 (shielded)
SHIELD	20	70 (normal), 35 (shielded)
WAIT	10	N/A
PROGN	0	N/A

Looking at all the available sensing information, there are 39 states the robot tank can be in (for example: not facing a wall, with enemy directly behind, within cannon range but not within shield radius). This number excludes impossible states like "facing a wall and enemy ahead". Excluding the possibility of compounding actions, there are six possible actions for each state (the six terminals). Therefore the range of possible behaviors of the tank is $6^{39} = 2.23 \times 10^{30}$. If we assume, conservatively, only the

Table 8.1
Tableau for "Artificial Death" Simulations

Objective:	Evolve efficient 'tank combat' programs.
Terminal set:	`FORW, TURNR, TURNL,` `WAIT, FIRE, SHIELD.`
Function set:	`IF_FB, IF_LR, IF_RADIUS, IF_RANGE,` `IF_WALL, RPROGN.`
Fitness cases:	One fitness case: random opponent (from within neighborhood), random initial facing.
Raw fitness:	A function of remaining energy of the victor.
Standardized fitness:	Equals raw fitness for this problem.
Hits:	Win or lose.
Wrapper:	None.
Parameters:	M = 500, G = 50 (usually).
Success predicate:	None.

possibility of a single **RPROGN** for each state, we get $(6x6)^{39} = 4.96x10^{60}$ - sufficiently large to ensure flexibility of evolved strategies. [Note: as a sanity check, for every strategy we observed, or were able to contrive, we were able to come up with a strategy to "break" it]

For illustration purposes, the following is one of the more complex tanks we designed, it uses an actively aggressive strategy:

```
(IF_RANGE  (IF_FB  (IF_LR  (FORW)   (FIRE)   (FORW))
                   (IF_LR  (TURNL)  (WAIT)   (TURNR))
                   (TURNR))
           (IF_FB  (FORW)
                   (TURNL)
                   (TURNL)))
```

Against a stationary target, this tank would behave as follows: as long as the opponent is not within cannon range, the tank will seek out its opponent. Note that it uses a fairly inefficient strategy to do so, only turning left. If the opponent is within range, the tank moves to where the opponent is directly in the line of fire, and shoots.

8.4 Method

One of the main features of our method is a new mechanism for fitness evaluation: individuals are evaluated two-by-two, by simulating their interactions in a fixed duration encounter in "the arena." The fitness evaluation function we are currently using is complementary-linear in the sense that the sum of the fitnesses of the two combatants (usually a "winner" and a "loser") will always be constant. As a consequence, our selection method, which utilizes the fitness-proportional selection function, is effectively a hybrid of fitness proportionate and tournament selection techniques.

Note that this is not a valid "overall" fitness evaluation, because the fitness of an individual is a measure of how it performed against its opponent in this generation. For comparative purposes we provide a suite of thirteen hand-crafted tanks that can be compared against every member of the population, basically creating an artificial set of thirteen "global" fitness functions that allows us to examine the performance of the population.

Because we use one-on-one combat as our evaluation technique, coupled with the semi-random initial positioning of the combatants in the arena, we have an effectively "high noise" fitness function. We postulate that this is analogous to "real life" situations and would expect the effects of this noise to average out over a number of generations. In an attempt to reduce this noise, we have also tested multiple-combat fitness evaluation functions (which also serve to effectively increase local interaction). However, we find that the increased computation time far outweighs its fairly small benefits.

Another major feature of our system is the addition of locality in terms of both combat opponent and mate selection. With respect to fitness evaluation, an individual is pitted against a opponent from the same geographic "neighborhood" with equal probability for all members of said set. Breeding mates are also selected from within the same neighborhood, and the resulting offspring are put back into the population within the same neighborhood. Neighborhood size may be varied to actively control the degree of local interaction desired in the run. Throughout our evaluation, both local and non-local simulations were studied to accurately gauge the effects of this introduction.

8.5 Discussion of Results

8.5.1 Population Seeding

A large number of trials were devoted to attempting to find parameter settings conducive to evolving viable strategies from a completely random initial population. While the evolution of many relatively passive strategies was fairly simple, we found that achieving a population with viable active as well as passive strategies was much more difficult. This was most evident with the addition of random initial combatant positioning to the simulation. The successful aggressive strategy must evolve not only an active attack mechanism (shield or fire) but must also locate and navigate to within striking distance of its prey. This proved to be too high a threshold for most of the runs we attempted from random populations and random positioning. However, in those runs where combatants were always started facing each other from across the arena, both active and passive strategies were found.

The problems in arriving at a decent population from scratch are most likely related to the large problem space and the comparatively small population size we use (500 individuals for most of the runs, run over 50 generations). Our choice of population size was limited by the computational resources at our disposal, because for each fitness evaluation the program has to do a combat simulation, which makes the runs quite slow. Additional resource would have allowed larger runs (in terms of populations size and number of generations) in order to examine the effect of locality on speed of evolution.

We decided that it would be more interesting from an ALife point of view to start off with a population that was seeded with viable basic strategies and then examine how they evolve further, rather than watch them struggle through the initial primitive stages of their evolution. To ensure the consistency of the initial populations across different runs we have provided a utility to reproducibly seed an entire population homogeneously from our test suite. With regard to the qualitative analysis of the effect of locality, seeding with well-defined strategies also facilitated explicit trait tracing throughout the generations. Furthermore, this allows us to examine locality in terms of which strategies are prevalent in various neighborhoods by comparing them to the behavior of the initial seed individuals.

Another available option would be to seed the entire population with a single individual and study the emergence of differentiation among the population as in Tom Ray's Tierra [Ray 1991] system; however, we are interested in studying a large variety of different strategies, and feel that this approach will prove less fruitful than the course we have followed.

8.5.2 Statistical Analysis

We use the thirteen seed tanks to evaluate the behavior of each individual in the final population. Every individual does combat with each of the seed tanks; and the results are combined to form a *behavior signature* of thirteen fitness values, yielding an easily quantifiable measure of phenotype. Ideally, we want this set of fitness functions to be orthonormal and complete, such that our signatures accurately classify all possible behavior variations. We can compare the behavior of two different individuals by checking how similar their signatures are. The correlation between two signatures gives us a measure of similarity. We calculate the correlation coefficients for every two individuals in the population, and use the overall average of these coefficients as a measure for diversity in terms of behavior, which we loosely call *phenotypical diversity*.

Figure 8.1 shows how the average correlation changes as a function of neighborhood size, for N = 10, 20, 50, 100, 200, 500 (because our population size is 500, a neighborhood size of N=500 means no locality). Whereas for most neighborhood sizes we use two runs with different random seeds, for N = 50 and 500, the graph shows the average over four runs for higher accuracy.

For N = 500 to N = 100, we clearly see that the average correlation goes down --and thus that the phenotypical diversity of the population improves -- as the neighborhood size gets smaller (increased locality). For even smaller neighborhood sizes, the average correlation seems to increase slightly, although it still stays well below the value for the non-locality runs.

We examine the genotypical similarity between individuals by observing the "gene frequencies" in each one. Calculating the relative frequency of occurrence for each of the six terminals and six functions in the S-expression representing the program for an individual, we get a *frequency signature* of twelve frequency values. Taking the overall average of the correlation between all these signatures provides a second measure of diversity of the population, referred to as *genotypical diversity*.

This measure of genotypical diversity is fairly crude, since it doesn't take any structural information into account. For example, it won't notice frequently recurring subtrees (so called *building blocks*), apart from the influence they have on the total gene frequencies. Despite these limitations, this approach does give us a much more nuanced view of the

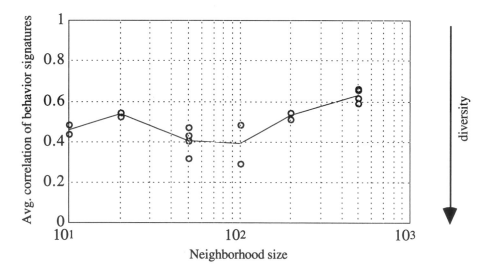

Figure 8.1
Measure of phenotypical diversity.

diversity in a population than, for example, the "variety" calculation used by Koza [Koza 1992], which is based purely on whether individuals are exactly identical (Lisp **EQUAL**).

Figure 8.2 shows the change in correlation of frequency signatures, as a function of neighborhood size, for N = 10, 20, 50, 100, 200, 500. (This figure, as well as figures 8.3 and 8.4, is generated from the same data as Figure 8.1)

This graph further shows, genotypically, how the addition of locality improves population diversity. Even for neighborhood sizes below 50, the average correlation in gene frequencies continues to decrease.

We have not yet found a satisfactory explanation for the apparent discrepancies in genotypical versus phenotypical diversity for small neighborhood sizes, although this is conceivably due to our small sample size of only two runs per neighborhood size. Another possible cause might lie in the fact that at N = 50 and below, deme effects start occurring which may affect the diversity (see section 8.5.3).

It should also be kept in mind that these neighborhood sizes of N =10 and N = 20 are rather small for a population size of M = 500, and only 50 generations. For example, with N = 10, the maximum speed at which the influence of an individual can propagate through the population (sometimes referred to as the *speed of light*) is 5 individuals per generation (to each side). This means that for 50 generations it is unlikely that an individual would manage to propagate through the entire population. It is possible that we are still seeing residual effects from the initial seeded population.

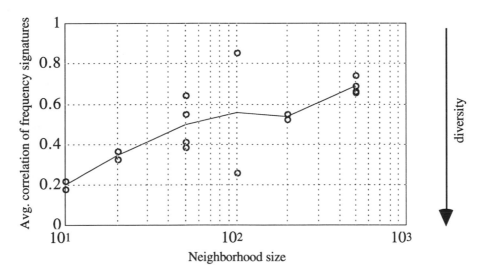

Figure 8.2
Measure of genotypical diversity.

We define generality of an individual in terms of uniformity of the points in its behavior signature. An individual with wildly varying fitness values in its behavior signature is said to be less general (more specialized) than an individual with a smooth behavior signature curve, though the two may have the same average fitness. While our test suite does not cover the entire range of possible behaviors, we do find this measurement to be useful for comparison between different populations. By seeding all of our populations identically with individuals from our general test suite, which incorporate a number of basic behaviors, we hope to reduce bias towards any one particular behavioral strategy.

Figure 8.3 shows the average generality of individuals (in terms of standard deviation over individual behavior signatures, lower standard deviation means a more general individual), with respect to varying neighborhood size.

The individuals in the population with N = 50 are slightly more general (about 5.6%) than those in the non-locality population (N = 500). However, because of the large variance between the individual runs, it is not clear whether the difference is sufficiently significant from a statistical point of view. The generality of the individuals seems to degrade again for smaller neighborhood sizes.

We can also examine the generality of an entire population by taking the average performance of each of the test tanks against each individual and thus generating a behavior signature for the population as a whole. As in the case of individual generality,

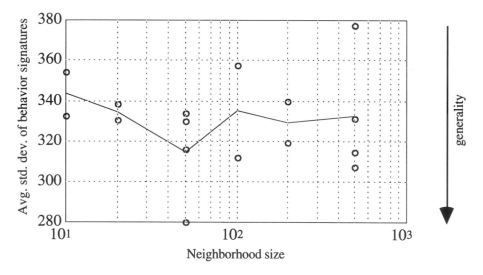

Figure 8.3
Measure of individual generality.

we quantify this by taking the standard deviation over the values of this signature.

Figure 8.4 shows the generality of the population (in terms of standard deviation over the behavior signature of the entire population, lower standard deviation is more general), with respect to varying neighborhood size.

This graph shows a large increase (approx. 30%) in generality between N = 500 and N = 50. This is probably due to the greater phenotypical diversity in populations using locality, in combination with the slightly larger generality of the individuals. Again, the generality of the populations seems to decrease for neighborhood sizes smaller than 50, although it stays better than the generality of the non-locality populations.

Because we do not have a measure of absolute fitness, it is difficult to determine whether or not the actual performance of the population increases with the addition of locality. One method of testing this would be to compare each individual of a population using locality with every individual of a second, non-local, population. Clearly the computational requirements of such testing make this impractical with the means at our disposal. For comparison, our normal runs of 500 individuals for 50 generations take approximately 8 hours to run. A full comparison alone, with only one random initial position of the opponents per combat, would take at least 10 times as long.

We have taken a simpler though less accurate approach to quantifying performance by taking the average of all behavior signature values over all individuals. While weak, this measure provides a crude mechanism for comparing the effect of locality and non-locality

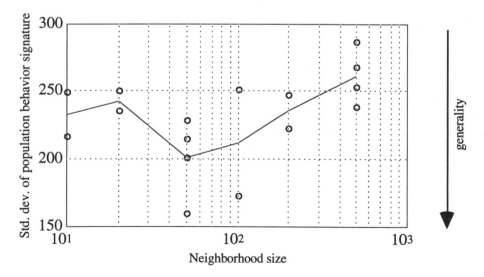

Figure 8.4
Measure of population generality.

in populations. According to this measurement, we see no statistically significant change in performance between the non-locally evolved population and that with neighborhood size 50.

8.5.3 Emergence of Demes

Although nothing in our model explicitly implements deme formation, we do find the emergence of deme-like structures in populations using sufficiently small neighborhood size. The Oxford English Dictionary defines a deme as "A local population of closely related plants or animals". We use this definition, as opposed to Wright's inter- and intra-subpopulation interaction criterion. Using our parameters, M=500, and G=50, the boundary at which this deme effect occurs is around neighborhood size 50.

Collins and Jefferson observed a similar effect [Collins and Jefferson 1991], when they were using local mating in a 2-dimensional population, to solve a graph partitioning problem. Because this problem has two complementary solutions, which are known in advance, they were able to construct a map of their population, showing which areas had converged to one solution or the other. The problem we are using has a whole continuum of strategies, which makes it more difficult to show the demes' location and ethnographic composition.

One way in which we could visualize the deme structures in the population, would be to take the average correlation of each individual's behavior or frequency signature with respect to all other individuals in the generation. Areas with a high correlation would then indicate demes. The problem with this technique is that demes with more common strategies would stand out more than demes with a rarer strategy, skewing the graph to appear to have more pronounced demes for the former.

Also, the contents of a typical population is very "noisy", because we are using a fairly large fraction of crossover and mutation. This creates a large number of individuals that are significantly different (in terms of their signatures) from their immediate neighbors, making it hard to distinguish any structure in the population.

We therefore adopted a local correlation measure to delineate the demes. First, we calculate for each individual a weighted average C^w of the correlations of its signature with the other individuals:

$$C^w_i = \frac{\sum\limits_{j \neq i} C_{i,j}/\text{distance}(i,j)}{\sum\limits_{j \neq i} 1/\text{distance}(i,j)}$$

This still does not take into account that individuals with a more "average" signature (i.e. whose correlation with respect to the rest of the population is higher on average), are also more likely to have a higher local correlation measure. We therefore subtract the average correlation, leaving just the *additional* correlation due to locality C^+:

$$C^+_i = C^w_i - \overline{C_i}$$

This measure also eliminates a lot of the noise caused by individuals created by mutation or crossover, because most of this noise appears both in C^w and in $\overline{C}$, and thus gets canceled out. The remaining noise can be filtered out with a simple smoothing function (Gaussian filter kernel with a spread of 10). The result is a graph of the entire population, in which the peaks represent demes, and the valleys represent boundaries between demes.

This technique can be applied to both the behavior signatures and the gene frequency signatures. Figure 8.5 shows these curves for one of the runs with $N = 20$ (None of the runs with $N = 100$, 200 and 500 show this effect: the curves for these just consist of some noise centered around zero). The bands in the graph are three typical demes that we selected by hand, based on the position of peaks in both curves.

Although they are by no means identical, the same structures are visible in the two curves. The curve derived from the frequency signatures is usually higher than the one derived from the behavior signatures. This is linked to the fact that a small difference in genotype can cause a large difference in phenotype, such that the phenotypic correlation between related individuals is generally not as high as the genotypic correlation.

When comparing the average performance (i.e. average over the behavior signature) of

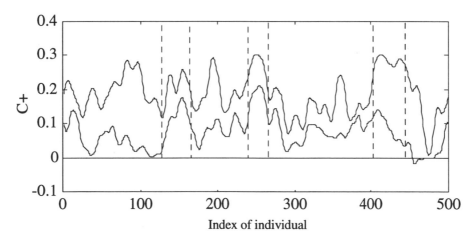

Figure 8.5
Occurrence of demes, measured based on phenotypic similarity (bottom curve), and genotypic similarity (top curve). Dotted regions indicated three hand-selected demes

individuals inside and between demes, we also find that the regions in between demes are characterized by a high percentage of poorly performing hybrid individuals. This result is similar to that observed by Collins and Jefferson [Collins and Jefferson 1991].

We can examine the dynamic behavior of the demes indicated in Figure 8.5. by tagging the members of each deme on a genetic level (see section 8.5.4) and letting the population evolve for another 50 generations. By plotting the gene frequencies of the genetic material that came out of these three demes, we can see how far their influence has spread. Figure 8.6 shows the demes tagged at generation 50 (the vertical bars), and the resulting gene frequencies (smoothed as above) after another 50 generations.

The genetic material for the demes we found at generation 50, from index 128-164, 240-265, and 404-444, has now shifted to respectively 494-89, 378-389, and 371-9. We observe the greatest shift for the left border of the first deme: 134 individuals in 50 generations. This is a lot slower than the "light speed" for this neighborhood size, which is 10 individuals per generation.

The genetic material from the second deme has all but disappeared from the population. In fact, taking into account that part of the remaining genes will probably be inside dead branches in the tank programs, only a small fraction of the original material is still being expressed in the population after 100 generations.

The material from the third deme on the other hand, has now spread to a region with more than three times its former area and makes up about 50% of the genetic material of the individuals in this area. At its extremities, the material from the third deme now mingles with material from the other demes, despite their initial geographic separation.

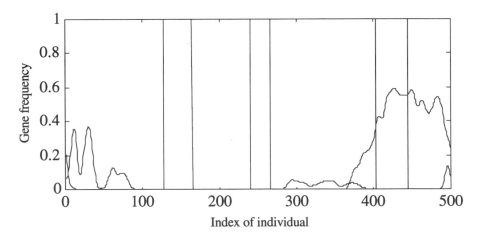

Figure 8.6
Drift of genetic material in three sample demes.

8.5.4 Structural Analysis of Individuals

The analysis of the members of a population involves detailed understanding of both the phenotypic and genotypic aspects of the individual. An understanding not only of how an individual functions, but the underlying mechanisms and origins of its behavior.

How a tank behaves has much to do with how its parents behaved. While part of later generation tanks are "traits" (functional blocks of code) copied from a parent, many develop modifications to existing behaviors, or entirely new behaviors, resulting from recombination with other members of the population (some of which may be identical to itself) and through mutation.

One mechanism used to study the evolution of programs is tagging. This technique is borrowed from populations studies in nature and from cellular biology. These fields have used this technique for years, marking individual entities and genes to explore their interaction with their respective environments over a period of time.

We use tagging in two ways. We mark the individual genes of each of our seed tanks with an indicator ID of that tank to facilitate recognition of the influence of seed tanks (by ID) and mutations (by the lack of ID) in future generation individuals. As described above, we also use tags to mark the individual genes of every member of a deme before allowing the simulation to continue evolving. This tagging allows us to trace the evolution, drift, spread, and inter-deme interactions of the deme members.

Phenotypic analysis is undertaken from a behaviorist viewpoint: rather than dwelling on the countless possible behaviors possible from the evolving individuals, we focus on the stimulus-response characteristics of each individual with respect to its surroundings. There are 39 viable sensory possibilities which can be distinguished by the tank, and to each, there may be a distinct response, which may be a small program in itself (utilizing **RPROGN**). From this simple model we can envision a graph of the world surrounding the tank and the response to stimuli of the tank in each region. Considering this as a phenotypic descriptor of the individual, comparing distinct population members

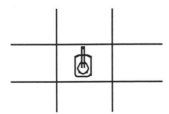

Figure 8.7
Basic layout of stimulus-response graph. Each of the 8 quadrants may contain a different response.

behaviorally is reduced to symbol matching the stimulus-response graphs. Figure 8.7 shows the basic layout of such a graph.

Genotypically, the origins of the pieces of a tank's genetic code are interesting in that they can explain the evolution of a tank's make-up. Much like the genetic code of most living organisms, much of the genetic information in our tanks is not utilized. For example, a routine for turning to face an individual and then fire may only be called only if an opponent is in front of the tank while it is facing the wall -- clearly an event which can never occur. This material is not wasted, however, in that it may be transferred to the offspring of the individual through crossover. Furthermore, a particular behavior trait might have multiple distinct implementations. Trait tracing (following functional blocks of code through generations of crossover) is done by following tag structures and comparing them with the original seed tanks. Interesting comparisons are made by comparing an individual with its genetic ancestors in both genotype and phenotype, as the genetic blocks passed from ancestor to individual do not always serve the same purpose in both tanks.

Figures 8.8 and 8.9 are two examples of tanks bred during our simulations. They are presented in reduced form such that all tags and much of the extraneous code has been removed. Appendix B shows the complete forms of these individuals including their tags (shown as integers following each function or terminal name) to allow full exploration of the genetic structure of the individuals. In addition to the code, a simplified stimulus-response phenotypic map is provided for each example in Appendix C.

The tank in figure 8.8 was taken from a simulation, using N=20, at position #244. It is an excellent example for a number of reasons. It has genotypic roots in seed tanks 1,10, and 13 (see appendix A), as well as pieces introduced through mutation. Phenotypically, it would seem that it takes portions from all 3 of its parent seeds. However, reduction analysis shows that most of its functionality comes from seed #10, and the modifications are from #13 (introduction of compound functions) and mutation

```
(IF_RANGE
    (IF_FB  (IF_LR  (TURNL)
                        (RPROGN (FIRE) (FIRE))
                    (TURNR))
             (RPROGN (FORW)
                        (IF_LR  (TURNL)  (WAIT)  (TURNR)))
             (TURNR))
    (WAIT))
```

Figure 8.8
Individual #244 (N=20) Reduced

```
(IF_FB
    (RPROGN (RPROGN (FORW)
                        (IF_RANGE (IF_LR (TURNL)
                                          (FIRE)
                                          (TURNR))
                        (FORW)))
            (IF_RANGE (IF_RADIUS (IF_LR (SHIELD)
                                          (RPROGN (FIRE) (FIRE))
                                  (IF_WALL
                                     (RPROGN (TURNL)
                                             (FORW))
                                     (SHIELD)))
                        (FORW))
            (FORW)))
    (FORW)
    (TURNR))
```

Figure 8.9
Individual #263 (N=50) Reduced

(adaptation of the **RPROGN**s to create a function which, when along side an opponent, allows the tank to move forward before turning to face -- thus stepping out of the line-of-fire). Thus, this example illustrates the combined results of multiple crossovers and mutation creating a new trait.

Another good example tank was found at position #263 during a run with N=50 (Figure 8.9). It has genotypic roots in seed tanks 11 and 12 (see Appendix A), as well as pieces introduced through mutation. This individual shows multiple crossovers with derivatives of seed #12, and through this method as well as mutation, greatly exploits the favorable feature **RPROGN**. By contrast to the previous example, this individual has very little unused genetic code, as comparison with its unreduced listing (Appendix B) will show. This example also serves to illustrate the limitations of simplistic representations of phenotypic reaction.

As illustrated in Appendices B and C, analysis of the evolved code for the tanks is an arduous task, even with the help of tagging facilities and application of reduction. The creation of a more complete tool kit with, for instance, automatic trait tracing and detection of functional blocks, would allow more concrete data to be collected on the phenotypic and genotypic aspects of the individuals.

8.6   Recommendations for Future Work

• Introduction of a novel beneficial feature to a population, in order to examine its spread throughout the population over the course of subsequent generations.

• Introduction of various neighborhood sizes throughout the population. Possible variation to modify the probability of selection within a neighborhood to be a function of distance from the neighborhood center.

• Introduction of "sex" or "species" to the population. For instance, allowing each sex to have a different size neighborhood and seeing how that effects population diversity, trait migration, etc.

• Separation of seeding population from signature population. It is possible that by generating our characteristic signatures of the test population from the same test individuals used to initially seed the population a bias in generality of individuals may have been introduced.

• Quantifying phenotypic description based on stimulus-response characteristics. For example, measuring phenotypic distance based on response in each of the 39 possible states.

• Comparison with larger populations and larger number of generations. Hopefully with a number of different population sizes the complete formula for deme emergence (from neighborhood size, population size, and number of generations) can be deduced.

• Seeding with less "complete" seeds. The introduction of intentionally damaged tanks, or seed tanks even more incomplete than the current ones, may provide insights as to the recombination capabilities of the evolutionary paradigm.

8.7   Conclusions

The tools introduced in this chapter (tagging, quantification of phenotypic behavior and genetic composition by means of signatures, and stimulus-response maps) have proved extremely useful in providing new insights into the mechanisms at work in our simulations. We feel that these same tools could be applied to an entire range of similar problems.

Our experiments show that the introduction of locality to the GP paradigm significantly improves population diversity, which, in conjunction with the slightly increased generality of individuals, yields a substantial improvement in the generality of the population as a whole. In addition we found no statistically significant decrease in overall average fitness (as measured by the average behavior signature).

Given the right conditions, demes spontaneously emerge despite the lack of explicit geographic isolation other than uniform isolation-by-distance. We consider the evolving deme structures to be more flexible than a static deme implementation. We encourage further comparisons of the two approaches.

Finally, we have shown how the neighborhood of an individual plays an important role in its evolution not only as a genetic pool but as evolutionary pressure toward specialization (through competition with similar individuals). The resulting pockets of different strategies may later diffuse through the population as a whole.

8.8 Appendix A: Seed Tank Code

```
; #1: "HUNTER 1"
(IF_RANGE (IF_FB (IF_LR (FORW) (FIRE) (FORW))
                 (IF_LR (TURNL) (WAIT) (TURNR))
                 (TURNR))
          (IF_FB (FORW) (TURNL) (TURNL)))

; #2: "HUNTER 2"
(IF_RANGE (IF_LR (IF_FB (FORW) (TURNR) (TURNR))
                 (FIRE)
                 (IF_FB (FORW) (TURNL) (TURNL)))
          (IF_FB (FORW) (TURNR) (TURNR)))

; #3: "WAITER 1"
(IF_RANGE (SHIELD) (WAIT))

; #4: "WAITER 2"
(IF_RADIUS (SHIELD) (WAIT))

; #5: "RUN-AWAY 1"
(IF_FB (IF_LR (TURNR) (TURNR) (TURNL))
       (FORW)
       (FORW))

; #6: "RUN-AWAY 2"
(IF_FB (IF_LR (TURNR)
              (RPROGN (TURNR) (FORW))
              (TURNL))
       (FORW)
       (FORW))

; #7: "RUN-AWAY 3"
(RPROGN (IF_LR (TURNR) (TURNR) (TURNL)) (FORW))

; #8: "CORNER-HIDER"
(IF_WALL (WAIT)
         (IF_FB (TURNL) (TURNL)
                (RPROGN (FORW) (IF_WALL (TURNL) (WAIT)))))
```

```
; #9: "LURKER 1"
(IF_RANGE (IF_FB (IF_LR (WAIT) (FIRE) (WAIT))
                 (TURNL) (TURNL))
          (WAIT))

; #10: "LURKER 2"
(IF_RANGE (IF_FB (IF_LR (TURNL) (RPROGN (FIRE) (FIRE)) (TURNR))
                 (TURNR) (TURNR))
          (WAIT))

; #11: "LURKER 3"
IF_RANGE (IF_FB (IF_LR (WAIT) (FIRE) (WAIT))
                (IF_LR (TURNL) (FIRE) (TURNR))
                (IF_LR (TURNL) (SHIELD) (TURNR)))
         (WAIT))

; #12: "LOOK-AHEAD"
(IF_FB (RPROGN (FORW) (IF_RANGE (FIRE) (FORW))) (TURNR) (TURNR))

; #13: "SHIELDER"
(IF_RADIUS (SHIELD)
           (IF_FB (FORW)
                  (RPROGN (FORW) (IF_LR (TURNL) (SHIELD) (TURNR)))
                  (TURNL)))
```

8.9   Appendix B: Original Example Tank Code

Unreduced code for individual from Figure 8.8 (N=20, Tagged Individual #244):

```
(IF_RANGE10 (IF_FB10 (IF_LR10 (IF_RANGE10 (IF_FB10 (IF_LR10 (TURNL10)
(RPROGN10 (FIRE10) (FIRE10)) (IF_WALL (IF_FB (SHIELD) (RPROGN (SHIELD)
(TURNL)) (FIRE)) (TURNR10))) (IF_RANGE (FORW13) (IF_WALL (TURNR10)
(IF_FB (FIRE) (TURNL) (FORW)))) (TURNR)) (WAIT10)) (RPROGN10 (FIRE10)
(FIRE10)) (TURNR10)) (IF_RANGE (IF_FB13 (FORW13) (RPROGN13 (FORW13)
(IF_LR13 (TURNL13) (WAIT) (IF_LR10 (TURNL10) (RPROGN10 (FIRE10)
(FIRE10)) (TURNR10)))) (FIRE)) (IF_WALL (IF_RANGE1 (IF_FB1 (IF_LR1
(TURNL) (FIRE1) (IF_WALL (IF_RANGE1 (IF_FB1 (IF_LR1 (TURNL) (FIRE1)
(FORW1)) (IF_LR1 (TURNL1) (IF_RANGE (SHIELD) (TURNL13)) (TURNR1))
(TURNR10)) (RPROGN (FORW) (IF_RADIUS (IF_FB10 (FORW) (IF_RANGE
(IF_RADIUS (TURNL13) (TURNL)) (IF_WALL (WAIT) (IF_LR1 (TURNL) (FIRE1)
(FORW1)))) (TURNR10)) (FORW)))) (FIRE))) (IF_LR1 (TURNL1) (SHIELD)
(TURNR1)) (TURNR10)) (RPROGN (FORW) (IF_RADIUS (FORW) (FORW)))) (IF_FB
(FIRE) (IF_LR10 (TURNL13) (RPROGN10 (TURNL1) (FIRE10)) (FORW13))
(FIRE)))) (TURNR10)) (WAIT10))
```

Unreduced code for individual from Figure 8.9 (N=50, Tagged Individual #263):

```
(IF_FB12 (RPROGN12 (RPROGN12 (FORW12) (IF_RANGE12 (IF_LR11 (TURNL11)
(FIRE11) (TURNR11)) (FORW12))) (IF_RANGE12 (IF_RADIUS (IF_LR (SHIELD11)
(RPROGN (FIRE)(FIRE)) (IF_RANGE (IF_WALL (RPROGN (TURNL)(FORW))
(SHIELD)) (FORW))) (IF_WALL (IF_RANGE (FORW) (FORW)) (FORW))) (FORW12)))
(FORW12) (TURNR12))
```

8.10 Appendix C: Stimulus-Response Maps of Example Tanks

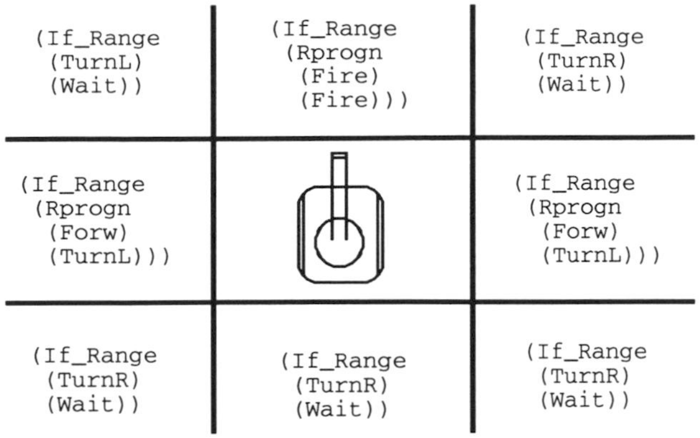

(If_Range (TurnL) (Wait))	(If_Range (Rprogn (Fire) (Fire)))	(If_Range (TurnR) (Wait))
(If_Range (Rprogn (Forw) (TurnL)))		(If_Range (Rprogn (Forw) (TurnL)))
(If_Range (TurnR) (Wait))	(If_Range (TurnR) (Wait))	(If_Range (TurnR) (Wait))

Figure 8.10
Simplified Stimulus-Response Map for individual from Figure 8.8 (N=20, Tagged Individual #244)

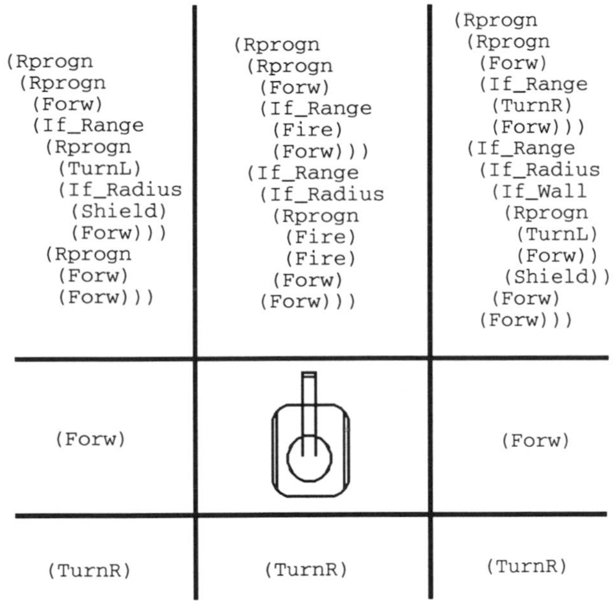

(Rprogn (Rprogn (Forw) (If_Range (Rprogn (TurnL) (If_Radius (Shield) (Forw))) (Rprogn (Forw) (Forw)))	(Rprogn (Rprogn (Forw) (If_Range (Fire) (Forw))) (If_Range (If_Radius (Rprogn (Fire) (Fire) (Forw) (Forw)))	(Rprogn (Rprogn (Forw) (If_Range (TurnR) (Forw))) (If_Range (If_Radius (If_Wall (Rprogn (TurnL) (Forw)) (Shield)) (Forw) (Forw)))
(Forw)		(Forw)
(TurnR)	(TurnR)	(TurnR)

Figure 8.11
Simplified Stimulus-Response Map for individual from Figure 8.9 (N=50, Tagged Individual #263)

Acknowledgments

We would like to thank John Koza, whose lectures on Genetic Algorithms and Artificial Life started all this, and the people on the genetic-programming@cs.stanford.edu mailing list for the high-frequency mailings and for pointing us to the theory by Wright. Also thanks to Jan Jannink for acting as a sounding board.

Our code is built upon our modified version of John Koza's "Little Lisp" code [Koza 1992].

Bibliography

Collins, R.J. and Jefferson, D.R. (1991) Selection in Massively Parallel Genetic Algorithms, in *Proceedings of the Fourth International Conference on Genetic Algorithms*, R. Belew and L. Booker, Eds., San Mateo, CA: Morgan Kaufmann.

Darwin, C. (1859) *On the Origin of Species by Means of Natural Selection*, London: Murray.

Koza, J. (1992) *Genetic Programming: On the Programming of Computers by Means of Natural Selection*, Cambridge, MA: The MIT Press.

Provine, W. B. (1986) *Sewall Wright and Evolutionary Biology*, Chicago, IL: University of Chicago Press.

Ray, T. (1991) An Approach to the Synthesis of Artificial Life, in *Artificial Life II*, C. G. Langton, C. Taylor, J. D. Farmer and S. Rasmussen, Eds., Redwood City, CA: Addison-Wesley.

Wright, S. (1977) *Evolution and The Genetics of Populations*, Chicago, IL: University of Chicago Press, Vol 3, Ch. 13.

Wright, S. (1932) The roles of mutation, inbreeding, crossbreeding and selection in evolution, in *Proceedings of the Sixth International Congress of Genetics* , The Congress.

9 The Evolution of Mental Models

Astro Teller

Most interesting problems do not have solutions that are simple mappings from the inputs to the correct outputs; some kind of internal state or memory is needed to operate well or optimally in these domains. Traditionally, genetic programming has concentrated on solving problems in the functional/reactive arena. This may be due in part to the absence of a natural way to incorporate memory into the paradigm. This chapter proposes a simple, Turing-complete addition to the genetic programming paradigm that seamlessly incorporates the evolution of the effective gathering, storage, and retrieval of arbitrarily complicated state information. A new environment is presented and used to evaluate this addition to the paradigm. Experimental results show that the effective production and use of complex memory structures can be evolved and that functions evolving the intelligent use of state quickly and permanently displace purely reactive and non-deterministic functions. These results may aid future research into the causes and constituents of mental models and are shown to open the field of genetic programming to include all learning strategies that are Turing-possible.

9.1 Introduction

In speaking about machine learning, we usually mean something more specific than the verb "to learn" in all its uses. Machine learning as a discipline often emphasizes problems of classification and the discovery of functions. These problems include tasks such as symbolic regression, broom balancing, and playing chess. All of these tasks are learning enterprises, but none of them requires the use of state. In general, an agent can neither act nor learn in an environment without some model of the environment. This model must come as a combination of those model features given to the agent and those it builds and maintains on its own. Learning, given an environmental model, is an important problem. Creating a model, however, is also a fundamental part of what it means to learn.

Imagine a small boy (or any small agent for that matter) riding a bicycle around his neighborhood. We might say that this agent has learned to ride the bicycle. This learning process is highly reactive. The ability to lean in different directions in order to remain upright requires no state information. We might also like to say that the agent has learned its way around the neighborhood. This learning is not only the developed ability to navigate but the construction of a mental map of the area. Genetic programming has been applied in the past to problems like the ability to balance on a bicycle but little work in evolutionary strategies has been done on creating and using a mental model of the environment.

The use of state to solve problems is not new to genetic algorithms or genetic programming. In both areas state has been used implicitly and explicitly. In no case, however, has it been a natural, integrated part of the process. In the field of genetic algorithms encodings have been done for both finite state automata and recurrent neural networks. [Jefferson *et al.* 1991] In both cases, binary encoding of lengths over 400 lead to 32 states. In an evolution of solutions to the Prisoner's Dilemma using state, the binary encoding used 2^m bits for an individual who could remember back **m** binary moves. [Lindgren 1991] In genetic programming, one of the techniques has been to use *Progn* as a way to string together side-effecting terminals. [Koza 1992] *Progn* is a more implicit use of state. It is not clear whether this use of state in genetic programming would be generally applicable in memory intensive problems and it does not lend itself to any kind of introspection (i.e. decision making based on a function's current state).

The organization and use of memory is often roughly divided into three classes: state information, iconic models, and sentential models. However distinguishing between these three categories is a very subjective business. State information can be thought of as "memory" that divides all possible states of the world into subsets. State information becomes an iconic model when these subsets of the possible world states are divided along boundaries that correspond to " simple features" of the environment. State information can be interpreted as a sentential model when there is a semantic assignment to the state information that shows the memory to contain recursively combinable statements concerning what is true about the environment (e.g. "A implies B"). For a more complete discussion of these issues try *Mental Models: Towards a Cognitive Science of Language, Inference, and Consciousness*. [Johnson-Laird 1983]

These categories of mental models may be useful for speaking about certain mental activities. However, the subjectivity of these divisions suggests that mental processes are a continuum and cannot be divided along any well defined boundaries. Even if these mental categories are interesting ranges along this continuum, there is no reason to believe that these categories, taken from how we think, will be of any relevance to how evolved agents build and use state information. Rather than force the results that follow into some predefined mental pigeon hole, this chapter hopes to help call into question the meaningfulness of categorizing the organization and use of memory.

The purpose of this chapter is threefold. First, it is an attempt to evolve individuals that use state effectively and that even build mental models of an environment given only temporally and spatially immediate sensory input. These mental model builders use an addition to the genetic programming paradigm called **indexed memory**. The second purpose of this chapter is to evaluate indexed memory with respect to the evolution of mental models and as a general, Turing-complete tool for using memory in genetic programming. And finally, this chapter presents the environment in which these agents operate, **Tartarus,** as a benchmark for the effective use of state. This environment condemns the population individuals, like Sisyphus, to spend forever fighting the entropy of their system. At the beginning of each new generation, the world is returned to its disordered state and the individuals must start again. In this spirit the test environment is named Tartarus after the underworld where Sisyphus spends an eternity pushing his boulder out of a global minimum.

Indexed memory was not designed to foster the emergence of mental models *per se*, but to be a general method for the use of memory in the field of genetic programming. The evolution of the intelligent use of state was an appropriate first trial of this paradigm; indexed memory facilitates a variety of uses for state and is simple enough not to overly bias the evolution of state use. This chapter, through results and discussions, will suggest that this new scheme for incorporating the use of memory into genetic programming will be practically applicable to a wide range of problems that have already been investigated and will open the door to new problem areas that are memory-critical. In addition, this attempt to produce agents that build up and effectively use state may be a small step into using computational evolution to study the nature of mental models.

9.2 The Method and The Model

Before indexed memory is described, motivations and criteria for its creation will now be summarized. The ideal state storage model should allow genetically evolved functions some atomic abilities to examine and change their state. The *Progn* model of state usage is an example of a strategy that does not allow the function to examine its state using the same mechanism used to evaluate its inputs. Any new state storage model should be paradigmatically neutral and should operate at a low level. There should be little bias on how evolved functions make use of their state and no data structures or "concepts" should be impossible because of the high level at which the state storage scheme operates. In genetic programming there is a constant tradeoff between the benefits and losses of constraining the search space of algorithms. The "lower" (more basic) the memory scheme we choose, the more of the algorithm space we can/have to explore. As we tackle increasingly difficult problems with genetic programming, our intuition fails us and it becomes increasingly important not to warp the fitness landscape, causing good solutions to become hard or impossible to obtain. For this reason, paradigmatic neutrality and basic functionality were important criteria in settling upon a memory scheme. In addition, and most importantly, any new state storage model should facilitate any computation (i.e. be Turing-complete).

In the general case of indexed memory, each population individual has not only a functional tree, but an array (**Memory**) of elements indexed from 0 to (M-1). Each of these M elements is an integer in the same range. The idealization of indexed memory is a memory array indexed over the integers whose element values are also integers. For practical purposes however, some M will always be chosen to approximate this idealization. Two non-terminals are added to any existing function set: **Read** and **Write.** Read takes one argument (Y) and returns Memory[Y]. Write takes two arguments: X and Y. It returns the old value in Memory[Y] and then puts X into Memory[Y]. Write could return the new value in Memory[Y] instead and indexed memory would work just as well. Both parameters can be any integer between 0 and (M-1) and both functions return an integer between 0 and (M-1). For the Tartarus world M was chosen to be 20. This may seem small but consider that the agents now have 20^{20} states available to them (approximately $1.05 * 10^{26}$). This number is significantly larger than the few dozen states available to genetic algorithm and genetic programming experiments done in the past. The choice of M=20 was largely for simplicity, but indexed memory with dynamically expandable M is Turing-complete as will be discussed later in the chapter. Also later in the chapter the evolutionary effects of significantly larger M will be discussed along with the ramifications of having so many states available.

There will be only two kinds of terminals: constants (between 0 and (M-1)) and inputs. The number and meaning of these inputs will be set when the environment is described. All functions were constrained to return integers between 0 and (M-1) so that any computed value was a legal memory index. This restriction could have been relaxed by taking the index modulo M before accessing memory, but was not for simplicity. The non-terminals selected for this problem were:

F = { OR,NOT,LESS,ADD,IF-THEN-ELSE,SUB,EQ,READ,WRITE,ADF }

(**OR** X Y) --> This returns the MAX(X,Y)
(**NOT** X) --> This returns 1 if X=0, otherwise it returns 0.
(**LESS** X Y) --> This returns 1 if X<Y, otherwise it returns 0
(**ADD** X Y) --> This returns X+Y (ceiling of M-1)
(**IF-THEN-ELSE** X Y Z) --> This returns Y if X>0, otherwise it returns Z
(**SUB** X Y) --> This returns X-Y (floor of 0)
(**EQ** X Y) --> This returns 1 if X=Y, otherwise it returns 0
(**READ** Y) --> This returns Memory[Y]
(**WRITE** X Y) --> This returns Memory[Y] and THEN sets Memory[Y] to X

In addition to the main genetically programmed functional tree, the agent is given an automatically defined function (ADF) tree with two arguments. In that subexpression the terminals and non-terminals already stated are legal. The ADF has two additional

Root or Subtree of Main Function   **Automatically Defined Function**

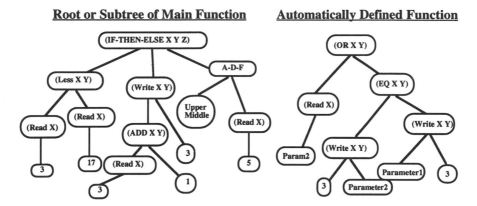

Figure 9.1

terminals: **Parameter1** and **Parameter2**.   The main function of the agent has the additional non-terminal (**ADF** X Y) which sets Parameter1 to the evaluation of the subexpression X and Parameter2 to the evaluation of subexpression Y and then evaluates **ADF**.   Like the other functions, **ADF** returns an integer in the range of 0 to (M-1).

The two trees shown above could be whole trees or subtrees of a main function and an ADF, respectively. In English the main branch above would translate into "If Memory[3] < Memory[17] then return Memory[3] and add 1 to it. Otherwise return the result of sending the value of the UpperMiddle sensor and the value of Memory[5] to the ADF." The returned value of Memory[3] is its value before the increment. Memory[3] is not always incremented because the tree is not fully evaluated. That means that the test branch of the IF-THEN-ELSE is evaluated and then, based on the result, either the THEN or ELSE subexpression is evaluated, but not both. The choice of whether to do full evaluation of an expression is an important one whenever there are side-effecting terminals or non-terminals. Because the simulator created for these experiments does not fully evaluate an agent's function, the agent is allowed some "choice" about what and when to write to memory. In full evaluation there are more total writes to memory but there is no way for a function to avoid writing to memory under certain conditions. Indexed memory works well, but differently, in the case of full evaluation.

9.3   The Environment

Tartarus is an NxN grid. Approximately $(N-2)^2/3$ boxes are randomly distributed on the inner $(N-2) \times (N-2)$ grid . In this arena the fitness of a particular agent is tested. The

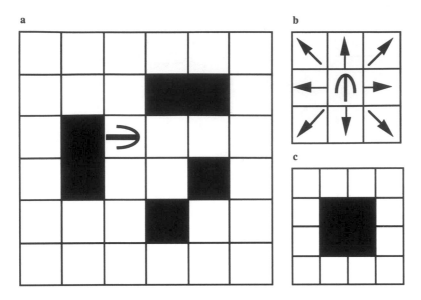

Figure 9.2

agent can be in any square that does not have a box and may face North, South, East, or West. N was chosen to be 6 in order to maintain computational tractability and to maximize certain criteria including the number of different possible initial configurations and low box density. This means that there are 6 randomly placed boxes on the inner 16 squares. At time step 0 of each fitness test, each agent finds itself on a random clear location facing in a random direction. In the discussion sections there will be mention of this choice and its ramifications. A typical trial might start as shown in figure 9.2.a.

The agent is given eight input sensor values for the eight nearest neighbors to the grid position it inhabits (figure 9.2.b). These eight inputs can take on one of three values: **Clear=0**, **Box=1**, or **Wall=2**. The **Wall** value (**2**) indicates that the position is outside the boundaries of the world. These eight input values can be used in the genetically programmed functions as eight terminals, called {UpperLeft, UpperMiddle, UpperRight, MiddleLeft, MiddleRight, LowerLeft, LowerMiddle, LowerRight}. Notice that these are relative to the heading of the agent; after an agent turns, while the environment will not change, one or more of the input values may have changed. This was chosen so that the agent could not figure out from the inputs which way it was facing. An agent has no knowledge of where it is or which way it is pointing unless it keeps track of that information using its indexed memory.

The agent's goal is to push all of the boxes out of the center onto the perimeter grid positions. At the end of a fitness test the agent is rewarded 2 points for every corner that contains a box and 1 point for every non-corner edge position that contains a box. Since no two boxes can occupy the same space there is a 10 point fitness test maximum.

The agent has only three possible actions: *MoveForward*, *TurnLeft*, and *TurnRight*. Because the agent is always on a clear square, turning left and right are always allowed. The agent is only allowed to move forward under two conditions: when the grid position in front is clear and inside the world boundary or when the position in front has a box but the space after that (in the same direction) is clear and inside the world boundary. In the first case the agent moves to the next square. In the second case the agent moves to the next square and the box is pushed one square forward. If the agent attempts to move forward into a wall or into a box which may not be pushed (either because it is against a wall or against another box) the action fails and the environment remains unchanged.

The main function for each agent does not return {*MoveForward*, *TurnLeft*, *TurnRight*}. It returns integers in the range 0 to (M-1). A filter is needed to map these numbers into one of the 3 possible actions. For M=20 the mapping was { Move (ReturnValue < 7), Left (6 < ReturnValue < 14), Right (14 < ReturnValue) }. Any partition with non-negligible ranges for each action would probably have worked.

The world is not always solvable by doing a random walk even in unbounded time, since there are many box configurations (e.g. figure 9.2.c) from which it is impossible to move any box to the wall. The simulator discards random initial environments with configurations like figure 9.2.c. Nothing, however, prevents agents from causing these configurations to occur during the fitness tests. Tartarus can, of course, be cleaned up more easily if the agent has a large amount of time. To keep the task challenging, turning left, right, and moving forward each take 1 time step (even when the move forward is unsuccessful). A complete tour of the board would take $N^2+2N-3 = 45$ and each agent is allowed only 80 time steps, somewhat less than two tours of the board.

Tartarus was designed to promote the intelligent use of state. The intention was to make a task with sufficient complexity in an environment with sufficiently random initial conditions so that agents who use memory would have a sizable advantage over agents who ignore their memory or who do not have memory. It cannot be overstressed how hard it is to achieve high fitness in the Tartarus environment without memory. Considerable effort has been put into writing a function that can reliably get more than 1 point every 80 time steps without using indexed memory. No human-generated or machine-generated program has yet materialized. As an example, imagine the set of conditions in which the agent should or should not move forward when there is a box in front of it and all its other sensors are clear. This could be the case when the box is against the wall, in which case moving forward would not change the environment and an agent without state information would spend the rest of the fitness test pushing the box against the wall. With the same input vector, the box could be against another box. Again, the agent that chooses to push in that situation runs the risk of spending the rest of the fitness test in futile pushing. Yet this input vector seems to be one in which pushing (moving forward into the box) is the best alternative.

It is, in fact, impossible to get better than minimal fitness without using state. Imagine an extreme case where the world is 1000x1000 and there is only 1 box somewhere on the inner 998x998 grid. Based only on the sensors it is, practically speaking, impossible to locate the box. The only other possibility is to do a pseudo-random walk and functional agents cannot do a pseudo-random walk unless they have access to a random walk action. This is a fundamental reason why Tartarus is a good environment in which to promote the use of state.

9.4 The Implementation

Table 9.1
Tableau for the Tartarus environment using Indexed Memory.

Objective:	Find an agent that performs significantly better than random or reactive agents competing in the Tartarus environment.
Fitness cases:	40 fitness cases in which the program has 80 time steps to move the 6 randomly distributed blocks to the edge and corners of the world.
Raw fitness:	Raw fitness is the sum of points earned in the 40 test cases. 1 point per box moved to the edge and an extra point per box moved into a corner.
Standardized fitness:	Standardized fitness: total possible points (400) minus the raw fitness.
Hits:	Same as raw fitness.
Wrapper:	{0 < MoveForward < 6} {7 < TurnLeft < 13} {14 < TurnRight < 19}
Parameters:	*Population* = 800. *Maximum Generation* = 100
Success predicate:	A program scores the maximum number of hits in all 40 test cases.
Overall program structure with automatic function definition:	One result-producing branch and one function definition called ADF which takes two arguments
Terminal set for the result-producing branch:	0...19 , UpperLeft , UpperMiddle , UpperRight , MiddleLeft , MiddleRight , LowerLeft , LowerMiddle, LowerRight
Function set for the result-producing branch:	NOT, OR, If-THEN-ELSE , LESS , ADD, SUB, EQ, READ, WRITE, ADF
Terminal set for the function definition ADF	0...19 , UpperLeft , UpperMiddle , UpperRight , MiddleLeft , MiddleRight , LowerLeft , LowerMiddle, LowerRight , Parameter1 , Parameter2
Function set for the function definition ADF	NOT, OR, If-THEN-ELSE , LESS , ADD, SUB, EQ , READ, WRITE

In most respects these experiments followed the standard genetic programming choices for various parameters. The population size used was 800. Empirically, any population size over 500 seemed to have sufficient diversity to keep the process going. The 800 initial individuals were created randomly. There was a 2 to 1 bias in the initial population randomizer in favor of the non-terminals READ, WRITE, and ADF and in favor of the input terminals. In the ADF there was a 2 to 1 bias in favor of using the terminals Parameter1 and Parameter2. Also, because 0 acts as False and all positive numbers as True, the constant 0 was as likely to come up as all the positive integers combined. The rationale for introducing this bias is that a solution to the Tartarus problem would probably contain more of these items (relative to others like the non-terminal AND or the terminal 6). This biasing does not warp the fitness space. At worst, it starts the population lower in the valleys. With effectively infinite diversity (i.e. population much larger than 800) this would have been unnecessary because there would have been no danger of "losing" a particularly useful terminal or non-terminal.

The maximum depth for the main function was set to 15 and the maximum depth for the ADF was set to 10. After the first 20 generations, the average number of nodes in a function stabilized at about 80 nodes for the main function and 30 for the ADF. The standard deviations were about 25 and 10 respectively. Fitness proportionate reproduction was used to create the mating pool. The crossover percent was 90. Forty-five percent were chosen in pairs and randomly selected subtrees were exchanged in their main functions. Another forty-five percent were similarly chosen and randomly selected subtrees were exchanged in their ADFs. One percent of each population was chosen, proportional to fitness, and mutated. If the mutated node needed a different number of subexpressions, the appropriate number of were added or subtracted. If subexpressions needed to be added, a subtree was created in the same manner that the original population was created. A non-elitist strategy was used, but the best of each generation was saved to a file for use in experimental results.

One fitness case is not sufficient to accurately determine the fitness of a particular individual. For each new generation, 40 environments were randomly generated and each member of the population tested on all 40 worlds. Given the large number of possible initial configurations, the chance that any agent was ever tried on the same environment twice is very close to zero. The number 40 was determined empirically as a reasonable trade-off between the need to get an accurate measure of the fitness for each agent and the need to keep down the run time for each generation. For each of the agent types described below, the runs each got about 10 hours of CPU time on a DEC5000/125. This usually achieved generation 80 for the mental agent with ADF and around 90 for the benchmark agents.

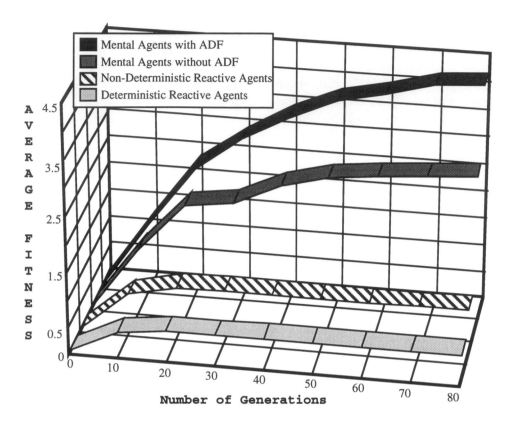

Figure 9.3

9.5   Experimental   Results

So far this chapter has concentrated on a new method for incorporating memory into the genetic programming paradigm. A type of population individual was described and the following results focus on the success of this kind of agent. In order to fairly asses the merit of this type of agent, three benchmark agent types are introduced that are used for comparison in the experiments. The graph above shows the average best of generation for each of these four types. This average was taken across 10 runs, each of which was allowed to go to generation 80.

Agent Type1: Mental agent with ADF ==>   This is the type already described.

Agent Type2: Mental agent with no ADF ==>   This type is like the one already described in every way except that it has no ADF non-terminal symbol (and hence no ADF). This agent type is used to measure the relative effect of having an ADF.

Agent Type3: Non-Deterministic Reactive ==>   This type has no state and no ADF. So READ, WRITE, and ADF are absent from its non-terminal set. Its filter divides the range of 20 up into four parts instead of three. The first three are still MoveForward, TurnLeft, and TurnRight. The fourth (ReturnValue > 15) causes one of the other 3 actions to be performed at random. This agent type is used to test whether the added ability to random walk is sufficient to solve the problem without state information.

Agent Type4: Deterministic Reactive ==>   This type is the real benchmark. This agent type does not have state (READ and WRITE are no longer non-terminals) nor does it have ADF. And, unlike agent type 3, its actions are limited to Move, Left, and Right. It is over this agent in particular that an improvement is sought.

The most important observation to take away from the graph above is that the agent types with indexed memory do much better than the agent types with no access to state. The benchmark agent type 4 (Deterministic Reactive), as expected, is the least successful. Without state this environment is too complex to allow even marginal fitness improvements, and ,as a result, after 80 generations, its best of generation is getting an average of about half a point per fitness test. The random agent type 3 (Non-deterministic Reactive) does almost twice as well, but after 80 generations it still does not consistently get even one point per fitness case.   While we can see that the ADF is very useful in this domain, it clearly is not the most dominant issue. The difference between the agent type 1 (Mental Agent with ADF) and agent type 4 is a factor of 8.5. The difference between agent type 2 (Mental Agent without ADF) and agent type 4 is about a factor of 6. But the difference between agent type 1 and agent type 2 is not even a factor of 2. The conclusion is that state information is critical to getting high fitness in this domain and that indexed memory is an effective evolutionary tool in achieving this use of state.

It is simple to determin if the agents that have evolved are using their memory; their memory can be switched off and their new fitnesses examined. Determining if these agents are using their memory in ways that correspond to features of the world is much more difficult, as we will see later in the chapter. In order to determine objective qualities of performance, it is necessary to perform experiments on a specific agent. The two following graphs show data taken for a typical agent type 1 Best of Generation 80. Its average fitness is 4.4 points per fitness test. This agent is shown directly after the graphs.

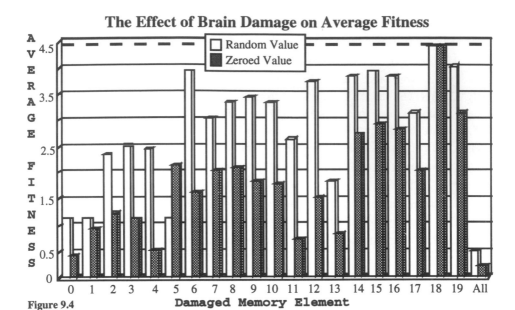

Figure 9.4

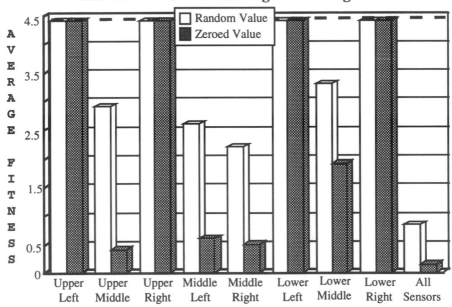

Figure 9.5

Automatically Defined Function

```
(WRITE (IF-THEN-ELSE (ADD (WRITE 0 Parameter1) (OR Parameter1 (READ
(WRITE (READ 0) (READ (ADD Parameter2 (ADD Parameter1 (READ 0))))))))
(ADD Parameter2 (ADD (ADD (READ (READ (READ 0))) (ADD Parameter1
Parameter2)) (ADD Parameter1 (READ MiddleRight)))) (WRITE (WRITE (WRITE
0 (WRITE (WRITE (OR (READ (WRITE Parameter1 Parameter2)) 0) Parameter1)
(WRITE Parameter1 Parameter2))) Parameter1) (WRITE (WRITE (READ 0) (READ
(WRITE Parameter1 Parameter2))) (WRITE Parameter1 0)))) (READ (READ
(READ (READ (EQ Parameter1 Parameter2)))))))
```

Main Function

```
(ADD MiddleLeft (WRITE (SUB 3 (NOT (OR UpperMiddle (EQ 2 (IF-THEN-ELSE
(SUB (SUB (READ 17) UpperMiddle) 8) (*A-D-F* (EQ UpperMiddle (IF-THEN-
ELSE (READ 17) UpperMiddle 10)) (WRITE 10 UpperMiddle)) (LESS (WRITE
UpperMiddle UpperMiddle) (*A-D-F* LowerMiddle (WRITE UpperMiddle
UpperMiddle)))))))) (WRITE 8 UpperMiddle)))
```

The first graph is an account, over many fitness tests, of how this particular agent fared with a form of brain damage. In each set of runs, some memory index **i** was selected and that memory index always returned a random number between 0 and 19 no matter what the actual value was. This was done for 100 random fitness tests, and an average fitness was found for the individual with that brain element damaged. This was done for each index **i** between 0 and 19. It is clear that this individual relies heavily on the values stored in certain memory positions. The same experiment was performed with the damaged memory index returning 0 independent of the true memory value. This caused greater reduction in the fitness of the individual. A possible explanation for why the constant misinformation was more deleterious to the performance than was random misinformation is that random numbers returned from memory may cause more random actions than the return of a constant. And in general, periods of random actions may help fitness, while periods of constant action usually lead to spinning in place or futile pushing against a box or the wall. When all of the memory indices were simultaneously subjected to the random or constant brain damage, the individual's fitness dropped to around 10% its normal fitness.

In the second graph we see the results of the complement to the brain damage study. This study subjects the same individual to sensory deprivation. Blinding each sensor input individually, 100 fitness tests were done and the histogram shows the resulting fitness averages. Each blinded sensor returned a random sensor value in the set {0,1,2}. The same tests were rerun and each blind sensor returned 0 regardless of the correct input. Here, as with the brain damage study, the sensor constant misinformation caused a greater loss of fitness than did the sensor random misinformation. A plausible explanation for the constant misinformation's greater effect on fitness may be that reacting to boxes is more important than reacting to clear spaces. When the "blind" sensor returns 0 all the time it is right an average of 63% of the time. When the "blind" sensor returns a random input it is right an average of 33% of the time. But the constant misinformation is never right in indicating there is a box (since 0 means Clear) and the random misinformation has a 33% chance of correctly indicating the appearance of a box at that sensor. When all the sensors were blinded simultaneously the individual's fitness dropped to 20% and 5% of the normal fitness (for the random and constant blinding respectively).

In this particular individual there are certain memory indices and certain sensors that are not used or, when damaged, do not adversely affect the individual's performance. For example, the particular individual shown above does not use any of its diagonal sensors {UpperRight, UpperLeft, LowerLeft, LowerRight}. This does not mean that these sensors are unnecessary. The UpperMiddle, MiddleLeft, and MiddleRight Sensors were the most popular among successful individuals from generation 80, but all of the other sensors were represented. Similarly, among successful generation 80 individuals positions 0,1,2,3, and 4 were the most heavily used memory positions, but all of the other memory indices were well represented.

9.6 Discussion of Indexed Memory

One of the advantages of the indexed memory strategy is its large number of possible states. Consider that for fewer than two dozen bytes of storage the agent gets over 10^{26} possible states (M=20) with constant access to 20 cross sections of the state space. Suppose we had a problem where there were 5 features to keep track of and each could be one of 4 values. With the indexed memory we could use 5 elements and store the values 0 through 3 in each one. Now this is wasteful in the sense that there are more compact representations, but the waste is a few states and this makes the use of state easier and less global in the following sense. Imagine a different memory strategy that has exactly 1024 different states (for example a recurrent neural network with 10 hidden units). There will be cases where a feature of the world changes and the agent switches from state 511 (0011111111) to state 512 (0100000000). This will require switching 9 of the 10 bits even though only one feature has changed. There are techniques, like the Grey code, which try to minimize the hamming distance between neighboring states. But if the order of state changes is not known ahead of time, it is hard to find a representation that will not sometimes have to change many or all of its elements just to effect a single change. Another way to say this is that redundancy brings with it a certain flexibility. This flexibility is the increased number of possible representations of a set of states. Since in using memory, the evolutionary process is trying to find some workable representation of the useful environmental features, more possible representations is better. Since we pay almost nothing for all the extra state in indexed memory, we effectively get the flexibility for free.

The efficacy of indexed memory is affected by the choice of M=20. Several other choices for M were investigated. For values of M less than 8 or 9 the performance noticeably decreased. For values of M larger than 40 or 50 there was also a slight decrease in the rate of increase of average fitness over generations. For low M the performance loss is probably attributable to insufficient memory space. When M is 5 there are only 3,125 possible states (5^5) and using all of them introduces the tight representation problem described above. A solution would be to have only 5 memory elements but to allow each element to store a larger range of integers. This solution would at least increase the number of available states. However, this solution would complicate indexing (Read X) by making the effective index be (X modulo M). Also, having a few large valued slots might still lead to the tight fit representation problem mentioned above. For large M the opposite problem occurs: there are too many memory elements. As a result, in early generations there is a lower likelihood of reading from the same memory element that useful information was written to. After a few generations certain memory elements become popular in sections of the population and this problem largely disappears. Part of the explanation for why certain memory elements become popular is that there is a bias in these experiments towards the numbers 0,1, and 2. All the boolean functions return 0 or 1 and the input values are {0,1,2}. For example, the typical agent shown above makes heavy use of memory positions 0,1, and 2. But they are not the only heavily used memory indexes for this individual; indices 3,4,10, and 13 are also particularly important. Though M was chosen to be 20, indexed memory is a general scheme which will work for any value of M, whether the element values have the same range as the memory index or not.

Along with the issue of the particular choice for M, the initial state of the memory elements at the beginning of each fitness is of some importance. For these experiments the memory elements of an agent were all set to 0 at the start of each new fitness test. Trials were done with other homogeneous constants and with fixed distribution of elements in the range 0 to M-1 (e.g. Memory[0] = 5, Memory[1] = 19, ...). As long as the configuration of the agent's memory was always the same on time step zero, the particular set of memory values made little difference. The future works section will mention experiments that go beyond these constraints.

The argument for the usefulness of an overabundance of available state is not the only important feature of the indexed memory paradigm. Another advantage is that for a non-trivial problem it is difficult to know how many states will be necessary. Picking some state size as a guess at the approximate amount of state that will be needed not only burdens the evolutionary process with the tight-fitting representation problem mentioned above, but runs the risk of having too few states and making high success at the task impossible. There will always be features of an environment whose values in a mental model are dubious, but by using indexed memory, part of the memory space can be used as a scratch pad for calculations and for those features whose values are only loosely tied to success in the domain.

Indirection is another critical aspect of indexed memory. While no additional constructs are necessary to facilitate (Read (Read X)), this adds a powerful ability to the memory system. Without indirection, data structures might evolve in the agents memory systems, but it would in general take linear time in the size of the memory to extract information from an element of a dynamic data structure. Think of a dynamic memory data structure as a variable sized area of memory containing closely related information which is larger than one memory element. With memory indirection, mental models that contain linked lists, queues, stacks, and other data organizations become not only possible, but relatively simple.

Indexed memory will work for a wide variety of problems requiring state or mental models. For example, the Prisoner's Dilemma problem has already been mentioned as a problem for which memory is useful. Indexed memory could easily accommodate keeping the last k moves as long as k < M and could probably find encoding that would work for k>M. As an example the function could maintain a queue by moving Memory[2] into Memory[1], Memory[3] into Memory[2], ... until Memory[k] was moved. Then it could put its most recent input (the past move) into Memory[k]. Another strategy could be to use a circular queue and leave out Memory[0] and Memory[1] as pointers to the head and tail of the queue. These two examples show that indexed memory is a sufficiently powerful tool for the problem. Undoubtedly the evolutionary process would come up with an equally valid technique that would not map into any familiar algorithm.

Indexed memory is not only powerful enough to solve the Prisinors Dilemna, it can be shown that indexed memory can support **any** data structure and carry out **any** computation. This proof, that a genetically evolved function with indexed memory is Turing-complete, is too long to include here in full. The intuition behind the proof, however, can be given in a few lines.

The essence of the action of a Turing machine is the **delta** function which maps the machine's current state and current tape position symbol to a new state, a new tape head position, and possibly a new symbol at the old tape position. Imagine that M is allowed to be any integer and its memory is expanded dynamically when a memory position is read or written which has not before been accessed. Now set aside two memory elements (e.g. Memory[0] and Memory[1]) and define all other elements with a strict ordering. Let Memory[0] hold the "state" of the Turing machine and Memory[1] hold the position of the "tape head." Now it is clear that the rest of the indexed memory can be used as a tape and that the genetically evolved function is exactly that **delta** function. The function maps the state (Read 0) and the current tape position symbol (Read(Read 1)) to a new state (Write NewState# 0), possibly a new symbol at the old tape position (Write NewSymbol# (Read 1)), and a new tape position. To move the tape head right or left the **delta** function will need to skip from -1 to 2 and 2 to -1 respectively. Here is how the **delta** function might move the tape head left:

```
(IF (EQ (Read 1) 2) THEN (Write -1 1) ELSE (Write (Sub (Read 1) 1) 1))
```

So now it remains only to show that the **delta** function can always be evolved using GP with indexed memory. There are a finite number of states, and a finite number of elements in **sigma** (the tape alphabet). So the mapping from <Current state, Current tape symbol> to <New state, New tape symbol, New tape head position> can always be done with a finite number of IF-THEN-ELSE's. So as long as the number of Turing machine states is bounded in advance or there is no limit on the size of the evolved function, **delta** can always be produced. Just like **delta** from a Turing machine, the GP function will be repeatedly called until it returns a reserved integer T (e.g. 1045047) which signifies termination.

Since indexed memory is Turing-complete, genetically evolved functions that use indexed memory can simulate any machine and run any algorithm. Furthermore, because these programs are being evolved in the complete space of algorithms, this means that, in principle, we can evolve any program we need without ever expanding upon the simple set of terminals and non-terminals described in this chapter.

9.7 Discussion of Mental Models

In the experiments section an agent was presented that attributed signs of a mental model. This agent's performance dropped significantly when particular sensors failed to accurately report the correct feature value of the environment. The agent's performance also suffered when it could not reliably recover the same value it had stored in particular places in its memory. These conditions are neither necessary nor sufficient criteria for showing that the agent has an iconic or a sentential model of the world. There is little reason, however, to look for these human-centric mental attributes. No part of the fitness function rewards "thinking" that falls into line with how we speak about thinking. Before we ask about some highly evolved agent, "what is it doing?" we must consider what kind of answer we expect. It seems possible, given the results above, that the current language of mental activities does not allow us to speak fruitfully about what that individual is doing. This research is, therefore, about mental models and not about iconic or sentential models specifically. Given that these agents are not human and that the human notions of iconic and sentential models have no context-independent justifications, it may be unjustified to ascribe iconic or sentential models to these agents' mental behaviors.

Whenever we try to find meaning in the actions of agents other than ourselves, we have to doubt what we see. There is in fact no perfect way of determining whether an agent has or is using a mental model of the world. During the course of the research and quite by accident, the agent shown below was noticed, whose average fitness is about 5 points per fitness test. This agent evolved under slightly different conditions than those described in this chapter, but its existence is instructive. The environment this successful generation 80 individual evolved in differed from the environment that this chapter details in that agents always started in the upper left corner of the world facing east.

Automatically Defined Function

```
(NOT (READ (WRITE (WRITE Param1 (WRITE (WRITE (WRITE Param2 Param1)
(WRITE Param2 Param1)) (NOT (READ (WRITE (WRITE (WRITE Param2 Param1)
(WRITE 12 Param1)) 0))))) 0)))
```

Main Function

```
(READ (OR (READ (READ (READ 0))) (WRITE (*A-D-F* 10 (READ (READ (OR
(READ (READ (READ (READ (READ (READ (READ 0))))))) 0)))) (SUB 0
(READ (READ (ADD 0 (READ (SUB (READ 0) (READ 0))))))))))
```

Notice that this agent never examines **any** of its sensors. This agent **does** make use of state and the results of the brain damage study on this agent are similar to the results shown in the previous section. However, the sensory deprivation tests have no effect on this agent. It scores this high level of fitness by tracing out a complicated, variable path that it has evolved to use. Increasing the size of the world to 7x7 drops its fitness to an average of 3.1 points per fitness case. This shows a certain robustness in this variable pattern method. Starting this agent in a random clear location with a random heading in the 6x6 world dropped this agents fitness to an average of 1.6 per fitness case, which is lower, but still high considering its apparent brittleness. This agent, using state very effectively to pace out its pattern, has evolved a sort of "built-in" model of the world.

The existence of this agent does not mean that this research did not succeed in producing agents with a mental model of the world. But it does remind us that the problem of determining whether an agent is "thinking" is hard enough that behavioral evidence is not sufficient and that there may, in general, be no conclusive way of determining the level of attention an agent pays to the world in which it is acting.

From a certain point of view this sensorless agent has "learned" more than some of its fellow best-of-80th generation agents (though in a simpler domain). It has evolved a general strategy that works well no matter what configuration of boxes is presented. When the starting location and heading for the agent were randomized, no reoccurrence of this type of successful sensor-insensitive agent was discovered. But we can never rule out the possibility of the appearance of such an agent. This is not the kind of mental model that this research was trying to induce, but this agent serves as a reminder that despite the complexity of an environment, there are often simple solutions quite unlike any human approach to the problem. And if these simple solutions do exist, they will probably evolve.

9.8 Future Work

The results described in this chapter lead toward at least two different lines of the research. The first is the use of techniques, like the one presented here to create or evolve mental models and to design experiments to study these models. An interesting step would be to make some distinction between simple use of state and a mental model based on phenomena observed in the memory units. Another would be to eliminate the types of environmental features that allowed the agent with a "built-in" model to appear. As was pointed out, though, this type of agent may be impossible to completely prevent in general. The configuration of an agent's initial state of memory as it begins each new fitness test also plays a role in these questions. Work has already begun to study various kinds of learning using the persistence of memory across fitness tests as well as the Baldwin effect, Lamarkian evolution, and the evolution of the initial memory state itself.

The second line is to continue to improve upon indexed memory as a paradigm in genetic programming or to expand this work in some natural way into genetic algorithms. The introduction of natural uses of memory in these environments may eventually lead to a broader range of problems that can be undertaken using evolutionary strategies.

To substantiate the claim that indexed memory is truly a general, practical solution to the problem of incorporating memory into genetic programming, it must be tried on a variety of other memory-critical problems. Work has already been started using indexed memory and the same non-terminal set discussed here to find solutions to the problem of learning a maze, the Prisoner's Dilemma, and a modified version of Simon-Says. The inputs are different for each problem and the filters map the integers returned into one of the possible actions. It is important to note that while GP with indexed memory is Turing-complete, there is no guaranty that it is an efficient method for evolving Turing machines. The crucial future research question will be "Does the generality of indexed memory justify the low level at which it operates ?"

9.9   Conclusions

An important goal of this research effort was to evolve agents that use state effectively and build mental models of their environment. As was shown using the brain damage and sensor damage graphs, the loss of particular memory elements or sensor elements results in significantly lower fitness. This means that memory and sensory inputs play a pivotal role in that individual's success. It was argued that a fitness dependence on inputs that correspond to the real feature values from the environment and a fitness dependence on the consistent ability to retrieve from memory the same value that was placed there, implies some non-trivial mental model. The fact that this use of memory cannot be described in terms of the familiar forms of mental activity suggests a need not only for further investigation of the agents, but also for a new way to talk about mental activities themselves.

The abilities of another successful individual were then discussed. This one used state very effectively but completely ignored its sensor values. The suggestion is that while this agent does not invalidate the conclusions just summarized, it brings into question our ability to reliably tell when an agent is "making a mental model" and when it is using its state in ways that are successful but have little or no correlation to the current feature values of the world. It was also posited that this second individual had a different kind of "built-in" model of the world and that it had found a "general" solution to a simplier version of the same domain.

These results were produced using the new technique of **indexed memory**. Indexed memory succeeds in this domain and will succeed in others for three main reasons. First, the large amount of state made available to the evolving agents makes it unnecessary for evolution to find compact representations for stored information. Second, the ability to have arbitrarily deep levels of memory indirection (i.e. (Read (Read (....))) facilitates the efficient creation of pointers and more complex data structures. And, most importantly, indexed memory is Turing-complete; it is theoretically sufficient for the evolution of any algorithm.

Work in progress includes the use of indexed memory to solve several familiar memory-based problems and suggestions were given about how indexed memory might be used by evolving systems to solve problems in other domains. The intuitions were given to a proof that index memory is Turing-complete and ramifications of this proof were discussed.

It has been argued that this research saw the emergence of mental models in agents that evolved from functional agents with access to indexed memory. These behaviors could not have come about without some type of simple, malleable, state information. As a representative of this type of state information system and as an addition to the genetic programming paradigm, indexed memory stands in a position to advance work done on the genesis, types, and structures of mental models and to open up new types of problems to the field of genetic programming.

Acknowledgments

I benefited greatly from discussions I had on this subject with John Koza, Nils Nilsson, and many others. The original motivation for this research came from a problem John Koza and Nils Nilsson brought to my attention which they call "the 7-layer Cake" problem. It was in quest of progress on this domain, the evolution of intelligent use of state, that this chapter came about.

Bibliography

Jefferson, David, Collins, Robert,. "AntFarm: Toward Simulated Evolution." In Langton, Christopher, et al. (editors), *Artificial Life II*. Addison-Wesley. 1991.

Jefferson, David, Collins, Robert, Cooper, Claus, Dyer, Michael, Flowers, Margot, Korf, Richard, Taylor, Charles, and Wang, Alan. "Evolution as a theme in artificial life: The genesys/tracker system." In Langton, Christopher, et al. (editors), *Artificial Life II*. Addison-Wesley. 1991.

Johnson-Laird, P.N. *Mental Models: Towards a Cognitive Science of Language, Inference, and Consciousness*. Cambridge, Mass. Havard University Press. 1983.

Koza, John R. *Genetic Programming: On the Programming of computers by means of natural selection*. The MIT Press. 1992.

Lindren, Kristian. "Evolutionary Phenomena in Simple Dynamics." In Langton, Christopher, et al. (editors), *Artificial Life II*. Addison-Wesley. 1991.

10 Evolution of Obstacle Avoidance Behavior: Using Noise to Promote Robust Solutions

Craig W. Reynolds

This chapter reports on investigations into the evolution of reactive control programs for obstacle avoidance in a "robot like" simulated vehicle for a corridor-following task. The focus of these experiments is the development of robust, general purpose controllers; in contrast to the brittle, overly specific controllers evolved in previous work. In these experiments, the evolved control program is invoked once per simulation step. Using proximity sensors, the controller inspects its environment and steers the vehicle to avoid collision with obstacles. Forward motion is constant and automatic.

The use of noise appears to discourage brittle solutions. Because those opportunistic solutions are easier to evolve, discouraging them makes the problem harder. In this sense, adding noise has an effect similar to increasing the number of fitness trials in the deterministic (non-noisy) case. It is possible to over-fit a finite deterministic training set. Appropriate use of noise should discourage over-fitting.

The side-effect of adding noise to the fitness test is that it inevitably produces noise in the fitness value determined for an individual. Fitness testing becomes stochastic and repeated fitness tests of identical programs yield differing results. This "fitness noise" and the associated variance in fitness values serves to mask the "true fitness" of an individual. When the variance due to noise is comparable to the variance due to genotype, the progress of evolution is markedly slowed.

10.1 Introduction

Previous work [Reynolds 1993b] has shown that the Genetic Programming Paradigm [Koza 1992] can be used to automatically create programs which enable a simple moving 2d vehicle to avoid collisions with obstacles by mapping sensory input (range data) into motor output (steering action). However, those evolved control programs were found to be quite "brittle." They do not solve the general obstacle avoidance problem. They can only guide the vehicle through one specific obstacle course. They will not even work in the *same* obstacle course if any of the vehicle's initial conditions (position or orientation) is slightly perturbed. These brittle controllers are a bit like a "house of cards" which stands only as long as absolutely nothing changes.

The crux of the problem is that evolution will discover solutions which capitalize on the deterministic, precisely repeatable nature of the fitness tests. Evolved controllers will come to depend on utterly insignificant coincidental properties of the vehicle's sensors, its actuators, and their interaction with the obstacle course.

The current work concerns one approach to avoiding this brittle behavior and seeks to evolve robust, general purpose control programs for the obstacle avoidance problem. The basic idea is to remove determinism from fitness testing by injecting noise into the system. This noise will tend to "jiggle" coincidental relationships between elements of the system and so tend to discourage evolution from capitalizing on them. Now the "house of cards" must be built on a shaky table. Clearly this is a much more difficult task, but if it can be accomplished, the evolved controller should be a robust solution to the problem.

Noisy fitness tests present new problems. Multiple trials are required to separate noise from underlying "true fitness." Scores from multiple trials must be combined into an overall fitness value. These experiments have included: taking the *minimum* of the trials (to encourage robust noise tolerance), taking the *average* of the trials (to encourage incremental improvement), and some hybrids of these approaches.

The GP function set for these experiments includes basic arithmetic operations, a sensor using ray-tracing to find the range to the nearest obstacle in a specified direction, and an effector which "steers" by a given angle. The vehicle always moves forward at a constant rate, so steering is its only means of avoiding collision. Raw fitness is simply the number of steps the vehicle moves before colliding with an obstacle.

These experiments have not yet produced a robust controller. But a host of lessons have been learned, and one by one, impediments to a solution have been eliminated. These issues and solutions will be presented along with results related to the interaction of noisy fitness functions with number of trials, methods of combining scores from multiple trials, elitism, and population sizes for Steady State Genetic Programming.

10.2 Previous Work

An earlier series of experiments described in [Reynolds 1993b] used Genetic Programming to create reactive controllers for a similar obstacle avoidance task. A successful run is shown in Figure 10.1. The vehicle has completed its entire run of 300 steps, using its sensors to guide its steering so as to avoid collisions with all obstacles. This control program was discovered after about 22000 new programs had been created and fitness tested. A population of 1000 programs was used, so this represents 22 "generation equivalents."

The fitness test used to guide the evolution of this control program employed a single, precisely repeatable simulation-based fitness test. This allowed evolution to take the easy path to discover a program which only solved this one specific control task. There was no "incentive," no survival advantage, to find a controller that could generalize to other related tasks. As a result, the controller it found was "brittle" and could not solve similar but slightly different problems.

Closely related to the experiments reported here is the work of the Evolutionary Robotics Group at the University of Sussex. While using a different model of evolution

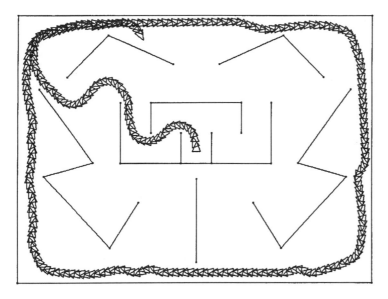

Figure 10.1:
A successful, but brittle, controller from earlier work.

(SAGA [Harvey 1992]) and a different model of controller architecture (dynamic, recurrent neural nets [Cliff 1993], [Harvey 1993]), they have investigated some closely related problems in evolution of robotic controllers. They were the first to investigate the role of noise in the evolution of robust controllers. The general approach used here, of evolving stimulus-response behavior based on simulated performance using simulated perception inside a closed simulated world, was originally inspired by [Cliff 1991a] and [Cliff 1991b]. Section 6.1 of [Harvey 1993] is an erudite rebuttal to the concept of evolving *programs* as reactive robotic controllers.

The experiments reported here have much in common with some of Koza's work, particularly the evolution of behaviors such as "wall following" and "box pushing" as reported in [Koza 1992]. See also Simon Handley's GP robotics work [Handley 1993].

The evolution of robust controllers is related to the larger problem of generalization in evolution. This issue of generalization is an active area of research in many branches of evolutionary computation. In GP, see for example [Tackett 1993] on the evolution of generality in a classification problem, and [Kinnear 1993a] on generalization in sorting.

Earlier work on obstacle avoidance behavior based on remote (distal) sensors (as opposed to touch sensors) for vehicles moving at "moderate speed" (see the section 10.4

Table 10.1
Tableau for the sensor-based corridor-following problem.

Objective:	Find a robust, noise tolerant, stimulus-response controller able to perform a corridor following task with a 2d vehicle.
Terminal set:	0, 0.01, 0.1, 0.5, 2
Function set:	`+, -, *, %, iflte, look-for-obstacle, turn`
Fitness cases:	For deterministic runs: 1, 2, or 4 simulation runs with unique combinations of initial orientation and obstacle course mirroring. For noisy runs: 16 or 32 simulation runs with randomized initial conditions.
Raw fitness:	For each trial: number of simulation steps (up to 50) taken before the first collision of vehicle and obstacle, divided by 50. Fitness is the average, minimum, or a hybrid combination (see text) of all trials.
Standardized fitness:	Not used.
Hits:	Number of collision-free runs.
Wrapper:	None.
Parameters:	M = 500, 1000, or 2000; G = 50
Success predicate:	Fitness = 1

for a definition) can be found in [Reynolds 1987], [Reynolds 1988], [Mataric 1993] and [Zapata 1993], among many others.

A classic reference on controllers and behaviors for this class of *vehicle* is [Braitenburg 1984] which is highly recommended.

10.3   Obstacle Avoidance as Genetic Programming

The Genetic Programming system used in these experiments is generally similar to the approach described in [Koza 1992]. Including, for example, the use of *roulette wheel* selection. Certain aspects differs slightly, for example, the fitness values used here are of the "bigger is better" polarity. The crossover operation used here produces a single offspring from two parents. In this system, constant expressions are replaced by a numerical constant. Perhaps most significantly, this system uses Steady-State Genetic Programming (SSGP) which, as described in [Reynolds 1993a] is derived from work reported in [Syswerda 1989], [Syswerda 1991] and [Koza 1992]. Table 10.1 describes the application of the Genetic Programming Paradigm to this problem

The functions `turn` and `look-for-obstacle` are specific to the obstacle avoidance problem. Each of them takes a single numeric argument which represents an angle relative to the current heading. Angles are specified in units of *revolutions*, a normalized angle measure: 1 revolution equals 360 degrees or 2π radians. These functions will be explained more fully below, but basically: `turn` steers the vehicle by altering its heading by the specified angle (which is returned as the function's value). `look-for-obstacle` "looks" in the given direction and returns a measure of obstacle proximity.

The Genetic Programming substrate used here, and the application-specific functions for the simulation, were originally developed on Symbolics Lisp Machines. For the current series of experiments, the software was ported (but not subsequently re-optimized) to Macintosh Common Lisp (version 2.0p2) and was run on Macintosh Quadra 950 workstations. As an example, to give a rough estimate of the rate of execution, for run S (see Results section) at 25400 individuals into the run, a typical fitness test took about 130 seconds to execute.

10.4   The Vehicle

The design of the simulated vehicle used in these experiments is kept intentionally vague and abstract. It could equally well represent an animal as a wheeled or legged robotic vehicle. The intent is to gloss over the low level details of locomotion and to concentrate instead on the more abstract issues of "steering" and "path determination." (This is not "path *planning*," since these reactive controllers neither plan nor learn.) In order to survive, the controllers need only steer along a clear pathway while avoiding contact with the walls surrounding it. The skill involved is similar to that required by a squirrel running along a tree branch, or by an automobile's driver, negotiating through a narrow alley.

The control programs being evolved by the Genetic Programming paradigm represent the vehicle's "thought process" for a single simulation step. During fitness testing, the evolved program is run at each time step of the simulation. Using its sensors, the program inspects the environment surrounding the vehicle from its own "point of view" (that is, relative to the vehicle's local coordinate space), performs some arithmetic and logical processing, and decides how to steer (adjust the heading of) the vehicle. The vehicle then automatically moves forward by a constant amount (half of its body length). The fitness test continues until the vehicle takes the required number of steps (50), or until it collides with one of the obstacles. The raw fitness score for each such fitness trial is the number of steps taken divided by the maximum number of steps, producing a normalized score between 0 and 1 (1 being best).

The goal of these experiments is a special case of *obstacle avoidance* behavior. Some might call this specific behavior *corridor following*. The term "obstacle avoidance" is often used to refer to moving around in wide open spaces while steering around the occasional tree or rock. That behavior is actually much less challenging than the corridor following task described here. The earliest experiments in this investigation, prior to the work reported in [Reynolds 1993b], involved a vehicle moving around a largely open environment which contained the occasional obstacle. Using the GP function set described here, evolution quickly plugged a forward-pointing proximity sensor into the steering primitive and produced a very simple, robust program that turned to the left by an amount proportional to the proximity of an obstacle in front of the vehicle. This trivial controller would easily avoid any isolated obstacle, as long as there was room on the left side to steer around it.

The "obstacle course" and restrictions placed on the vehicle in these experiments are designed to present a much more demanding task. The vehicle is forced to steer precisely through a crowded environment. It is surrounded by obstacles and must make its way down narrow corridors where simply steering away from the obstacles is not an option. It is forced to turn in both directions around "U turns" and right angles. The radius of curvature must be carefully selected to avoid collisions on both the inside and outside of the turns.

The vehicles in these experiments have a fixed minimum turning radius. This serves two purposes. As a practical consideration, it prevents a class of uninteresting and degenerate solutions to the fitness test described above. If the vehicles could turn very sharply, they could simply spin in place. If no obstacle is hit the first time around, they can keep spinning collision-free until the simulation is over. In the absence of a limitation on turning radius, evolution will discover this trivial solution very quickly and it will soon dominate the population. The turning rate limitation used here is implemented by simply clipping the absolute value of the change in heading accumulated each simulation step. The evolved control program makes arbitrary calls to the `turn` function, its argument is summed into a per-simulation-step "virtual steering angle" accumulator, which is clipped and then used to actually alter the vehicle's heading at the end of the simulation step. The maximum per-simulation-step turn allowed in these experiments is ±0.05 revolutions (18 degrees or 0.31 radians). This limitation implies a minimum turning circle which is somewhat larger than the width of the corridors of the obstacle course. As a result, the vehicle cannot spin in place, it cannot turn around in the corridor, and its only choice is to travel along the corridor.

The other purpose of a limitation on turning radius is that it produces a model of a vehicle moving at "moderate speed." This qualitative description is intended to capture a relationship between the vehicle's momentum and its available turning acceleration. At "low speed" a vehicle has relatively little momentum, it might be able to bring itself to a stop, or make an abrupt change of heading in a single time step. At "moderate speed" momentum begins to dominate acceleration and changes of heading require many time steps. The upshot is that in this speed regime, maneuvers are less abrupt and paths tend

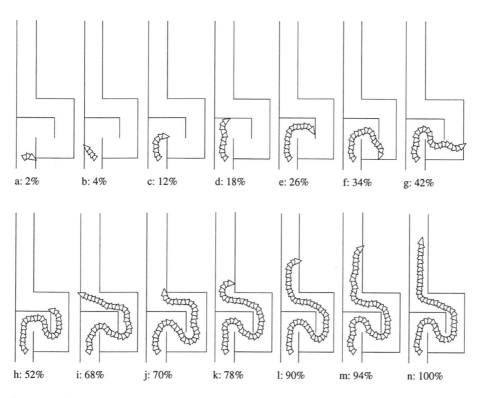

a: 2%   b: 4%   c: 12%   d: 18%   e: 26%   f: 34%   g: 42%

h: 52%   i: 68%   j: 70%   k: 78%   l: 90%   m: 94%   n: 100%

Figure 10.2
The obstacle course used in these experiments, and an assortment of typical trials. The course consists of two opposing U-turns followed by two opposing right angle turns and then a straight-away. These examples were collected from several different evolved control programs. Listed below each image is the raw score earned for that trial. Note that scoring is based on the number of steps taken, not the distance covered.

to curve more gently than at low speeds. Effectively we have a model more reminiscent of running than of crawling.

10.5   The Obstacle Course

The design of the obstacle course used in these experiments was intended to force the control programs to perform "corridor following obstacle avoidance" behavior and to quickly reject controllers which were completely unsuited to the task. This approach allows the majority of the computational effort to go into testing higher fitness individuals. A fitness function based on this kind of simulation has the desirable property that execution time is roughly proportional to the individual's fitness.

Figure 10.2 shows the obstacle course and an assortment of typical trials. The vehicle is initialized in the center of the corridor and pointing toward one wall or the other. Initial random headings range from 0.1 to 0.15 revolutions (36 to 54 degrees) off of the corridor's midline. A vehicle whose controller turns continuously (see Figure 10.2(a)) or does not steer at all (see Figure 10.2(b)), will run into a wall almost immediately. To survive more than a few steps the controller must develop the skill of turning away from a glancing collision and proceeding down the corridor.

As the vehicle travels a few steps further into the obstacle course, it finds a wall directly ahead, blocking its path. It must turn to avoid the wall ahead, or end up as in Figure 10.2(d). It can not start turning too soon or it will clip the inside of the U-turn, as in Figure 10.2(c). Immediately after there is another U-turn in the opposite direction. This ensures that the controller knows how to make both right and left hand turns. The vehicle follows a short straight-away, then two opposing right-angle turns and then runs out the remainder of its steps in a long straight-away.

Note that it is during this final leg that the controller first sees an opening in the obstacles ahead of it. Many controllers seem to be confused by this "light at the end of the tunnel" and will turn into the wall as they first approach it. This is may be due to the fact that until they reach relatively high fitness they have spent their entire evolutionary history in a realm where they are always surrounded by obstacles. They have probably developed an "expectation" of being surrounded and must now evolve an alternate strategy for dealing with a gap in the wall.

An obstacle course used in earlier experiments had a longer straight-away before and after the first turn. This arrangement seemed to promote premature convergence, particularly in small populations. Some individuals would discover how to make it to the first turn (and sometimes past it) but they would then so dominate the population that diversity would be lost and the population would never discover how to get around the second turn. In the subsequent experiments described here these problems were addressed in two ways. First, the long straight-aways were moved from the beginning of the course to the end. This forced the turning problem to be addressed sooner, and weeded out the non-turners sooner. Second, the entire obstacle course was periodically mirrored about the axis of the first corridor. In noisy runs the obstacle course was flip-flopped at random, in deterministic runs it was flopped every other fitness test. This procedure required that evolving controllers could turn equally well in both directions and so prevented convergence towards right-turners.

10.6 Sensors

The `look-for-obstacle` function serves as the vehicle's sensors and is the controller's only source of information about its world. All adaptive, time-varying behavior must be derived somehow from the variations of these sensor readings as the vehicle moves through its world.

The sensors measure proximity to a distant object. They are *distal* range sensors as opposed to touch sensors. They could represent sonar range finders, visual perception through fog, or photo sensing of reflected light from a vehicle-mounted light source.

When `look-for-obstacle` is called, a *ray-tracing* operation is performed. That is, a geometric ray ("half line") is constructed from the vehicle's position in the direction specified by the sum of the vehicle's heading and the argument to `look-for-obstacle`. The intersection (if any) of that ray with each obstacle is calculated. The obstacle whose ray intersection is closest to the vehicle's center is deemed the *visible obstacle*.

The value returned from `look-for-obstacle` is a number between zero and one. A value of one would indicate that the obstacle is coincident with the vehicle. (This does not occur in practice since this would mean a collision had already occurred.) As the distance between the vehicle and the obstacle increases, the proximity signal drops off quadratically in strength. At a certain threshold value (15 body lengths in these experiments) the signal will have reached zero. Hence a value of zero returned from `look-for-obstacle` indicates that the closest obstacle in the given direction (if any) is more than 15 units away.

In the original conception of this project, it was imagined that evolution would design a sort of *retina* consisting of a series to calls such as (`look-for-obstacle 0.1`) with various constant values used for the argument. Hence evolution would select the "field of view" and density distribution of the receptors in the retina. As always, evolution turned out to have its own ideas. There was nothing in the problem statement that said the argument to `look-for-obstacle` had to be a constant, and so the controllers soon evolved bizarre mechanisms based on calculating a dynamic value and using that to specify the direction in which to look. For example one species of obstacle avoider seemed to be based on the puzzling expression (`look-for-obstacle` (`look-for-obstacle 0.01`)). In hindsight it becomes clear that the correct model is not that of a retina but rather a simple form of *animate vision* [Cliff 1991b] where the controller aims its visual sensor at the area of interest.

10.7 Noise

The goal of these experiments was to investigate the ability of noise to disrupt the functioning of brittle control programs, and so to encourage the evolution of robust, noise-tolerant programs. For a control program to be useful in real-world applications it needs to be very robust, and in particular it needs to be able to deal with real-world noise. If we seek to evolve control programs for eventual application to real world tasks, we must take noise into account. Our fitness criteria should include noise to force evolved programs to implement noise-tolerant solutions to their task.

Inman Harvey argues in [Harvey 1993] for just such an approach regarding evolution of connectionist (dynamic, recurrent neural net) robotic controllers. He also suggests that this desire to incorporate noise implies that neural nets are a more appropriate media for evolved controllers than are programs. He says that it is "difficult to justify" the injection of noise into high level programming languages.

In contrast, this work assumes that injection of noise is a natural part of the model. The programs themselves are based on a traditional deterministic von Neumann model. The noise is presumed to come from the interaction with the real world. Hence it is the sensors and actuators used by the deterministic program which are modeled as being noisy. See [von Neumann 1987]. A simple example of the kind of system this is intended to represent might be a digital thermostat. The inherently noisy temperature sensor provides input to the deterministic controller, and its output is an inherently noisy heater. Because the thermostat is based on the robust principle of negative feedback, its operation is noise-tolerant and stable.

In these experiments the arithmetic and logical primitives of the Genetic Programming function set (+, -, *, %, and iflte) are deterministic and exact. The sensor and actuator primitives (look-for-obstacle and turn) are considered to be in contact with the real world and so are modeled as noisy. In order to simulate a noisy actuator in the steering system, the value passed to turn is added to a random number before being used to adjust the vehicle's headings. Similarly, random numbers are added to the parameter and returned value of the look-for-obstacle function. This simulates noise in both the aiming of the sensor, and the value it measures. In all cases, the random numbers used are ±2.5% of full range of the sensor or actuator (for example, angular noise is ±0.025 revolutions). This implies a signal-to-noise ratio around 20, about 13 dB.

10.8   Measuring Fitness in the Presence of Noise

While injecting noise into fitness testing has benefits, it also introduces complications. Noise in fitness tests creates variations in fitness values. There is no longer a way to measure the one "true fitness" value of an individual. Each single fitness evaluation will produce a different fitness value. Fitness must now be treated as a stochastic variable. In order to characterize this value we must perform a series of fitness tests.

The number of fitness trials used to analyze an individual is a trade-off between computational efficiency and accuracy. A smaller number of trials makes fitness faster to compute. Increasing the number of trials will generally give a more representative value and reduces the variance in fitness measurement. In these experiments either 16 or 32 trials have been used in noisy fitness tests. Even at these rates, significant variation is seen when a given individual is fitness tested multiple times (multiple sets of 32 trials). This variation is affected by method of combination, see below. While it may be desirable to reduce this variance by increasing the number of fitness trials, these simulation-based fitness trials are already *very* time consuming and so practicality argues against more trials.

Having run a series of fitness trials, the list of fitness values obtained must be combined somehow into a single representative value. Several statistics might be used. Certainly the *average* (*mean*) of the samples is an obvious candidate. The mean is a very intuitive statistic which captures the gist of the sampled values. However the goal here is not merely to summarize a list of fitness values, but to provide the best information to guide the evolutionary process.

This consideration has lead the Sussex group to use the *minimum* of the fitness trials as the overall fitness score [Cliff 1993], [Harvey 1993]. Since the goal is to evolve controllers which work well *despite* the effects of noise, it seems reasonable to ask which have the best "worst case" performance. Doing so seems to encourage noise tolerance. The fact that an individual might occasionally achieve a very high score is less significant than its ability to reliably achieve a certain level of performance.

In the current series of experiments both the mean and minimum methods have been tried, and both have demonstrated shortcomings. One undesirable aspect of using the average value of fitness trials is that it does not encourage consistency, particularly in the early part of a run. Consider two evolved control programs, each of which has an average fitness of 50%. Controller A scores 25% on half of its trials and 75% on the other half. Controller B consistently gets a score of 50% on all trials. While on the average these two controllers are equally good at their task, the reliability of B is more appealing. Note that A's fitness is sensitive to the exact mix of good and bad trials. Since these are random trials, the good-to-bad ratio will vary, and so the average fitness will potentially have a large variance. In order to make efficient progress, evolution needs to have a reliable measure of quality for each individual. If the fitness variance due to inconsistent performance is comparable to the fitness variance due to quality of solution, evolution will be unable to distinguish good solutions from bad ones, and it will be unable to progress. With too much variation in fitness, high fitness individuals are indistinguishable from those who happened to get a "lucky break." The symptom of this problem is that the fitness of the population will increase to a certain point and then never improve.

Using the minimum fitness combination method fixes this problem, but causes another. Because controllers are judged to be only as good as their worst case performance, controller A would get a fitness score of 25% and B would get 50%. This clearly captures the superiority of B with regard to consistency. It also seems to encourage solutions to reduce their fitness variance, since there is a selection pressure against even occasional low scores. However another kind of masking effect was observed while trying to use the minimum method in these experiments. Consider another controller C which gets 75% on "almost all" of its fitness trials, but which on occasion slips up and scores only 50%. Intuitively we would consider C to have better performance than B, yet typically they will both get a fitness of 50%, their minimum fitness value. In these experiments the symptom of this was typically that the population average never progressed past a value corresponding to a collision with the "sharp edge" on the inside of the first U turn (see Figure 10.2(c)). Most controllers would collide with this point at least occasionally, causing them all to have the same fitness. The controllers which occasionally got higher scores seemed to be equivalent, they just happened to get a lucky break. When these high scorers were re-tested by hand, they achieved scores down near the population average.

These two techniques for combining fitness trials (average or minimum) seem to have complementary strengths and weaknesses. Using the minimum fitness encourages noise-tolerant solutions, but masks the difference between an always-bad and a usually-good-but-occasionally-bad solution. Using the average fitness encourages incremental improvements, favoring a good-but-occasionally-great solution over a merely good solution, but masks the benefit of consistency. These tradeoffs suggest that an optimal strategy might fall somewhere between these two alternatives. Current experiments use a weighted average to combine individual fitness trials. It emphasizes the minimum score over the average score (4/5 minimum + 1/5 average). This combination method appears to work well, but in the absence of overall success it is hard to evaluate individual components.

Table 10.2
Summary of the 25 runs described in this chapter.
Notes:
1. variable number of trials, up to 32
2. evolved program *return* the steering angle, `turn` removed from function set
3. size limitation (no more than 50 atoms in each program)
4. initial random population given constant fitness value (in lieu of fitness testing)

run	population	noise?	trials	combine method	individuals processed	best fitness	average fitness	average size	notes
A	200	no	2	min	3686	42%	28%	303	
B	500	no	2	min	11678	42%	30%	1049	
C	2000	yes	32	min	26279	28%	19%	48	1
D	4000	yes	32	min	13334	28%	8%	20	1
E	2000	yes	32	min	108944	28%	12%	130	1
F	500	yes	16	min	38400	28%	10%	75	
G	500	yes	16	ave	28300	70%	47%	100	
H	200	yes	32	ave	10600	64%	47%	143	
I	200	no	4	ave	20200	76%	65%	263	
J	500	yes	32	ave	25000	58%	40%	304	
K	1000	yes	32	ave	21900	50%	30%	96	
L	500	no	1	...	6000	100%	36%	26	
M	500	no	2	ave	19870	100%	57%	34	
N	500	no	4	ave	15000	52%	39%	52	
O	1000	no	4	ave	61960	100%	73%	48	
P	2000	yes	32	ave	22384	56%	35%	42	
Q	2000	yes	16	ave	11900	48%	17%	30	
R	2000	yes	32	m+a	50430	30%	19%	101	
S	2000	yes	32	4m+a	57400	25%	14%	87	
T	2000	yes	16	4m+a	11900	21%	7%	24	
U	1000	no	4	ave	26779	86%	67%	86	2
V	2000	yes	32	4m+a	28200	20%	13%	49	2
W	2000	yes	32	4m+a	20700	20%	12%	30	2, 3
X	1000	no	4	ave	40900	95%	69%	36	2, 3
Y	10000	yes	32	4m+a	45200	23%	4%	7	2, 3, 4

10.9   Results

The most fundamental result to report based on these experiments is that so far, **no** robust, noise-tolerant controller has been evolved for this corridor-following problem. The results of all runs are summarized in Table 10.2. Of particular significance is that none of the runs with noise have attained 100% fitness. Two possible explanations suggest themselves. One possibility is that the evolution of such a controller is

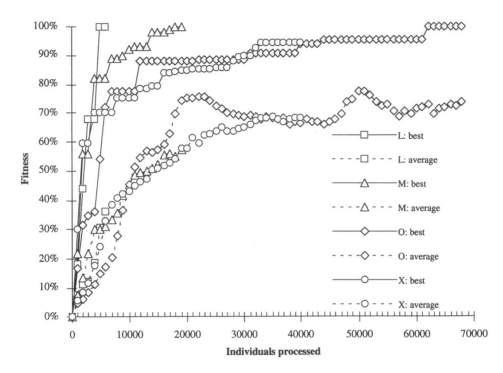

Figure 10.3
Results from four runs of the deterministic case. This chart shows fitness plotted against the number of new individuals processed (created through crossover and then fitness tested). Run L used a single deterministic fitness case. Run M used two cases. Runs O and X both used four deterministic fitness cases. Run X differs from the others in that it merely *computes* the steering angle.

impossible for fundamental reasons, given the experimental design used here. The other explanation, the one to which the author continues to cling, is that the evolution of this sort of controller is possible, but that it is a hard problem and it lies beyond the "evolution power" that has been brought to bear on it so far.

Some would argue (see section 6.1 in [Harvey 1993]) that the model used here, specifically the use of evolved programs instead of evolved dynamic recurrent neural nets, is inappropriate for this problem domain. Certainly nothing in these results can be used to refute that contention. However in each of the evolution runs in this series of experiments, an ability to partially solve the given control task was clearly seen. The symptom of failure has been that the spontaneous self-organized improvement due to evolution in the population of controllers eventually falls off. As a result the population fitness rises for a while and then flattens out asymptotically at a level below the global

```
(iflte 0.5
       (turn (look-for-obstacle 1.03999996))
       (turn (look-for-obstacle (* (turn 0.1) 2.0)))
       (+ (+ (* 0.05 (look-for-obstacle 0.04))
             (look-for-obstacle (turn 0.1))
             (turn (look-for-obstacle
                       (+ 2.00999999
                          (* 0.05
                             (look-for-obstacle 0.2))))))
          (look-for-obstacle 0.04)
          (look-for-obstacle (look-for-obstacle (turn 0.1)))))
```

Figure 10.4:
The most compact of several programs from run O which attained perfect scores. This run was based on a
fitness function which took the average of four deterministic trials. This program has not been simplified.

maximum. While this could be due to a fundamental problem, it seems much more
reminiscent of a more ordinary problem: the application of an underpowered GA to a
hard problem, or the use of a fitness function which is unable to properly guide the
evolutionary process.

One argument which suggests that the problem is merely a lack of "evolution power" is
the trend seen in the series of experiments with deterministic (non-noisy) runs.
Deterministic runs L, M and O each eventually attained 100% fitness. See Figure 10.3.
Run L used one fitness trial and required about 6000 individuals to succeed. Run M used
two fitness trials and required about 20000 individuals to succeed. Run O used four
fitness trials and required about 62000 individuals to succeed. While these numbers have
limited significance because they are based on a sample size of one for each situation,
there is a clear trend towards increased difficulty as the number of trials goes up.
Increasing the number of trials is analogous to requiring increasing generality.

There is a temptation, which is certainly not justified by this tiny collection of
datapoints, to see a tripling of effort required for each doubling of the number of
deterministic fitness test cases. There is little reason to believe this relation would hold
up as the number of fitness cases increased. Presumably the difficulty-versus-fitness-
cases curve flattens out at some point. It seems unlikely that a fitness test with 2,000,000
cases is really much harder than one with 1,000,000 cases. Further research is required to
characterize the relationship between the number of deterministic fitness cases and the
difficulty of evolving a solution.

While deterministic runs L, M and O each succeeded, deterministic runs A, B, I and N
did not. (Note that in these experiments, "did not succeed" implies that the author made a
subjective judgment that the run had "topped out" due to premature convergence or loss

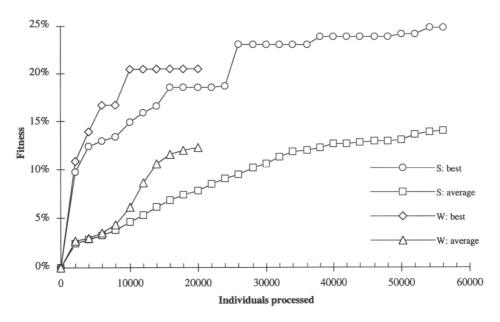

Figure 10.5
Two noisy runs, both with a population of 2000. Run S has `turn` in its function set and no restriction on program size. Run W does not use `turn` and limits its programs to size 50.

of information due to noise. Generally this was based on an observation that the population's average fitness had remained essentially unchanged for too many generations.) In runs A, I and N the failure appears to be related to the use of an undersized population. That is, in the two fitness trials case, A failed with a population of 200 but M succeeded with a population of 500. Similarly for four fitness trials, run I failed with 200, N failed with 500, and O succeeded with 1000. (Note that B failed, perhaps because of the unusual experimental multiple crossover it used which created very large programs.)

Figure 10.4 shows the most compact of the 9 perfect solutions found after 68000 individuals were tested in run O. This corresponds to 68 generations of population 1000. Note that it is a "perfect solution" only for the four cases of the fitness test used in run O. It is moderately brittle, with only a limited ability to solve the general noisy corridor-following problem. When this program was evaluated (for example) by the more demanding fitness test for run R, it got a score of only 21%. Of the 32 trials which made up the R fitness test, this program's highest trial score was 100%, but its lowest trial score was just 2%. The average of the 32 trials was about 40%.

```
(iflte (look-for-obstacle 0.1)
       (look-for-obstacle 0.0)
       (- (turn (+ 0.11
                   (turn 0.1)
                   (look-for-obstacle (+ 2.01999998
                                          (look-for-obstacle 0.0)))))
          (turn (look-for-obstacle -0.1))
          (turn 0.2))
       (turn (turn (look-for-obstacle 0.1)))))
```

Figure 10.6
Best-so-far individual from run S, its fitness is 25%. The evolved program has been hand-simplified from its
original size of 67 down to this size 25 version.

Of the 13 noisy runs, none ever reached 100% fitness. None of them even got very
close. Certainly none of them were capable of reliably making it most of the way down
the corridor without colliding with a wall. This lack of success was consistent across a
wide variety of parameters. Populations ranged from 500 to 4000. The number of
individuals processed ranged up to 109,000. The number of generation-equivalents
ranged up to 80. The number of trials per fitness test ranged up to 32. Five different
methods of combining the scores from multiple trials were used. While the fitness
measures were different in most cases (and so it is not necessarily meaningful to compare
fitness values between runs) it is worth noting that the highest fitness value attained by
any individual in the noisy case was in run G which used only 16 trials. This high-
scoring individual apparently just had 16 lucky breaks in a row, because retesting it
produced a fitness value down near the population average.

As of this writing the noisy runs shown in Figure 10.5 are still in progress. Both have a
population of 2000, 32 trials per fitness test, and a hybrid method for combining fitness
cases: a weighted combination that emphasizes the minimum by a factor of four:
$((4 \bullet m)+a)/5$. The best individual from run S is shown in Figure 10.6. Run W differs
from S in two respects. First, W uses a reduced function set, it does not use the turn
function. Instead the evolved program computes and returns the steering angle. The
fitness test then calls turn on this computed angle. This may simplify the problem
somewhat and so provide better results. Run W also differs in that it limits programs to
be of size 50 or less. This prevents "bloat" and may help focus the effect of crossover on
relevant portions of code.

One of the enjoyable aspects of this work is watching the parade of candidate
controllers which go by during fitness testing. For a given task, one can easily imagine
expert and inept behavior. But between those there is a surprisingly rich middle ground
of buggy behavior. Some controllers will make a perfect nimble run through the corridor

on one trial, and will inexplicably crash into a wall then on the next trial. Some controllers steer smoothly, some jitter back and forth nervously. Occasionally there are even programs that seem to exhibit anti-expert behavior such as "wall seeking."

10.10 Future Work

There are many directions in which this work may proceed. Certainly one goal is to continue current ongoing experiments to see where they will lead. Another tack will be examine the assumptions that have been held constant in the current series of experiments and investigate what happens when they are changed. For example all of the runs described here used "roulette wheel" selection and Steady-State Genetic Programming. It would be worthwhile to try tournament selection and "generational" GP. One suspicion about the current methodology is that because of its *greedy* nature, SSGP holds on to high-scoring individuals for too long. This could be especially troublesome in the presence of noisy fitness tests with wide variance. If an individual happens to get several lucky breaks it will get a high score. It may be detrimental to let the greedy SSGP mate that lucky individual over and over again allowing it to dominate the population based on an unrealistically high fitness value. Using a generational GP would not prevent high scores due to lucky breaks but it would limit the lifetime of any individual and so limit the amount of influence it could have on the population.

Given how often these runs exhibit symptoms of having merely "pooped out", an obvious direction to go would be toward more powerful GP techniques. Using a larger population is one approach, but it is fairly daunting given the already tediously long runtimes for these experiments. (A new run Y has been started with a population of 10000, this may terminate early due to memory limitations.) Another possible direction would be to use Automatic Function Definition which has been demonstrated [Koza 1993], [Kinnear 1993b] to significantly improve the power of GP for hard problems. The use of *demes* and *geographically-based mating* also appear to add significant power to Genetic Algorithms by preserving diversity through the use of geographically separated subpopulations [Collins 1992]. This technique has recently been applied to the evolution of behavioral controllers in [Ngo 1993] for the dynamically-balanced locomotion of 2d articulated figures.

Because of the sharp demarcation in solvability seen here between the deterministic runs and the noisy runs, it would be interesting to investigate whether this is a qualitative or quantitative effect. Is it the mere presence of noise that makes evolution harder, or does it have to do with the signal-to-noise ratio? Does turning down the "noise magnitude knob" make the problem easier and if so, does difficulty fall off smoothly with noise, or is there a sharp phase transition?

When the one-collision-and-you're-out model used in this work was first described to John Koza he dubbed it the "electrified fence" model. While this approach has the desirable property that bad programs are rejected early (hence saving on execution time) it has the potential drawback that it is harsh in its judgment of a controller which is almost perfect except for one little scrape early in the obstacle course. Instead we could allow each controller to run the entire number of steps, *constraining* it to stay inside the walls, and making the raw score be the number of steps *without* wall contact. This may offer a "smoother" fitness landscape that might be easier to ascend.

The limitation on sharp turns may be considered too *ad hoc* by some. It would be possible to use other techniques to discourage controllers from lapsing into the degenerate fitness sink of spinning endlessly in place. For example, introducing a *predator* as in [Reynolds 1993a] would make spinning in place much less attractive. A gazelle which spins in place has little chance of evading a hungry lion. Similarly a vehicle-eating predator could be introduced into the simulated world to motivate the vehicle to move in an efficient distance-covering manner. The predator might start a few steps behind the vehicle and then simply move toward it at the same speed. Any hesitation by the vehicle would put it in peril.

It would be interesting to investigate ways to fold other constraints into the fitness function, allowing control over additional qualities of the evolved solution. For example many controllers evolved here call the computationally expensive ray-tracing operation `look-for-obstacle` and then ignore the returned value. This is wasteful both at the conceptual level and at the practical GP-execution-time level. One way to address that would be to assign a cost to each member of the function set. (Representing, for example, real or simulated execution time.) A cost metric would be accumulated during the fitness test. When a preset cost threshold was exceeded, the fitness test would end, the individual would get only the "fitness points" it had collected up to that point. This would force the evolved programs to not just solve the problem, but to do so while using no more than a certain fixed amount of the costly resource. Similarly, the controllers could be allowed only a certain amount of turning angle to be used throughout their run, this could be used to force the vehicles to use a very smooth, low-energy path through the obstacle course.

As mentioned in [Reynolds 1993b] but never pursued, it is tempting to think about *coevolving* obstacle-avoiders and obstacle-courses. In the specific task examined here, this would mean the coevolution of corridor-followers and corridors. As in these experiments, the vehicles would be graded on their ability to avoid collisions with the walls. The coevolving corridors would be graded on their ability to "trick" the vehicles into having fatal collisions.

Acknowledgments

I wish to thank my employer Electronic Arts for providing the hardware and software facilities used for this work. I owe special thanks to my supervisor Steve Crane, and to his supervisor, Luc Barthelet, who have allowed me to pursue my research interests "in the background," despite the lack of any immediate applicability to the current needs of the company. I am deeply indebted to John Koza first for having invented the Genetic Programming Paradigm, and then for being so helpful and supportive to those who use it. Thanks to James Rice for his role in organizing the GP community and for being the un-moderator of our mailing list. Thanks to Kim Kinnear, editor of this volume, and organizer of the Workshop from which it came, for his many hours of service to the community. Thanks to Kim, James, Pete Angeline and Walter Alden Tackett for helpful discussions about this work. Thanks to the Evolutionary Robotics group at the University of Sussex for pursuing their thoughtful investigations which inspired aspects of this research. Thanks to my coworkers at Electronic Arts who have contributed their workstations on nights and weekends for the cause of Genetic Programming: Emmanuel Berriet, Steve DiPaola, Frank Giraffe, David Rees, and Randy Moss. Finally, heartfelt thanks to my wife, Lisa Berson Reynolds, who bravely encourages me to pursue my research interests--knowing that to do so means I will always be in a deadline panic and never have any free time.

Bibliography

Braitenburg, V. (1984) *Vehicles*, MIT press, Cambridge, Massachusetts.

Cliff, D. (1991a) Computational Neuroethology: A Provisional Manifesto, in *From Animals To Animats: Proceedings of the First International Conference on Simulation of Adaptive Behavior* (SAB90), Meyer and Wilson editors, MIT Press, Cambridge, Massachusetts.

Cliff, D. (1991b) The Computational Hoverfly; a Study in Computational Neuroethology, in *From Animals To Animats: Proceedings of the First International Conference on Simulation of Adaptive Behavior* (SAB90), Meyer and Wilson editors, MIT Press, Cambridge, Massachusetts.

Cliff, D. (1993) P. Husbands, and I. Harvey Evolving Visually Guided Robots, in *From Animals to Animats 2: Proceedings of the Second International Conference on Simulation of Adaptive Behavior* (SAB92), Meyer, Roitblat and Wilson editors, MIT Press, Cambridge, Massachusetts, pages 374-383.

Collins, R. J. (1992) *Studies in Artificial Evolution*, Ph.D. thesis, University of California at Los Angeles..

Handley, S. (1993) The Automatic Generation of Plans for a Mobile Robot via Genetic Programming with Automatically defined Functions, in *Advances in Genetic Programming*, K. E. Kinnear, Jr., Ed. Cambridge, MA: MIT Press.

Harvey, I. (1992) Species Adaptation Genetic Algorithms: The Basis for a Continuing SAGA, in *Toward a Practice of Autonomous Systems: Proceedings of the First European Conference on Artificial Life*, Varela and Bourgine editors, MIT Press/Bradford Books, pages 346-354.

Harvey, I. (1993) P. Husbands, and D. Cliff, Issues in Evolutionary Robotics, in *From Animals to Animats 2: Proceedings of the Second International Conference on Simulation of Adaptive Behavior* (SAB92), Meyer, Roitblat and Wilson editors, MIT Press, Cambridge, Massachusetts, pages 364-373.

Kinnear, K. E. Jr. (1993a) Generality and Difficulty in Genetic Programming: Evolving a Sort, in *Proceedings of the Fifth International Conference on Genetic Algorithms*, S. Forrest, Ed San Mateo, CA: Morgan Kaufman, pages 287-294.

Kinnear, K. E. Jr. (1993b) Alternatives in Automatic Function Definition: A Comparison of Performance, in *Advances in Genetic Programming*, K. E. Kinnear, Jr., Ed. Cambridge, MA: MIT Press.

Koza, J. R. (1992) *Genetic Programming: on the Programming of Computers by Means of Natural Selection*, ISBN 0-262-11170-5, MIT Press, Cambridge, Massachusetts.

Koza, J. R. (1993) Automatic Function Definition as a Means for Automatic Function Decomposition, in *Advances in Genetic Programming*, K. E. Kinnear, Jr., Ed. Cambridge, MA: MIT Press.

Mataric, M. J. (1993) Designing Emergent Behaviors: From Local Interactions to Collective Intelligence, in *From Animals to Animats 2: Proceedings of the Second International Conference on Simulation of Adaptive Behavior* (SAB92), Meyer, Roitblat and Wilson editors, MIT Press, Cambridge, Massachusetts, pages 432-441.

Ngo, J. T. (1993) and J. Marks, Spacetime Constraints Revisited, Proceedings of SIGGRAPH 93 (Anaheim, California, August 1-6, 1993), in *Computer Graphics Proceedings*, Annual Conference Series, 1993, ACM SIGGRAPH, New York, pages 343-350.

Reynolds, C. W. (1987) Flocks, Herds, and Schools: A Distributed Behavioral Model, in *Computer Graphics*, 21(4) (SIGGRAPH '87 Conference Proceedings) pages 25-34.

Reynolds, C. W. (1988) Not Bumping Into Things, in the notes for the SIGGRAPH '88 course *Developments in Physically-Based Modeling*, pages G1-G13, published by ACM-SIGGRAPH.

Reynolds, C. W. (1993) An Evolved, Vision-Based Behavioral Model of Coordinated Group Motion, in *From Animals to Animats 2: Proceedings of the Second International Conference on Simulation of Adaptive Behavior* (SAB92), Meyer, Roitblat and Wilson editors, MIT Press, Cambridge, Massachusetts, pages 384-392.

Reynolds, C. W. (1993) An Evolved, Vision-Based Model of Obstacle Avoidance Behavior, in *Artificial Life III*, Santa Fe Institute Studies in the Sciences of Complexity, Proceedings Volume XVI, C. Langton, Ed. Redwood City, CA: Addison-Wesley.

Syswerda, G. (1989) Uniform Crossover in Genetic Algorithms, in *Proceedings of the Third International Conference on Genetic Algorithms*, pages 2-9, Morgan Kaufmann Publishers.

Syswerda, G. (1991) A Study of Reproduction in Generational and Steady-State Genetic Algorithms, in *Foundations of Genetic Algorithms*, G. J. E. Rawlins, Ed. San Mateo, CA: Morgan Kaufmann, pages 94-101.

Tackett, W. A. (1993) and A. Carmi, The Donut Problem: Scalability, Generalization, and Breeding Policy in the Genetic Programming, in *Advances in Genetic Programming*, K. E. Kinnear, Jr., Ed. Cambridge, MA: MIT Press.

von Neumann, J. (1987) Probabilistic Logics and the Synthesis of Reliable Organisms from Unreliable Components, in *Papers of John von Neumann on Computing and Computer Theory*, W. Aspray and A. Burks Eds. Cambridge, MA: MIT Press.

Zapata, R. (1993) P. Lépinay, C. Novales, and P. Deplanques, Reactive Behaviors of Fast Mobile Robots in Unstructured Environments: Sensor-based Control and Neural Networks, in *From Animals to Animats 2: Proceedings of the Second International Conference on Simulation of Adaptive Behavior* (SAB92), Meyer, Roitblat and Wilson editors, MIT Press, Cambridge, Massachusetts, pages 108-115.

11 Pygmies and Civil Servants

Conor Ryan

A new selection scheme, the Pygmy Algorithm, which uses disassortative mating, taken directly from population genetics, to reduce premature convergence in Genetic Algorithms and Genetic Programming is introduced. This reduced premature convergence is shown to allow stronger selection pressure and the use of small population at very little extra computational cost and with little risk of losing genetic material.

11.1 Introduction

In applying Genetic Programming, and Genetic Algorithms in general, to problems with variable length solutions, it is often desirable to evolve a solution which uses as few instructions as possible. It is also desirable to preserve a population which, although evolving in a directed manner to some solution, maintains a degree of diversity to prevent premature convergence on a non-optimal solution. The degree to which a population is directed tends to reduce the diversity in it, and for this reason elitism, the use of only the top percentage of the population for breeding, tends to be overlooked when searching for a method of avoiding premature convergence.

Unfortunately, most previously used methods for avoiding premature convergence, such as Crowding [DeJong 1975], Sharing [Goldberg 1987], restricted mating, spatial mating etc. all involve extra computation, while methods which reward parsimony [Koza 1992] run the risk of losing genetic material.

It is the aim of this chapter to present a new method for avoiding premature convergence with as little cost as possible. A secondary aim is to defend the use of elitism in even relatively small populations to enable them to enjoy the benefits brought about by elitism.

11.1.1 The Problem Space

To demonstrate some of the more interesting methods mentioned above, the work described here evolves a minimal sorting network [Knuth 1973]. Sorting networks present a twofold and possibly threefold problem; a K-sorting network must correctly sort any combination of K numbers, but it must also do so in as few exchanges as possible. One can also consider the parallelism of a network, as this can have a profound effect on the speed of a network if implemented in circuitry. However, this work will not consider parallelism.

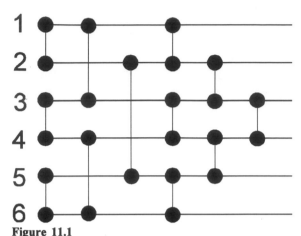

Figure 11.1
Network sorter for 6 numbers.

Sorting networks, not to be confused with sorting algorithms, are a list of fixed instructions which sort fixed length lists of numbers. Sorting networks have a convenient diagrammatic representation as in Figure 1; each input line represents one number and each connection, or "comparator module", compares the numbers at those positions and swaps them if necessary. For instance, the top left hand connection would be coded as

(if (> K[1] K[2]) (swap K[1] K[2]))

11.1.2 Genetic Programming Or String GA?

The simulations presented in this chapter were originally implemented using a variable length String GA rather than the Genetic Programming paradigm as an aid for comparison to other methodologies. Fortunately, variable length string GA is quite similar to GP due to its evolving genome which allows of possible solutions to grow and contract. Because of this, the results yielded by the GA experiments are very applicable to GP. For this reason the Pygmy algorithm described below is also applied to GP.

11.1.3 Implementation Notes

Sorting networks can be generated for any amount of numbers. In fact, this is not the first time artificial evolution has been applied to their generation. The most notable effort was that of Hillis [Hillis 1989] in which he evolved sorting networks for 16 inputs.

As the amount of inputs to a sorter increases, the amount of compares required increases exponentially. This results in an exponential increase in the time to evolve a sorter due to increased testing time. It was considered more important to be able to examine a large quantity of results for relatively easier problems than a small quantity of results for extremely difficult ones, e.g. a run to evolve a 9-number sorting network requires a much larger population than a say, 6-number sorting network. One experiment to evolve a 9-number sorting network with 5000 individuals running for up to 500 generations took up to 48 hours of CPU time on Decstation 5000/125.

To avoid having to wait inordinate amounts of time before any useful data was obtained, the experiments in this chapter were scaled down to concentrate on generating 6-number sorting networks, again over 500 generations, but with populations <=750. On the same machine, depending on population size, several hundred runs of these experiments could be carried out in a day, facilitating greatly the collection of empirical data.

As with most GP and GA experiments, the smaller the population, the more difficult it is to produce a satisfactory solution. Because of this, the population size in the experiments below is varied to vary the difficulty. The relationship between population size and difficulty was not linear, but in general, the smaller the population, the more difficult the problem.

It was also found that none of the experiments derived any great benefit from the amount of generations that they were run for. One of stopping conditions was if there was no change in the breeding pool for 20 generations. It was found that in the runs that failed, this invariably happened before 500 generations. It was also found that if a particular run was going to produce a perfect individual it would appear relatively early on, usually in the first 75-100 generations and always before the 300th, echoing results obtained by [Collins 1991] .

11.1.4 The Benefits of Elitism

An elitist breeding strategy differs from traditional, roulette-wheel strategies in that only the top 10-20% performing individuals are permitted to mate. A potential problem with this, especially with small populations, is that a group selected in this way may not be representative of the population at large. Genes which are common in the population may not appear at all in the elite group, and one cannot discount the possibility that those genes may be useful at some future time.

The obvious advantage of this is that poorly performing individuals do not hold back the rest of the population; elitism tends to be used when the implementor is more concerned with quick convergence [Eshelman 1991] on a solution than maintaining variety in a population. A method which permits elitism while still maintaining diversity could well vindicate elitism.

11.2 Traditional Methods

Any suggested improvements to genetic algorithms can only be appreciated when viewed beside existing methods. To aid this understanding, a simple implementation of String GA to the problem of evolving minimal sorting networks is presented using the following strategy. Individuals were of variable length, up to a maximum of 50 instructions, all of which were of the form **(Compare X Y)** which compared the numbers at positions X and Y and exchanged them if necessary. All the other experiments in this chapter use instructions such as these.

An elitist strategy was used with a breeding pool of 20% of the total population maintained. In the event of a tie in the fitness score, the individual with the fewest instructions was rated higher; this ensured that as the population evolved, the genes of shorter individuals propagated through the population.

As stated in the abstract, it is the intention to produce an algorithm that can preserve diversity under very strong selection pressure. To exaggerate the selection pressure associated with elitism further, and thus increase the difficulty in maintaining diversity, a bias in favour of the higher performing individuals was included.

11.2.1 The Fitness Function - Punish or Reward?

It has been shown [Knuth 1973] that a K-number sorting network that will correctly sort every K combination of 0s and 1s will sort every K combination of numbers, for 6-number sorters this involved 2^6 (64) fitness cases. To measure how well a set of numbers is sorted, the means usually employed is the counting of what Knuth called "inversions", which are considered to be numbers out of order. In a list sorted in ascending order an inversion is considered to be each number with a number greater than it on its left. Thus the sequence 1 2 3 4 5 6 has no inversions while the sequence 1 2 3 4 6 5 has one inversion.

A problem with this method is that it is more suited to traditional methods of sorting which move through the list, so a number can be counted as being out of order several times, the result of which the sequence 6 2 3 4 5 1 has 9 inversions with only two numbers needing reordering, while the sequence 1 4 2 5 3 6 has three inversions yet has *four* numbers in the incorrect positions.

This is not very suitable for sorting networks which do not move through the list but which arbitrarily select two numbers to put in the correct order. Measuring inversions seems to take the approach of punishing an imperfectly sorted list rather than rewarding a partially sorted list. To ensure the latter occurred the following scoring method was used :

```
score=0;
for (i=1;i<=5;i++)
if (num[i]<=num[i+1]) score++;
```

This means that it is possible for an individual to score five points for each correctly sorted test case so that an individual which correctly sorts all 64 cases will receive a score of 320. This *partial fitness* was extremely useful early on in each experiment as it was not uncommon for every individual in the initial population to fail to sort every test case. The score derived in this way is referred to as an

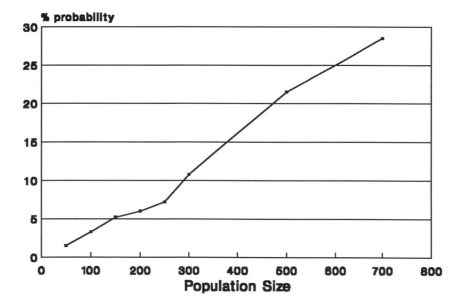

Figure 11.2
Probability of a perfect individual appearing

individual's **reward**; for the initial experiments an individual's fitness is equal to it's reward.

11.2.2 Early Results

The above experiment was applied to a number of different population sizes, varying from 50 to 750 individuals, the results of which can be seen in Figure 11.2. Each experiment was repeated on 1,000 different populations; and, to aid comparison, each population size started with identical seeds when generating the initial population in each case. A perfect individual is a sorting network which will correctly sort any combination of 6 numbers in 12 exchanges. Twelve exchanges has been shown [Knuth 1973] to be the minimum needed for such a sorting network.

High performing individuals were kept in the breeding pool until dislodged by a better individual; because a constant training set was used these individuals did not need to be tested again, so the entire population could be replaced each generation. No mutation was permitted.

It was found that every experiment produced a 100% correct individual, but not all experiments produced a perfect individual, i.e. 100% correct using only 12 exchanges.

As can be seen, the adjusting of population size has a significant effect on the probability of a perfect individual appearing. This is because of the effect a population's size has on the variety of individuals contained within it. A population which does not produce a perfect individual is said to have prematurely converged; such a population does not contain sufficient variety of potential parents to evolve further.

The remainder of this chapter is concerned with the avoidance of premature convergence.

11.3 Maintaining Diversity In Artificial Evolution

Due to their very nature, GA and GP actively encourage a population to converge on the same fitness level; however, premature convergence does not seem to be a problem for many of the species in nature which have survived to this day, a system which apparently works on the same principles. There are countless life forms on the planet which have evolved from humble beginnings and show no signs of being trapped on some local minima. If they have become stationary in evolutionary terms it is either because it suits the species at this current time or it forces the species to become extinct. Obviously, artificial evolution would prefer the former.

In natural evolution it is rare, if ever, that a species develops needing to perform only one task; even in the relatively primitive world of plants, a species requires several abilities. It must extract nutrients from soil, its leaves must be efficient at photosynthesis, its flowers must be attractive enough for insects to visit and collect pollen. Few species exist in very varied environments however, and the requirement of several abilities to survive enables plants to fill different environmental niches. Depending on the local environment, plants with extremely productive leaves may flourish, or a plant may need to excel at nutrient extraction in order to survive in a different area.

This leads to adaptation as different ecotypes of the same plant or species exploit different niches; they are different in their phenotypic expression, and are similar enough at genotypic level to mate. A result of different ecotypes is the maintenance of diversity in the breeding pool, a prerequisite for avoiding premature convergence.

11.3.1 Sharing And Crowding

A simple method which encourages niche formation, introduced by De Jong [DeJong 1975], is the *crowding* scheme. This scheme used overlapping generations with new individuals replacing not the individual of the lowest fitness as one might expect, but individuals that are genotypically similar to themselves.

However, to prevent an impractical number of comparisons when an individual is created, newly-born individuals are not compared to every other individual. Instead, a *Crowding Factor*(CF) is decided upon; that is, the number of individuals which a new individual will be compared to. This has been used with some success by De Jong with a crowding factor of 2 and 3. Crowding was originally designed with multimodal functions in mind, i.e. functions with several peaks in the fitness landscape. It has been found [Goldberg 1987] that crowding does not prevent the population from ending up on one or two peaks within the fitness landscape, despite which it is not really suitable for putting pressure on unimodal functions such as the problem we are currently concentrating on.

Another niche method, that of *sharing*, was introduced by Goldberg and Richardson [Goldberg 1987]. Sharing is based on the maxim that an environment contains limited resources and that phenotypically similar individuals must share these resources, an idea that seems very reasonable from a biological point of view. Similar individuals suffer penalties to their fitness depending on their similarity to others within the population. Again, sharing was designed with multimodal functions very much in mind, and, like crowding above, would not put pressure on a single solution.

11.3.2 Isolation by Distance

The mating schemes looked at thus far are *Panmictic* schemes, i.e. any individual may mate with any other in the breeding pool. For some populations, particularly large populations, it is unreasonable to assume that any individual may mate with the individual of its choice; even the concept of elitism becomes difficult to rationalise, both from a practical point of view and from biological analogy. In the relatively small populations used here, i.e. <=750, it is reasonable to assume the global knowledge necessary for the use of the elitism strategy, but, as populations become larger, >5000, the effort required to maintain this centralised control becomes unwieldy. In addition, large populations require prohibitively long runs, even for a few generations, so are commonly implemented on parallel machines, where centralised control is avoided as much as possible.

To avoid this excessive centralisation, the notion of *isolation by distance* is used, a notion widely held in natural evolution. There are two widely used techniques for the implementation of Isolation by Distance; Spatial Mating [Hillis 1989] [Collins 1991] and the *Stepping Stone* or *Islands* model. The Islands model divides the population into several *demes*, each of which evolves at its own rate and in its own direction with some amount of emigration between islands, while the Spatial Mating model has individuals placed in a toriodal grid and allows mating only between neighbours. This method implicitly creates dynamically sized demes which grow and contract as their fitness varies relative to neighbouring demes.

While both methods, in particular Spatial Mating, have been shown to maintain diversity, they are not very comparable to other methods examined in this chapter, which permit the use of elitism and high selection pressure. For this reason they will not be considered further.

11.3.3 Steady State Genetic Algorithms

SSGA, as discussed by Syswerda [Syswerda 1989], has shown itself to be a good method for the maintenance of diversity. SSGA differs from traditional GA/GP in that in each generation, very few, as few as one, individuals are created, using the current population as potential parents. A new individual created in this way is permitted to remain in the population only if it is fitter than another individuals chosen to be replaced and is unique with respect to all other individuals. The uniqueness requirement involves comparing the individual to all others in the population rather than the 20% required by elitism. 20% is required because clones only become a problem when they are potential parents.

SSGA can be looked upon as a special case of elitism [Baeck 1991] with the only difference being the elite group, which numbered 20% of the population in the above experiments, is a much larger proportion of the total population. However, the more *extinctive* a population is, i.e. the fewer individuals kept from generation to generation, the more directed the population will be.

As this chapter is concerned to a large extent with highly directed populations, SSGA will not be implemented. However, because of its close resemblance to elitism, any strategies which use elitism can also be implemented in SSGA, simply by increasing the number of individuals preserved from generation to generation.

11.3.4 Restricted Mating

An investigation of the gene pool in the runs that failed to produce a perfect individual above would reveal it to be choked with clones - identical individuals. It has been suggested [Eshelman 1991], and is used as a matter of course by many, that the prevention of clones helps avoid premature convergence in Artificial Evolution. In fact, the first extension to the experiment described above was the disabling of

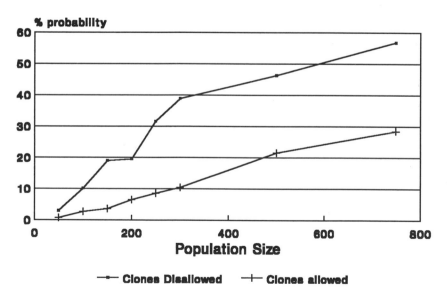

Figure 11.3
The effect on performance of disabling clones

clones, an act which normally requires the comparison of each individual to every other individual in the population as in SSGA.

One of the advantages of elitism over traditional roulette-wheel breeding strategies, like that used in GP, is that it is necessary to ensure only that the breeding pool contains no clones. To ensure this in the current simulation one must compare an individual to 20% of the total population in the worst case, reducing the required computation by a factor of 5. This reduction in computation was a deciding factor in the choice of strategy for selecting a breeding pool. The considerable benefit derived from the disabling of clones can be seen in Figure 11.3.

Eshelman went further than this and suggested using *incest prevention*, where only genetically different parents are allowed to mate, to prevent a population from becoming fixated. However, this strategy does have some problems. How different should parents be? How does one decide what is different? Eshelman's strategy is to calculate the Hamming distance between two individuals chosen to become parents, then, if the difference is above a threshold the individuals are allowed mate. However, as a population converges, parents become more and more alike so it becomes increasingly likely that potential parents will be rejected. If a situation

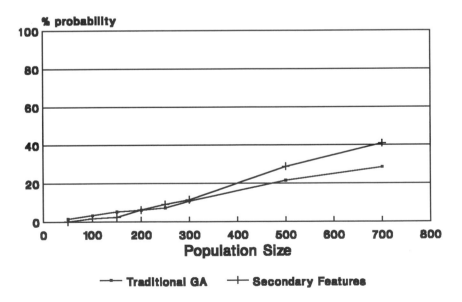

Figure 11.4
Comparison of traditional GA against breeding for secondary features.

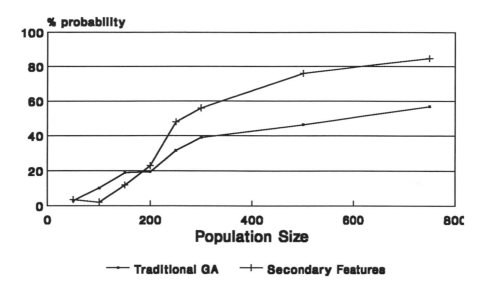

Figure 11.5
The effect on the relative performances with clones disabled.

where no parents are allowed to mate arises the threshold mentioned above is relaxed slightly.

In GP, and the high level string GA in this chapter, calculation of Hamming distances is not possible. Any calculation of differences between high level structures would mean choosing in advance the difference between every operator. Clearly, this would require much prior knowledge of each operator and is to be avoided. It is also quite likely that parents will be of different lengths, leading to some uncertainty as to the classification of difference. One faces a dilemma when two individuals contain identical instructions except for some extra instructions in one. Clearly, they are different, but how different?

Eshelman's method for incest prevention, although shown to drastically reduce clones at a lower cost than the method for clone prevention described above, aims to prohibit inbreeding to as great an extent as possible. It is a commonly held belief of those in the world of Artificial Life that inbreeding of any description is a recipe for disaster; [Brindle 1981] this is simply not the case. It is true that, in general, inbreeding, or sib-mating, can be detrimental; that said, however, inbreeding can have the effect of exaggerating certain traits of the parents, not necessarily a bad thing when individuals are being evolved with one aim in mind. This exaggeration could well be considered an advantage in the case where a high selection pressure is being used.

Clearly then, what is needed is a method which does not attempt to explicitly measure genetic differences, for this leads to much difficulty when defining exactly what constitutes difference. Also useful is a method which prevents inbreeding to *some extent* to keep some degree of diversity in the population; however, one must not lose sight of the fact that the primary purpose of GA and GP is to converge on a solution, *not* solely to maintain diversity.

11.3.5 Breeding For Secondary Features

A method which can be implemented cleanly and easily is that suggested by Koza [Koza 1992] which involves the inclusion in the fitness function of a weight for some secondary feature; in this case the feature would be length. The inclusion of this secondary feature permits the survival not only of efficient(fit) individuals but also that of short individuals.

The raw fitness is now modified by the adding of a size factor(sf) similar to that used by Kinnear [Kinnear 1993] to the fitness function which now becomes

reward + ((maxsize-size)*sf).

The results of this are now compared to the original scheme, *without disabling clones* in Figure 11.4.

As can be seen, at higher population levels, this outperforms the traditional method, and this outperformance is even more impressive when clones are disabled, see Figure 11.5. The cost of disabling clones is identical to the cost in the previous simulation.

This method was not intended to maintain diversity, but to aid the evolution of functions with more than one requirement, and therefore it is not unreasonable to expect that individuals tend to trade off one part of the fitness measurement against the other. For example, the above fitness function gives an efficiency weighting of 90% and weighting of 10% to shortness(parsimony), but in the case where the *best-so-far* individual has a fitness function with a score of 85 in efficiency and 8 in length, which is easier to evolve, an individual which is more efficient or an individual which is shorter? Of course, one cannot know.

Neither can one know whether a population would find it easier to lose an instruction that contributes relatively little to the fitness and thus gain a score in the parsimony measure than to utilise an extra instruction which might add to the fitness at a cost to parsimony.

The possibility of "malicious" loss of genetic material gives rise to concern. In certain cases, i.e [Kinnear 1993], where the size factor is used more as a nudge in a particular direction than a significant portion of the fitness function, the problem does not arise, but where the size, or indeed any secondary feature, is critical to the performance of any solution, difficulties may arise. It was found that at lower populations, where traditional GA outperformed this method, that genetic material was more prone to loss.

11.3.6 Pygmies And Civil Servants

The methods for avoiding premature convergence examined so far fall into two categories. The first being niching, where similar individuals tend to congregate together, and the second is the breeding for secondary features described above, where individuals try to maintain a balance between two different goals. The problems with these are the extra computation needed, difficulty in defining

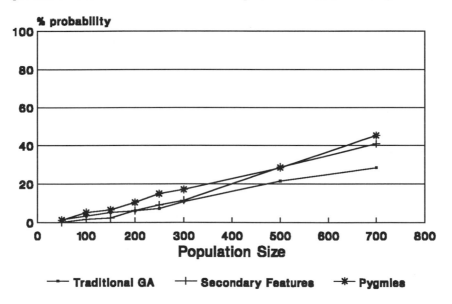

Figure 11.6
A comparison of the three methods with clones allowed.

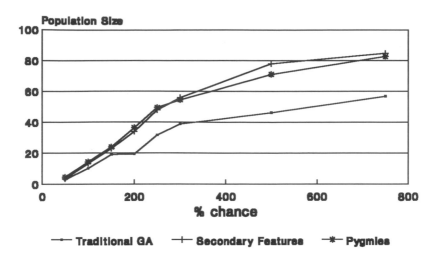

Figure 11.7
A comparison of the three methods with clones disabled.

difference/similarity, the trading off of fitness for length etc. To overcome these problems, a new method, which uses disassortative mating, is now suggested. Disassortative mating, the breeding of phenotypically different parents, is known to occur in some natural populations, and is sometimes used by plant breeders to maintain diversity.

Plant breeders do not have nearly as much information about the genetic make up of their populations as genetic programmers, and it is not uncommon to make decisions based on the appearance of parents rather than to over-analyse their genetic structure. In the world of Artificial Life, many have fallen prey to the problem of trying to over control their populations only to find that the resulting pay off was not worth the effort [Goldberg 1989].

The implementation of disassortative mating introduced below requires very little computational effort yet still provides a diverse array of parents. Rather than explicitly select two very different parents, the method presented here merely *suggests* that the parents it selects are different. No attempt is made to measure how different.

This implementation of disassortative mating, the *Pygmy Algorithm*, uses a strategy rather like the elitist method, in that a sorted list of the top performing programs is maintained. The Pygmy Algorithm, however, requires the maintenance of two lists of individuals, each with its own fitness function. For the purpose of this chapter, individuals in the first list are referred to as Civil Servants, the fitness function of which is simply their performance(efficiency at sorting); ties are resolved by rating the shorter individual higher as before. Individuals that do not qualify for the "Civil Service" are given a second chance by a slight modification of their fitness function to include a weighting for length, i.e. the shorter the better. Such an individual will then attempt to join the second list, members of which will be referred to as Pygmies.

When selecting parents for breeding, the Pygmies and Civil Servants are analagous to differing genders. One parent is drawn from the Pygmy list and one from the Civil Servant list, with the intention of a child receiving the good attributes of each and resulting in a short, efficient program. The presence of the Civil Servants ensure that no useful genetic material will be lost, and the presence of the Pygmies increase the pressure for the shortening of programs. The difference between genders in the Pygmy Algorithm and those in nature is that the gender of an individual is not adopted until after the fitness function(s) are evaluated; whereas in nature, individuals are born with a particular gender.

By analysing the data during runs it was found that each group influenced the new members of the other. Pygmies ensured that new Civil Servants became progressively shorter while the Civil Servants maintained a relatively high efficiency among the Pygmies. While the length of the Civil Servants never became shorter than the optimum length, the length of the Pygmies was frequently shorter as they tried to trade off efficiency for length.

The ancestors of this method are plain to see, for its roots are firmly entrenched in niche and speciation; elitism with its powerful selection pressure is also involved as is the strategy of breeding for secondary features.

It must be stressed that no effort is made to calculate how different Civil Servants are from their short cousins, the fitness function automatically decides which type an individual is. When selecting for breeding it is possible that two close relatives, even siblings, may be chosen; but this will only ever happen when they have complementary features, i.e. efficiency and shortness.

11.3.6.1 Implementation

To give a fair comparison, the Pygmies and Civil Servants were split evenly among 20% of the population. In the previous experiments an elite group of 20% was maintained. The fitness of the Civil Servants is simply their efficiency, while the fitness of the Pygmies is much like that above reward + ((maxsize-size)*sf).

but with a higher size factor, in this case 20% of the total fitness.

When running the Pygmy implementation it was found that the Pygmies tended to be smaller by several instructions and about 90-95% efficient. Soon into each run, the Civil Servants were all 100% efficient, but somewhat taller than the Pygmies. As each population progressed, the size of the Civil Servants followed that of the Pygmies, the size of which, in turn, decreased as that of the Civil Servants approached.

As can be seen from the graph of performance in Figure 11.6, the Pygmy method consistently outperforms each of the other two methods. It was expected that under easier conditions, i.e. higher populations, that the breeding for secondary features strategy would approach the Pygmy method, but it never outperformed it. It was mentioned previously that the secondary features strategy makes the possibility of losing genetic material more likely, and it was found that the more difficult the problem was, the higher the probability of this happening was and the worse the relative performance of this strategy became.

As described in Section 11.1.3, the problem was scaled down to obtain a large

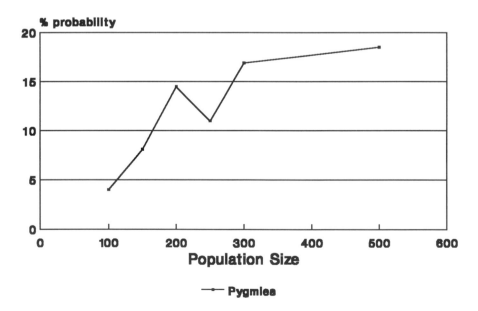

Figure 11.8
The Pygmy Algorithm applied to GP, clones enabled.

quantity of results, and populations were similarly scaled down to increase the difficulty of a problem.

The size of a population is directly related to the diversity of possible solutions within it and it is significant that the relative performance of the Pygmy Algorithm is best under the more difficult conditions.

11.3.6.2 Extending the model

It was pointed out earlier, in section 11.3.4, and was a focus of much of Eshelman's work, that while the disabling of clones maintains diversity to a great degree, there is much effort involved, an effort greatly reduced by the use of elitism. The use of the Pygmy model effectively *halves* the effort required by elitism : no Pygmy will ever be a clone of a Civil Servant, and vice versa.

When clones are disallowed, the results are as figure 11.7. From these results it appears that the Pygmy algorithm and the strategy of breeding for secondary features have an almost identical performance, except that at higher populations the Pygmy algorithm fares slightly worse. However, as mentioned above, the disallowing of clones in the Pygmy algorithm is done at half the cost.

11.3.6.3 Generalising the model

The Civil Servants tend to be 100% efficient at the primary feature and their fitness follows that of the Pygmies in regard to the secondary feature. There is no reason why the features need be efficiency and length; length could be substituted for a number of features, variety of instructions, parallelism, etc.

11.4.0 Pygmies and Genetic Programming

The elitist strategy is rarely implemented in GP, the closest thing being the Greedy Overselection model which is recommended only for populations >1000. The reason for this is that cited earlier: too much selection pressure leads to premature convergence. However, use of the Pygmy method allows elitism yet still prevents premature convergence. Because copies of the best performing individuals are guaranteed to be kept, the reproduction operator need not be used and crossover can be performed with a probability of 1.

The disabling of clones is not used in GP. The reproduction operator is relied upon to maintain a balanced population, at the risk of permitting clones. If, however, clones were to be disabled, the overhead would be enormous, every individual would have to be compared to every other individual; ten times the effort required by the Pygmy method. It is the intention of this chapter to compare empirical results with existing methods, but there is no empirical evidence to suggest that clones should be disabled when using GP. To present such evidence, or indeed, to prove that none exists, is beyond the scope of this chapter and the disabling of clones with regard to GP will not be considered further. Suffice to say that any such disabling can be done at a much lower cost using the Pygmy method than any other selection strategy.

The Pygmy method was compared to an implementation of traditional GP (.90 crossover, .10 reproduction etc.) on populations varying from 100-500 with the results as below. It must be stressed that the problem GP faced was slightly more difficult as much effort was put into the shuffling of PROGNX commands also. The PROGNX commands were *not* counted as instructions when calculating the parsimony measure, only swap commands. The swap commands were indivisible commands of the form **(SwapAB)** which would compare the numbers in positions A and B, swapping them if necessary. The swap commands were then linked together by the **PROGN2** and **PROGN3** connectives.

When running the simulations under the GP implementation, there were several surprise results. Although it was expected that GP would find some difficulty maintaining a balance between length and efficiency, it failed to find any "perfect" individual under any population size. The best individual it found was a perfectly efficient sorter, but required 13 swaps. To make the jump from 13 to 12 swaps would obviously require a larger population.

The Pygmy Algorithm fared much better, finding 13-instruction individuals as a matter of course, and every set of experiments produced a perfect sorter.

As with all the experiments in this chapter, as the population increased, the probability of a perfect sorter appearing tended to increase also. One unexpected result was that with a population of 250, there was a drop in performance. No reason for this occurence could be determined and the other results were as expected.

Although the results for GP seem slightly inferior to those in the string GA presented in Figure 11.6, GP faced the additional problem of rearranging the **PROGNX** commands to suit the **Swap** commands. Such performance shows howsuitable GP can be even to problems requiring fitness functions that are difficult to define and those that need a relatively large amount of direction.

Table 11.1
Tableau for the evolution of network sorters

Objective:	Evolve a computer program to emulate a minimal 6 input sorting network.
Terminal Set:	(SwapAB) where 1<=A<B<=6. Swap numbers at positions A and B if necessary.
Function Set:	PROGN2, PROGN3.
Fitness Cases:	Every combination of 0 and 1 six digits long.
Raw fitness:	Amount of correctly sorted numbers plus weighting for length.
Standardised fitness:	Total score plus maximum length minue raw fitness.
Hits:	Same as raw fitness.
Wrapper:	None.
Parameters;	M=100-500. G=100.
Success Predicate:	An S-expressions scores 320 hits using 12 swaps.

11.5.0 Conclusion

The Pygmy algorithm was born out of a reaction to the overcontrol of genetic algorithms. It can be implemented easily and can be adapted to virtually any problem involving the evolution of individuals needing more than one feature. Use of the Pygmy algorithm brings several benefits.

Elitism and high selection pressure can be used without causing premature convergence. It can maintain diversity much more efficiently than traditional methods under difficult conditions, in particular, when using small populations.

There is no uncertainty over the contribution of each feature to the fitness function, i.e. the question "Which is better, a short, reasonably correct individual or a long, perfectly correct individual?" never arises. It is not possible for the Pygmy method to lose genetic material by the trading off of efficiency for size, the best individual in any run will *always* be proficient in that part of the test case with the greatest reward.

Another useful contribution is that which arises when disabling clones, the Pygmy method can do so at half the cost of traditional methods and at just 10% of the cost in GP.

Finally, the Pygmy Algorithm achieves what was set out to do : in the face of difficult circumstances it can maintain a diverse breeding pool, using elitism and small populations, to prevent premature convergence to a greater degree than comparable methods through the use of multimodal fitness.

11.6 Future Work

The Pygmy approach works on the basis that the mating of two relatively different parents enough times will eventually yield an offspring with the good attributes of both. In a problem that requires three features to be evolved there is no reason why there could not be three parents. Theoretically, there could be as many parents as there are instructions. Obviously there would be some law of diminishing returns applicable to this, the more parents involved the less of an influence each will have on the child produced. Where this law of diminishing returns begins to take effect has yet to be determined.

While this chapter considers the Pygmies and Civil Servants to be distinct genders, work is currently under way which looks upon each group as being a *race*. Individuals in each race can be given a *Racial Preference Factor*(RPF), which is a measure of its inclination to breed with an individual from the other race. Use of the RPF can permit inbreeding within each group in a manner similar to that of demes and thus allow exploration of many more solutions.

The relationship between GP and clones has yet to be established, the effect of disabling clones and its implications for the reproduction operator is worthy of further investigation. All the experiments with string GA suggested that disabling clones is the route to higher performance and it has been shown that the Pygmy method can easily be modified to prevent clones at a low cost.

11.7 Appendix: The Pygmy Algorithm

The Pygmy Algorithm can be implemented with a minimum of effort on virtually any GA/GP system. The algorithm used to obtain the empirical results in this chapter is as follows.

• Initialise population as normal, without testing for uniqueness.

• Repeat for each individual in current population :

>   - Test individual against main fitness function.

>   - If fitter than least fit Civil Servant insert into Civil Servant list.

>   - Else calculate secondary fitness function and if fitter than least fit Pygmy insert into Pygmy list.

- Repeat until new population created :
    - Randomly select one individual from Civil Servant list.
    - Randomly select one individual from Pygmy list.
    - Create new offspring using crossover only.
- Repeat above two steps until terminating criterion is fulfilled.

Acknowlegements

Thanks to Gordon Oulsnum, Peter Jones and Dermot Ryan for their comments and suggestions. Thanks also to Kim Kinnear for encouraging me to submit this paper in the first place.

Bibliography

Brindle, A. (1981) *Genetic Algorithms for function optimisation.* Doctoral Dissertation, Computer Science Department, University of Alberta.

Collins, R. (1992) *Studies in Artificial Life.* Doctoral Dissertation, Department of Computer Science, University of California, Los Angeles, CA.

De Jong, K. (1975) *An analysis of the behaviour of a class of genetic adaptive systems.* Doctoral Dissertation, Department of COmputer and Communication Services, University of Michigan, Ann Arbor, MI.

Eshelman (1991) Preventing premature convergence in Genetic Algorithms by preventing incest, in *Proceedings of the 4th International Conference on Genetic Algortihms,* R. K Belew, L.B. Booker, Eds. San Mateo, CA : Morgan Kaufmann.

Goldberg, D and Richardson, J (1987) Genetic algorithms with sharing for multimodal function optimisation. *Proceedings of the 2nd International Conference on Genetic Algortihms.,* J. J. Grefenstette, Ed. San Mateo, CA : Morgan Kauffmann.

Goldberg, D (1989) Zen and the art of Genetic Algorithms. *Proceedings of the 3rd International Conference on Genetic Algortihms.,* J.D. Schaffer, Ed. San Mateo, CA : Morgan Kauffmann.

Hillis, D (1989) Coevolving parasites improve simulated evolution as an optimisation procedure, in *Artificial Life II,* Santa Fe Institute Studies in the Sciences of Complexity, C. G. Langton, Ed. Addison-Wesley.

Kinnear, K (1993) Generality and difficulty in GP : Evolving a sort, in *Proceedings of the 5th International Conference on Genetic Algortihms,* S. Forrest, Ed. San Mateo, CA : Morgan Kauffmann.

Knuth, D (1973) *The art of computer programming.* Reading, MA : Addison-Wesley, .

Koza, J (1992) *Genetic Programming.* Cambridge, MA : MIT Press.

Syswerda (1989) Uniform Crossover in Genetic Algorithms. in *Proceedings of the 3rd International Conference on Genetic Algorithms,* Editor?, Ed. San Mateo, CA : Morgan Kauffmann.

12 Genetic Programming Using a Minimum Description Length Principle

Hitoshi IBA, Hugo de GARIS, and Taisuke SATO

This paper introduces a Minimum Description Length (MDL) principle to define fitness functions in Genetic Programming (GP). In traditional (Koza-style) GP, the size of trees was usually controlled by user-defined parameters, such as the maximum number of nodes and maximum tree depth. Large tree sizes meant that the time necessary to measure their fitnesses often dominated total processing time. To overcome this difficulty, we introduce a method for controlling tree growth, which uses an MDL principle. Initially we choose a "decision tree" representation for the GP chromosomes, and then show how an MDL principle can be used to define GP fitness functions. Thereafter we apply the MDL-based fitness functions to some practical problems. Using our implemented system "STROGANOFF", we show how MDL-based fitness functions can be applied successfully to problems of pattern recognitions. The results demonstrate that our approach is superior to usual neural networks in terms of generalization of learning.

12.1 Introduction

Most of Genetic Programming (GP) computation time is taken up in measuring the fitness of GP trees. Naturally, the amount of time depends directly upon the size of the trees [Koza 1992]. The size of GP trees is usually controlled by user-defined parameters, such as the maximum number of nodes or maximum tree depth. We tried Koza-style GP to obtain a solution tree to the 6-multiplexer problem [Higuchi et. al. 1993], (see equation (12.1)), using the function set [AND, OR, IF, NOT], and using the address and data variables as terminal set. A typical solution tree is shown in Figure 12.1 However, due to the lack of a fitness function to control the explosive growth (i.e. depth) of trees, the tree growth got out of hand and the result was unsatisfactory in terms of computational time.

To overcome this difficulty and to get better results with much lesser population, we introduce a Minimum Description Length (MDL) principle [Rissanen 1978] to define fitness functions in GP, so as to control the growth of trees. We choose a "decision tree" representation for the chromosomes, and show how an MDL principle can be used to define the GP fitness functions. We then apply the MDL-based fitness functions to some practical problems. In earlier papers [Iba et. al. 1993a, 1993b], we presented a software system called "STROGANOFF" for system identification. This system integrates both analog and digital approaches, and its fitness evaluation was based upon an MDL principle. Using STROGANOFF, we show how our MDL-based fitness functions can also be applied successfully to problems of pattern recognitions.

12.2   GP using an MDL principle

In this section, we describe how decision trees can be evolved with GP to solve Boolean concept learning problems. Boolean concept learning has recently been developed in an attempt to formalize the field of machine learning [Anthony 1992]. We use Boolean concept learning as a means to treat the validity of MDL-based fitness functions. The reason why we chose a decision tree representation for the GP chromosomes is discussed later in Section 12.4.

12.2.1   Decision Trees and Genetic Programming

Decision trees were proposed by Quinlan for concept formation in machine learning [Quinlan 1983, 1986]. Generating efficient decision trees from preclassified (supervised) training examples, has generated a large literature. Decision trees can be used to represent Boolean concepts. Figure 12.2 shows a desirable decision tree which parsimoniously solves the 6-multiplexer problem. In the 6-multiplexer problem, a_0, a_1 are the multiplexer addresses and d_0, d_1, d_2, d_3 are the data. The target concept is:

$$output = \overline{a_0}\,\overline{a_1}d_0 + a_0\overline{a_1}d_1 + \overline{a_0}a_1d_2 + a_0a_1d_3 \qquad (12.1)$$

A decision tree is a representation of classification rules, e.g. the subtree at the left in Figure 12.2 shows that the output becomes false (i.e. 0) when the variable a_0 is 0, a_1 is 0, and d_0 is 0.

Koza discussed the evolution of decision trees within a GP framework and conducted a small experiment called a "Saturday morning problem" [Koza 1990]. However, as we shall see in later experiments (section 12.2.3), Koza-style simple GP fails to evolve effective decision trees because an ordinary fitness function fails to consider parsimony. To overcome this shortcoming, we introduce fitness functions based on an MDL principle.

12.2.2   MDL-based fitness functions

We introduce MDL-based fitness functions for evaluating the fitness of GP tree structures. This fitness definition involves a tradeoff between the details of the tree, and the errors. In general, the MDL fitness definition for a GP tree (whose numerical values is represented by "mdl") is defined as follows:

$$mdl = (Tree\_Coding\_Length) + (Exception\_Coding\_Length) \qquad (12.2)$$

The "mdl" of a decision tree is calculated using the following method [Quinlan 1989]. Consider the decision tree in Figure 12.3 for the 6-multiplexer problem (X, Y and Z

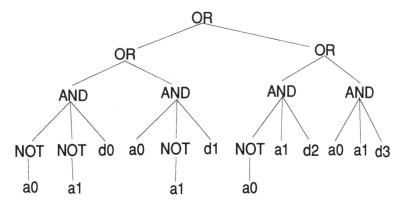

Figure 12.1
A Typical Tree Produced by non-Parsimonious GP

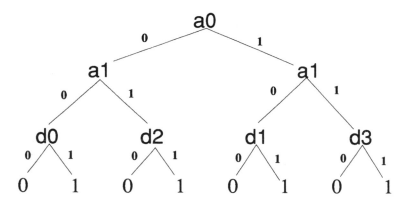

Figure 12.2
A More Desirable and Parsimonious Decision Tree

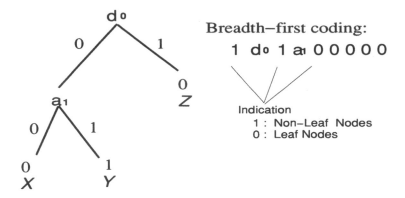

Figure 12.3
A Decision Tree for the 6-Multiplexer Problem

Table 12.1
Classified Subsets for Encoding Exceptions

Name	Attributes	# of elements	# of correct cl.	# of incorrect cl.
X	$d_0 = 0 \land a_1 = 0$	16	12	4
Y	$d_0 = 0 \land a_1 = 1$	16	8	8
Z	$d_0 = 1$	16	12	20

notations are explained later).
Using a breadth-first traversal, this tree is represented by the following string:

$$1\ d_0\ 1\ a_1\ 0\ 0\ 0\ 0\ 0\ 1 \tag{12.3}$$

Since in our decision trees, left (right) branches always represent 0 (1) values for attribute-based tests, we can omit those attribute values. To encode this string in binary format,

$$2 + 3 + 2 \times log_2 6 + 3 \times log_2 2 = 13.17\ bits \tag{12.4}$$

bits are required since the codes for each non-leaf (d_0, a_1) and for each leaf (0 or 1) require $log_2 6$ and $log_2 2$ bits respectively, and $2 + 3$ bits are used for their indications. In order to code exceptions (i.e. errors or incorrect classifications), their positions should be indicated. For this purpose, we divide the set of objects into classified subsets. For the tree of Figure 12.3, we have three subsets (which we call X, Y and Z from left to right) as shown in Table 12.1.

For instance, X is a subset whose members are classified into the leftmost leaf ($d_0 = 0 \wedge a_1 = 0$). The number of elements belonging to X is 16. 8 members of X are correctly classified (i.e. 0). Misclassified elements (i.e. 4 elements of X, 8 elements of Y, and 20 elements of Z) can be coded with the following cost:

$$L(16, 4, 16) + L(16, 8, 8) + L(32, 20, 32) = 65.45\ bits, \qquad (12.5)$$

where

$$L(n, k, b) = log_2(b + 1) + log_2\left(\binom{n}{k}\right). \qquad (12.6)$$

$L(n, k, b)$ is the total cost for transmitting a bit string of length n, in which k of the symbols are 1's and b is an upper bound on k. Thus the total cost for a decision tree in Figure 12.3 is 78.62 ($= 65.45 + 13.17$) bits.

In general, the coding length for a decision tree with n_f attribute nodes and n_t terminal nodes is given as follows:

$$Tree\_Coding\_Length = (n_f + n_t) + n_t log_2 Ts + n_f log_2 Fs \qquad (12.7)$$

$$Exception\_Coding\_Length = \sum_{x \in Terminals} L(n_x, w_x, n_x) \qquad (12.8)$$

Where Ts and Fs are the total numbers of terminals and functions respectively. In equation (12.8), summation is taken over all terminal nodes. n_x is the number of elements belonging to the subset represented by x. w_x is the number of misclassified elements of n_x members.

In order to use this MDL value as a fitness function, we need to consider the following points. First, the optimum MDL value is found by minimization, i.e. the smaller, the better. Second, although MDL values of decision trees are non-negative, the general definition of "mdl" (equation (12.2)) allows arbitrary values, i.e. range from $-\infty$ to ∞. Since the range is not known beforehand, a scaling technique must be used to translate the MDL value into a form suitable for a fitness function. For this purpose, we use a "scaling window" [Schraudolph and Grefenstette 1992]. Let $Pop(t)$ be a population of chromosomes at the t-th generation, and let N be its population size, i.e.

$$Pop(t) = \{g_t(1), g_t(2), \cdots g_t(N)\}. \qquad (12.9)$$

Each individual $g_t(i)$ has a data structure consisting of three components, i.e. its genotype, its MDL value, and its fitness.

$$g_t(i) : \qquad \textbf{Genotype} \qquad Gtype_t(i) \qquad (12.10)$$

Table 12.2
GP Algorithms with MDL-based fitnesses

Step 1 (Initialization) Let $t := 1$ (generation count).

Step 2 Call **GP** and generate $Pop(t)$.

Step 3 Calculate the MDL value $\{mdl_t(i)\}$ for each genotype $\{Genotype_t(i)\}$.

Step 4 $Worst(t) := \max_{i=1...N}\{mdl_t(i)\}$.

Step 5 $Wmdl := \max\{Worst(t), Worst(t-1), \cdots Worst(t-Wsize)\}$.

Step 6 For i:=1 to N do $fitness_t(i) := Wmdl - mdl_t(i)$.

Step 7 $t := t + 1$. Go to Step2.

Table 12.3
GP Parameters

Population size:	100
Probability of Graph crossover:	0.6
Probability of Graph mutation:	0.0333
Terminal set:	$\{0,\ 1\}$
Non-terminal set:	$\{a_0, a_1, d_0, d_1, d_2, d_3\}$

$$\text{mdl \quad value} \qquad mdl_t(i) \tag{12.11}$$

$$\text{fitness \quad value} \qquad fitness_i(t), \tag{12.12}$$

then our GP algorithm using MDL-based fitnesses works as shown in Table 12.2. $Wsize$ designates the size of scaling window. With this technique, the MDL value is translated into the windowed fitness value, so that the smaller an MDL value is, the larger its fitness is.

12.2.3 Evolving decision trees

In this section, we present results of the experiments to evolve decision trees for the 6-multiplexer problem using GP. Table 12.3 shows the parameters used. Where 1 (0) value in the terminal set represents a positive (negative) example, i.e. true (false) value. Symbols in the non-terminal set are attribute-based test functions. For the sake of explanation, we use S-expressions to represent decision trees from now on. The S-expression $(X\ Y\ Z)$ means that if X is 0 (false) then test the second argument Y and if X is 1 (true) then test the third argument Y. For instance, $(a_0\ (d_1\ (0)\ (1))\ (1))$ is a decision tree which expresses that if a_0 is 0 (false) then if d_1 is 0 (false) then 0 (false) else 1 (true), and that if a_0 is 1 (true) then 1 (true).

Figure 12.4 shows results of experiments in terms of correct classification rate versus generations, using a traditional (non-MDL) fitness function (a), and using an MDL-based fitness function (b), where the traditional (non-MDL) fitness is defined as the rate of correct classification. The desired decision tree was acquired at the 40th generation when using an MDL-based fitness function. However, the largest fitness value (i.e. the rate of correct classification) at the 40th generation when using a non-MDL fitness function was only 78.12%. Figure 12.5 shows the evolution of the "mdl" values in the same experiment as Figure 12.4(b). Figure 12.4(a) indicates clearly that the fitness test used in the non-MDL case is not appropriate for the problem, because there just is not a strong enough evolutionary pressure. This certainly explains the lack of success in the non-MDL example. The acquired structure at the 40th generation using an MDL-based fitness function was as follows:

```
(A0
  (A1
    (D0
      (A1 (D0 (0) (0)) (D2 (0) (A0 (1) (1))))
      (1))
    (D2 (0) (1)))
  (A1 (D1 (0) (1)) (D3 (0) (1))))
```

whereas the typical genotype at the same generation (40th) using a non-MDL fitness function was as follows:

```
(D1
  (D2
    (D3 (0)
      (A0 (0)
        (D3 (A0 (0) (0))
          (D0 (D0 (1) (D1 (A0 (0) (1)) (D0 (D0 (0) (0)) (A1 (1) (1)))))
            (D1 (0) (0))))))
    (A0
      (A0
        (D1 (1)
        (D1 (A0 (D1 (1) (1)) (A0 (D0 (D2 (1) (D1 (D1 (0) (0)) (1))) (0)) (1)))
          (0)))
        (A0
          (A0
            (D2
              (A1 (1)
                (D0
                  (D3 (0)
                    (A1 (D0 (A0 (1) (1)) (D3 (D0 (1) (0)) (A0 (0) (1))))
                      (A1
                        (D2
                          (D2 (D0 (D2 (A1 (D3 (0) (D3 (0) (0))) (A1 (0) (1)))
                            (D3 (D3 (0) (0)) (1)))
                            (D2 (1) (D0 (1) (0))))
```

```
                          (0))
                      (D0 (D3 (D2 (A0 (D3 (0) (0)) (1)) (1)) (1)) (1)))
                      (1))))
                  (D0 (D3 (A0 (1) (A0 (0) (A1 (0) (A0 (1) (1))))) (D3 (1) (0)))
                      (1))))
              (1))
          (A0 (A0 (0) (0)) (A1 (0) (D0 (1) (1)))))
      (A1
          (A0 (1)
          (D3 (D1 (D2 (D1 (0) (A0 (1) (D3 (D1 (0) (1)) (1)))) (0)) (1)) (1)))
          (0))))
      (0)))
  (D2 (A1 (A1 (1) (0)) (A1 (1) (0))) (1)))
```

As can be seen, the non-MDL fitness function did not control the growth of the decision trees, whereas using an MDL-based fitness function led to a successful learning of a satisfactory data structure. Thus we can conclude that an MDL-based fitness function works well for the 6-multiplexer problem.

12.3   Evolving trees with an MDL-based fitness function

This section describes the application of MDL-based fitness functions to more practical problems, using our implemented system called "STROGANOFF' (i.e. STructured Representation On Genetic Algorithms for NOn-linear Function Fitting). This system uses a hierarchical multiple regression analysis method (GMDH, Group Method of Data Handling [Ivakhnenko 1971]) and a structured GA-based search strategy [Iba et. al. 1993a]. The aim of our system is to establish an adaptive learning method for system identification problems. System identification techniques are used in many fields to predict the behaviors of unknown systems, given input-output data. This problem is defined formally in the following way. Assume that the single valued output y, of an unknown system, behaves as a function f of m input values, i.e.

$$y = f(x_1, x_2, \cdots, x_m) \qquad (12.13)$$

Given N observations of these input-output data pairs as below:

input				output
x_{11}	x_{12}	$\cdots$	x_{1m}	y_1
x_{21}	x_{22}	$\cdots$	x_{2m}	y_2
		$\cdots$		$\cdots$
x_{N1}	x_{N2}	$\cdots$	x_{Nm}	y_N

the system identification task is to approximate the true function f with $\overline{f}$. Once this

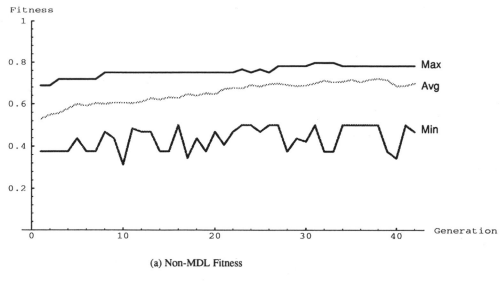

(a) Non-MDL Fitness

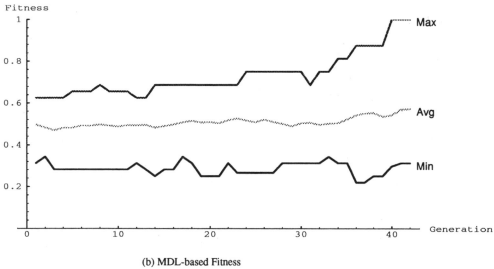

(b) MDL-based Fitness

Figure 12.4
Experimental Result (Fitness vs. Generations)

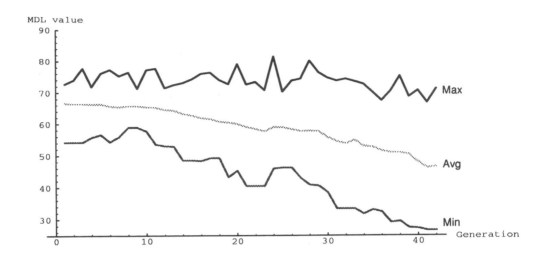

Figure 12.5
MDL vs. Generations

approximate function $\overline{f}$ has been estimated, a predicted output $\overline{y}$ can be found for any input vector $(x_1, x_2, \cdots, x_m)$, i.e.

$$\overline{y} = \overline{f}(x_1, x_2, \cdots, x_m), \tag{12.14}$$

This $\overline{f}$ is called the "complete form" of f.

GMDH is a multi-variable analysis method which is used to solve system identification problems [Ivakhnenko 1971]. This method constructs a feedforward network (as shown in Figure 12.6) as it tries to estimate the output function $\overline{y}$. The node transfer functions (i.e. the Gs in Figure 12.6) are quadratic polynomials of the two input variables (e.g. $G(z_1, z_2) = a_0 + a_1 z_1 + a_2 z_2 + a_3 z_1 z_2 + a_4 z_1^2 + a_5 z_2^2$) whose parameters a_i are obtained using regression techniques. GMDH uses the algorithm in Table 12.4 to derive the "complete form" $\overline{y}$.

The "complete form" (i.e. $\overline{y}$) given by the GMDH algorithm can be represented in the form of a tree. For example, consider the GMDH network of Figure 12.6, where the "complete form" $\overline{y}$ is given as follows:

$$z_1 = G_{x1,x2}(x_1, x_2) \tag{12.16}$$

$$z_2 = G_{x3,z1}(x_3, z_1) \tag{12.17}$$

and

$$\overline{y} = G_{x4,z2}(x_4, z_2) \tag{12.18}$$

Table 12.4
GMDH Algorithms

Step 1 Let the input variables be $x_1, x_2, \cdots, x_m$ and the output variable be y. Initialize a set labeled VAR using the input variables, i.e. $VAR := \{x_1, x_2, \cdots, x_m\}$.

Step 2 Select any two elements z_1 and z_2 from the set VAR. Form an expression G_{z_1, z_2} which approximates the output y (in terms of z_1 and z_2) with least error using multiple regression techniques. Regard this function as a new variable z,

$$z = G_{z_1, z_2}(z_1, z_2). \tag{12.15}$$

Step 3 If z approximates y better than some criterion, set the "complete form" (i.e. $\overline{y}$) as z and terminate.

Step 4 Else $VAR := VAR \cup \{z\}$. Go to Step2.

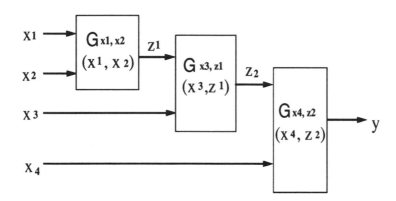

Figure 12.6
GMDH Network

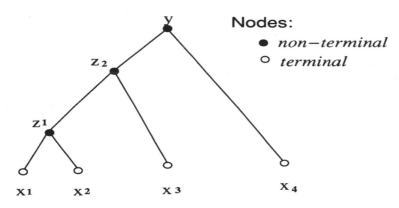

Figure 12.7
Equivalent Binary Tree

where equations (12.16) and (12.17) are subexpressions given in Step2 of the GMDH algorithm. By interpreting these subexpressions as non-terminals and the original input variables $\{x_1, x_2, x_3, x_4\}$ as terminals, we get the tree shown in Figure 12.7. This tree can be written as a (Lisp) S-expression,

```
(NODE1
  (NODE2
    (NODE3 (x₁) (x₂))
    (x₃)
  (x₄)))
```

where NODE1, NODE2 and NODE3 represent intermediate and output variables $\overline{y}, z_2, z_1$ respectively. Each internal node records the information corresponding to its G expression.

In this paper, each G expression takes the following quadratic form.

$$G_{z1,z2}(z_1, z_2) = a_0 + a_1 z_1 + a_2 z_2 + a_3 z_1 z_2 + a_4 z_1^2 + a_5 z_2^2 \qquad (12.19)$$

If z_1 and z_2 are equal (i.e. $z_1 = z_2 = z$), G is reduced to

$$G_z(z) = a'_0 + a'_1 z + a'_2 z^2 \qquad (12.20)$$

The coefficients a_i are calculated using a least mean square method. Details are described in [Iba et. al. 1993a].

We used an MDL-based fitness function for evaluating the tree structures. As mentioned in Section 12.2, this fitness definition involved a tradeoff between the details of the tree, and the errors (see equation (12.2)). The MDL fitness definition for a GMDH tree is defined as follows [Tenorio and Lee 1990]:

$$Tree\_Coding\_Length = 0.5 k log N \qquad (12.21)$$

$$Exception\_Coding\_Length = 0.5 N log S_N^2 \qquad (12.22)$$

where N is the number of data pairs, S_N^2 is the mean square error, i.e.

$$S_N^2 = \frac{1}{N} \sum_{i=1}^{N} |\overline{y_i} - y_i|^2 \qquad (12.23)$$

and k is the number of parameters of the tree, e.g. the k-value for the tree in Figure 12.7 is $6 + 6 + 6 = 18$.

STROGANOFF was applied to several applications such as pattern recognition and time series prediction [Iba et. al. 1993a]. We briefly present the results of an experiment called "SONAR" (i.e. classifying sonar echoes as coming from a submarine or not) [Gorman and

Table 12.5
GP Parameters (2)

Population size:	60
Probability of Graph crossover:	0.6
Probability of Graph mutation:	0.0333
Terminal set:	$\{(1),(2),(3),\cdots(60)\}$

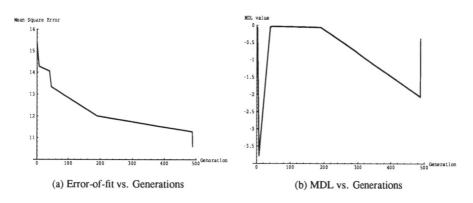

(a) Error-of-fit vs. Generations   (b) MDL vs. Generations

Figure 12.8
Pattern Recognition (Sonar Data)

Sejnowski 1988]. In the "SONAR" experiment, the number of data points was 208 and the number of features was 60. Table 12.5 shows the parameters used. (i) in the terminal set signifies the i-th feature. The final expression $\bar{y}$ was then passed through a threshold function.

$$\textbf{Threshold}(\bar{y}) = \left\{ \begin{array}{lll} 0, & \bar{y} < 0.5, & \text{negative example} \\ 1, & \bar{y} \geq 0.5, & \text{positive example} \end{array} \right. \tag{12.24}$$

Experiment 1
We randomly divided 208 data points into two groups, 168 training data and 40 testing data.

Figure 12.8 shows the results, i.e. mean square error and MDL vs. the number of generations. The elite evolved structure at the 489th generation classified the training data with 92.85% accuracy, and the testing data with 78.94% accuracy. This result is comparable to that of [Gorman and Sejnowski 1988].

The resulting elite structure evolved at the 489th generation is shown below with node coefficients (Table 12.6):

```
(NODE35040
  (NODE35041 (54)
    (NODE35042♠
      (NODE35043 (34)
        (NODE35044 (22)
          (NODE35045
            (NODE35046
              (NODE35047 (46) (NODE35048 (40) (NODE34446 (39) (32))))
              (NODE34650
                (NODE34623
                  (NODE34624
                    (NODE34625 (36)
                      (NODE34626
                        (NODE34627
                          (NODE34628 (35)
                            (NODE34629 (NODE34630 (42) (20))
                              (NODE34631 (48)
                                (NODE34632 (NODE34633 (52) (23)) (48)))))
                          (NODE34634 (NODE34635 (NODE34636 (28) (24)) (41))
                            (NODE34637 (9) (17))))
                        (NODE34638
                          (NODE34639
                            (NODE34640 (17) (43)) (NODE34641 (26) (38)))
                          (NODE34642 (1) (30)))))
                    (31))
                  (40))
                (26)))
              (NODE34667 (NODE34668 (44) (NODE34453 (52) (52)))
                (NODE34454
                  (NODE34455
                    (NODE34456
                      (NODE34457
                        (NODE34458 (60)
                          (NODE34459 (2)
                            (NODE34460
                              (NODE34461 (59)
                                (NODE34462 (4) (NODE34463 (36) (42))))
                              (4))))
                        (NODE34464 (NODE34465 (54) (NODE34466 (28) (14)))
                          (NODE34467 (48)
                            (NODE34468 (NODE34469 (NODE34470 (51) (22)) (13))
                              (NODE34471
                                (NODE34472
                                  (NODE34473 (34) (NODE34474 (46) (47))) (57))
                                (20)))))
                        (NODE34475 (NODE34476 (52) (NODE34477 (53) (20)))
                          (NODE34478 (42) (3))))
                      (NODE34479
                        (NODE34480
                          (NODE34481 (29)
                            (NODE34482 (NODE34483 (NODE34484 (58) (58)) (19))
```

Table 12.6
Node Coefficients

Node	NODE34152	NODE33549	NODE33548
a_0	1.484821	0.89203346	0.27671432
a_1	-3.2907495	2.631569	0.79503393
a_2	-12.878513	-14.206657	-0.66170883
a_3	1.5597038	-256.3501	0.95752907
a_4	3.294608	275.04077	-0.395895
a_5	55.61087	111.96045	0.11753845

```
                    (NODE34485 (58) (35))))
                (NODE34486 (57)
                  (NODE34487 (16)
                    (NODE34488 (41)
                      (NODE34489 (NODE34490 (NODE34491 (31) (43)) (37))
                          (NODE34492 (18) (16)))))))
                (NODE34493 (47) (14))))
              (NODE34494 (23) (NODE34495 (57) (23))))))))))
      (6)))
  (NODE34151 (NODE33558 (NODE33548 (NODE33549 (50) (49)) (25)) (56))))
```

Experiment 2

To compare the performance of STROGANOFF with those of neural networks, we conducted an experiment using the same conditions as described in [Gorman and Sejnowski 1988]. The combined set of 208 cases was divided randomly into 13 disjoint sets so that each set consists of 16. For each experiment, 12 of these sets were used as training data, while the 13th was reserved for testing. The experiment was repeated 13 times in order for every case to appear once as a test set. The reported performance was obtained by averaging the 13 different test sets, with each set run 10 times.

The results of this experiment are shown in Table 12.7. The table shows averages and standard deviations of accuracy for test and training data. NN (i) indicates a neural network with a single hidden layer of i hidden units. The neural data was reported by Gorman and Sejnowski [Gorman and Sejnowski 1988].

The following points from the table should be noticed.

1. STROGANOFF is not necessarily as good as neural networks at learning training data.

2. Judging from the neural testing data results, neural networks may suffer from an overfitting of the training data.

Table 12.7
STROGANOFF vs. Neural Networks

	Training data		Testing data	
	Average	Standard	Average	Standard
STROGANOFF	88.9	1.7	85.0	5.7
NN (0)	89.4	2.1	77.1	8.3
NN (2)	96.5	0.7	81.9	6.2
NN (3)	98.8	0.4	82.0	7.3
NN (6)	99.7	0.2	83.5	5.6
NN (12)	99.8	0.1	84.7	5.7
NN (24)	99.8	0.1	84.5	5.7

3. STROGANOFF established an adequate generalization resulting in good results for the testing data.

Thus we conclude that the MDL-based fitness functions can be used for effective control of tree growth. These methods appear to achieve good results for the testing data by not fitting the training data too strictly. This generalizing ability of STROGANOFF is desirable not only for pattern recognition tasks but for other applications as well [Iba et. al. 1993a, 1993b].

12.4   Discussion

Previous sections have shown the validity of MDL-based fitness functions for Genetic Programming. However MDL cannot be applied to every kind of problem to be solved by GP. Trees evolved in the previous experiments had the following characteristics in common.

Size-based Performance The more the tree grows, the better its performance (fitness).

Decomposition   The fitness of a substructure is well-defined itself.

The first point is a basis for evaluating the tradeoff between the tree description and the error. Decision trees and GMDH networks are typical examples.

The second point claims that the fitness of a subtree (substructure) reflects that of the whole structure, i.e. if a tree has good substructures, its fitness is necessarily high. For instance, when applying an original Koza-style GP to the 6-multiplexer problem, the function set [AND, OR, IF, NOT] and the terminal set of address and data variables were

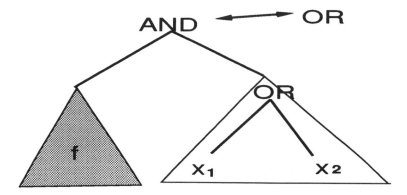

Figure 12.9
Koza-style Tree

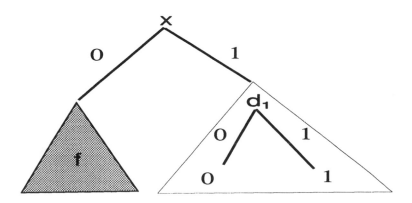

Figure 12.10
Decision Tree

used. Consider the substructure $(OR\ x_1\ x_2)$ of the tree shown in Figure 12.9. The whole value of this tree is heavily dependent on the root node as shown in the following table:

f	$x_1 \vee x_2$	returned value	
0	0	0	0
0	1	0	1
1	0	0	1
1	1	1	1
root label		AND	OR

This table shows that changing the root node from an AND to an OR causes a vast change in the truth value of the whole tree. Hence the fitness of the substructure cannot be evaluated itself. It is affected by the whole structure, especially by the upper nodes. On the other hand, consider the decision tree in Figure 12.10. The truth table is as below:

x	d_1	returned value
0	0	f
0	1	f
1	0	0
1	1	1

As can be seen, the truth value of the substructure $(d_1\ (0)\ (1))$ is kept unchanged whatever the root node x is, i.e. if x is 1 then the returned value is always d_1. In this way, the semantics of a substructure are kept, even if its upper nodes are changed. It follows that the fitness of a substructure is well-defined itself. The same argument is true of a GMDH network (Figure 12.11). In a GMDH network, if subnetworks z_1 and z_2 are combined into x with $x = G(z_1, z_2)$, then the fitness of x is better (i.e. no worse) than both z_1 and z_2 because of the layered regression analysis. Hence a GMDH tree which has good substructures is expected to have high fitness.

MDL-based fitness functions work well for controlling tree growth when the above two conditions are satisfied. The first point guarantees that MDL can be used to evaluate the trade off between the tree description and the error. The second point claims that good subtrees work as building-blocks for MDL-based fitness functions. Hence these points should be considered carefully before applying MDL-based fitness functions to GP. Representation of chromosomes is essential when designing effective GP techniques.

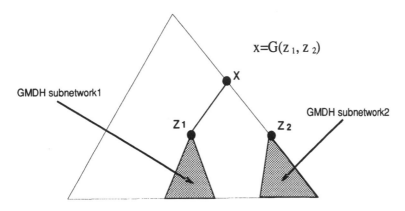

$x = G(z_1, z_2)$

GMDH subnetwork1

X

GMDH subnetwork2

Z_1

Z_2

Figure 12.11
GMDH Network

12.5   Conclusion

This paper has proposed an MDL principle to define fitness functions for GP. A decision tree representation was chosen for the chromosomes. We then showed how the MDL principle could be used to define GP fitness functions. Using our STROGANOFF system, we showed that MDL-based fitness functions can be applied successfully to practical problems such as pattern recognition. We discussed two conditions necessary to design effective representations for GP, i.e. "Size-based Performance" and "Decomposition". Extending an MDL principle to more general applications, in which the above two conditions are not necessarily satisfied, remains a challenging topic in GP. We are currently researching this.

Bibliography

Anthony, M. and Biggs, N. (1992), Computational Learning Theory, Cambridge Tracts in Theoretical Computer Science 30, Cambridge.

Gorman, R. P. and Sejnowski, T. J. (1988) Analysis of Hidden Units in a Layered Network Trained to Classify Sonar Targets, *Neural Networks*, vol.1.

Higuchi, T., Iba, H. de Garis, H. Niwa, T. Tanaka, T. and Furuya, T. (1993) Evolvable Hardware: Genetic-based Generation of Electronic Circuitry at Gate and Hardware Description Language Levels, ETL-TR93, April.

Iba, H., Kurita, T., deGaris, H. and Sato, T. (1993) System Identification using Structured Genetic Algorithms, ETL-TR93-1, to appear in *Proc. of 5th International Joint Conference on Genetic Algorithms*.

Iba, H., Higuchi, T., deGaris, H. and Sato, T. (1993) A Bug-Based Search Strategy for Problem Solving, ETL-TR92-24, to appear in *Proc. of 13th International Joint Conference on Artificial Intelligence*.

Ivakhnenko, A. G. (1971) Polynomial Theory of Complex Systems, *IEEE Tr. SMC*, vol.SMC-1, no.4, pp.364-378.

Koza, J. (1990) Genetic programming: A Paradigm for Genetically Breeding Populations of Computer Programs to Solve Problems, Report No. STAN-CS-90-1314, Dept. of Computer Science, Stanford Univ..

Koza, J. (1992) Genetic Programming, On the Programming of Computers by means of Natural Selection, MIT Press.

Quinlan, J. R. (1983) Learning Efficient Classification Procedures and their Application to Chess End Games, in *Machine Learning*, (eds. Michalski, R.S., Carbonell, J.G. and Mitchell, T.M.), Springer-Verlag.

Quinlan, J. R. (1986) Induction of Decision Trees, *Machine Learning*, vol.1, no.1, pp.81-106.

Quinlan, J. R. (1989) Inferring Decision Trees using the Minimum Description Length Principles, *Information and Computation*, vol.80, pp.227-248.

Rissanen, J. (1978) Modeling by Shortest Data Description, *Automatica*, vol. 14, pp.465-471.

Schraudolph, N. N. and Grefenstette, J. J. (1992) A User's Guide to GAucsd 1.4, CS92-249, CSE Dept., UC San Diego.

Tenorio, M.F. and Lee, W. (1990) Self-organizing Network for Optimum Supervised Learning, *IEEE Tr. Neural Networks*, vol.1, no.1, pp.100-110.

13 Genetic Programming in C++: Implementation Issues

Mike J. Keith and Martin C. Martin

The purpose of our current research is to investigate the design and implementation of a Genetic Programming platform in C++, with primary focus on efficiency and flexibility. In this chapter we consider the lower level implementation aspects of such a platform, specifically, the Genome Interpreter. The fact that Genetic Programming is a computationally expensive task means that the overall efficiency of the platform in both memory and time is crucial. In particular, the node representation is the key part of the implementation in which the overhead will be magnified. We first compare a number of ways of storing the topology of the tree. The most efficient representation overall is one in which the program tree is a linear array of nodes in prefix order as opposed to a pointer based tree structure. We consider trade-offs with other linear representations, namely postfix and arbitrary positioning of functions and their arguments. We then consider how to represent which function or terminal each node represents, and demonstrate a very efficient one to two byte representation. Finally, we integrate these approaches and offer a prefix/jump-table (PJT) approach which results in a very small overhead per node in both time and space compared to the other approaches we investigated. In addition to being efficient, our interpreter is also very flexible. Finally, we discuss approaches for handling flow control, encapsulation, recursion, and simulated parallel programming.

13.1 Introduction

In this chapter we explore the lower level implementation issues surrounding what we call the *Genome Interpreter*. Provided is example code from 5 test programs which were used to evaluate performance. Section 13.8 summarizes the results of these tests and discusses the trade-offs involved with the various implementations.

For the upcoming discussion, what we call an interpreter specifies the following lower level aspects of the design:

- the raw node representation
- how a tree of nodes is represented
- the method for evaluating an individual node
- the method for evaluating the tree as a whole
- the methods for (or methods to assist) those genetic operators which are dependent on the node or tree representation.

A key point is that the interpreter specifies the node implementation which is the particular part of the platform-coding in which the overhead will be magnified. Therefore, the interpreter is the most crucial component in the overall design with respect to space/time efficiency.

In order to illustrate the extremity of the node magnification, consider an application which uses a population size M of 2000 programs in which the average size P_{ave} of each individual program tree consists of 200 nodes. Also, consider a typical scenario that, in order to establish the fitness of each program, it must be executed over 20 test cases (let C be the number of test cases required). Therefore, for each generation, the total number of nodes which must be processed N_p and stored N_s is:

$$N_p = M P_{ave} C = 8,000,000$$

$$N_s = M P_{ave} = 400,000 \tag{13.1}$$

For tougher problems, the magnification factor increases at least quadratically since both the program size and the population size must increase.

13.2 Pointer Based Implementations

Koza [1992] and Tackett [1993] offer pointer based implementations for use in genetic programming in which each program is a parse tree and each node contains a pointer to each child or input. This "traditional" approach for representing the tree structure is typically coded in C as shown in Figure 13.1.

Here, **RETURN_TYPE** stands for the data type that the node evaluates to. Note that the **Args** parameter is an array of pointers to other **Node**s and allows a collection of such **Node** structures to be connected into a tree. A memory management mechanism is required to allocated and de-allocated memory for these nodes on demand. One can typically execute such a tree structure as shown in Figure 13.2.

This recursive routine will execute the genome in a post-order fashion (evaluating children before parent) unless the **Type** of the node is **MACRO**. In this case, the node must call **Eval()** to evaluate whichever of it's own functions it needs.

```
struct Node {
   unsigned char Type ;
   unsigned char Arity ;
   RETURN_TYPE Value ;
   struct Node *Args[MAX_ARGS] ;
   RETURN_TYPE ArgValues[MAX_ARGS] ;
   RETURN_TYPE (*EvalFunc)();   /* pointer to function that
         evaluates the subree rooted at this node. */
}
```

Figure 13.1
A traditional pointer-based tree structure coded in C.

```
RETURN_TYPE Eval (struct Node *N)
{
    if (N->Type == TERMINAL) return N->Value ;
    if (N->Type == MACRO) return (*N->EvalFunc)(N);
    else {
        for (i=0; i<N->Arity; i++)
            ArgValues[i] = Eval(N->Args[i]);
        return (*N->EvalFunct)(N);
    }
}
```

Figure 13.2
This routine recursively evaluates a subrtee; in the results section it is referred to as the "if statement tree". If the subtree is a macro, we simply call the evaluation function. Otherwise, we recursively call **Eval()** on each

```
class Node { public: virtual RETURN_TYPE Eval()=0; };

class AddNode : public Node
{
    Node *Arg[2];
public:
    RETURN_TYPE Eval() {return Arg[0]->Eval() + Arg[1]->Eval();}
};

class VarNode : public Node
{
    int Index;
public:
    RETURN_TYPE Eval() {return VarTable[Index];}
};
```

Figure 13.3
An example of a pointer based tree representation in C++; in the results section it is referred to as the "virtual function tree". Inheritance and virtual functions are used to ensure that the appropriate evaluation function is called and so that each node takes up only the amount of memory necessary.

The overhead in both time and memory for the example code above is significant. In many cases, the interpreter spends more time parsing the **Type** information than it takes to execute the node itself. In fact, because evaluating a typical GP token (like **ADD**, **OR**, **const**, etc.) is usually accomplished in a single machine instruction, it is imperative that we omit the type parsing completely. We can do this in C by making all functions macros and removing the **Type** field, or more elegantly in C++ using virtual functions as in Figure 13.3.

Obviously, this is a superior approach and reduces the node time-overhead. Note that the tree structure is executed in a pre-order traversal. A minor disadvantage to this coding is that the **Eval** method in each of the derived classes is slowed down by the virtual function mechanism used in C++ [Eckel 1990]. Therefore, we coded this approach without the inheritance in C which resulted in the fastest implementation in this chapter. The only overhead required to evaluate the child of a node is two pointer indirections, an array reference and a function call.

The minimum average memory used to represent a node in a pointer based representation can be calculated using a small trick. Each node, although it may have any number of children, has exactly one parent (except for the root). By associating each edge with it's child instead of it's parent, we see that in a tree of N nodes, there are $N - 1$ edges. For large N, then, there is on average one edge per node and hence one **Arg** pointer per **Node**. The linear schemes below show how to represent a tree with no memory overhead at all used to record the topology of the tree. Overall, then, the efficiency of pointer-based representations is quite poor in comparison to the linearized implementations discussed below. This C++ approach above allows each class derived from **Node** to uniquely define its parameters thus minimizing the memory overhead and achieving the lower bound of one pointer per node to store structure. Therefore, this is our preferred pointer-based approach. We should also point out that all of the pointer-based implementations require a memory management service which adds some additional speed overhead and coding complexity to their implementations.

13.3   A Postfix, Stack-Based Approach

We now introduce a stack based approach in which each function and terminal is responsible for getting its own arguments (if any) by popping them from a stack and pushing its single output on a stack. This is similar to the FORTH programming Language (see Winfield [1989]). Consider the 7-node program below:

```
Postfix: a b + c d - +            Infix: (a+b) + (c-d)
```

The stack convention just described provides an "implicit connection" between nodes. For example, terminal node **a** will push its value onto the empty stack, then **b** will do the same. At this point the stack contains the two values with **b**'s on top. The **+** node then pops two values (in this case **a** and **b**), adds them and puts the result back on the stack. This implicit connection allows us to omit the **Arg** pointers needed in the pointer based implementation and store the entire program as an array of nodes, as shown in Figure 13.4

Each program entity can be implemented as a record in which the program's tree is stored, not with pointers, but with a simple linear array. To execute such a program, we can simply index through the array and execute each node using a **case** statement as shown in the Figure.

```
class Program
{
    int Size;
    Node Genome[MAX_GENOME_SIZE];
    Stack stack;
public:
    RETURN_TYPE Eval() ;
};

RETURN_TYPE Program::Eval () {
    for (int i=0; i<Size; i++) { // For each node in order:
        switch (Genome[i].GetType()) {
        case ADD: { stack.PUSH(stack.POP() + stack.POP()) ; break; }
        case MULT: { stack.PUSH(stack.POP() * stack.POP()) ; break; }
        case VAR: { stack.PUSH(genome[i].GetVar()); break; }
        }
    }
}
```

Figure 13.4
The postfix program representation and an example tree evaluation function. There is no memory used to represent the tree topology.

Thus the node execution overhead in this implementation consists of 1 **case** lookup per node as well as 1 or 2 increments/decrements per node to maintain the stack, and an increment to move along the genome. Note that the **GetType** and **GetVar** routines are methods encapsulated in an unspecified **Node** class. A potential advantage of this scheme is that the base functions can be in-line thus avoiding the function call overhead. However, we found the **case** statement overhead in addition to the stack overhead made this approach relatively slow.

Stack-based implementations in general do not require the linear string of opcodes to represent a syntactically correct parse tree. This is a potential advantage of the stack-based approach since the other interpreters we considered require the syntax to be maintained.

We can see that our first linear approach meets many of the goals which we set out to address. We can directly access any entry in the genome-array which allows many manipulations of the genome to be simplified, especially choosing a random node. Furthermore, the array representation also allows the memory management mechanism to be completely omitted. The primary advantage of this approach, however, is that it provides large savings on the node size overhead since the implicit-connection feature uses no memory to represent the tree's topology.

13.3.1   Memory Efficiency

Using fixed-sized arrays to hold variable-sized Genomes can obviously result in a certain amount of space being unused. Therefore, although each node in an implicit-connection scheme can only take up 2 bytes, the effective node size (S_e) is greater than the actual node size (S_a) in relation to the average unused array space (U_{ave}) and average genome size (G_{ave}):

$$S_e = S_a + \frac{U_{ave}}{G_{ave}} \tag{13.2}$$

However, the fact that individuals tend to always use up as much space as they can, tends to minimize this effect and cause the array size to implicitly provide parsimony.

There is a way to further minimize this overhead and to allow the population to explore a variety solutions which have different space requirements on the genome. The idea is to simply provide a distribution of array sizes in the population. Each program structure has a **TraitIndex** parameter which specifies a set of characteristics for that individual or perhaps all individuals in a given region:

```
class Program { int Size; char TraitIndex; /* others...*/};
class Trait { int MaxSize, Minsize, MaxLoopDepth, /* others...*/};
```

Thus the **TraitIndex** is an index into an array of **Trait**s which allows us to reference the array size (**MaxSize**) of the individual. So if a given problem requires an expected genome size of N, we create a set of Individuals whose array sizes are distributed between *N-d* and *N+d* where *d* is the configurable array-size deviation. Note that the genetic operators must not perform an alteration on an individual which violates its **MaxSize** constraint. Also note that this still does not require us to use a memory manager since our variable-sized arrays remain static once created. In general, this trait-table approach allows us to explore a range of possibilities for any characteristic we chose.

13.3.2   Manipulating Postfix Programs

In order to faithfully implement the traditional GP operators, we must ensure that after initialization, crossover and mutation we have a valid representation of a tree. The basis of the following algorithms is to use the arity of every node to allow the syntax to be adhered to. At this point it should be mentioned that since none of these operations are performed while calculating fitness, their efficiency in general is not of prime concern.

13.3.2.1   Postfix Initialization

We can initialize a linear array of nodes as an implied tree in the same recursive manner that pointer based trees are initialized [Koza 1992]. One difference is that in addition to a **MaxDepth** constraint, one needs to also constrain the size of the implied tree to be no greater than the genome array size. This can be accomplished by having a parameter which keeps a count of the number of open branches and using it as follows:

```
if ((NumOpenBranches + CurrentSize) == TargetSize) Resolve = TRUE;
```

Where **Resolve** is a flag which forces all subsequent nodes to be terminals thus forcing all open branches to be resolved. A second difference associated with initializing a postfix tree is that we need to initialize from right to left so that we can begin with function nodes and end with terminal nodes. We conclude the initialization process by copying all of the nodes left so they start at the zero position in the array.

Note, however, that the above method will tend to produce lopsided trees. To avoid this, we can allocate enough memory to hold the biggest tree for the depth we want, and then recursively initialize the array, as show in Figure 13.5

However, the fact that we have a linearized implementation should motivate us to explore alternative techniques for various genetic operators. Based on the current scheme, one can initialize a valid postfix expression by simply requiring that the ongoing **Stack-Count**, as we scan from left to right, is never negative and that the final count is 1. Note that the **StackCount** for a token equals the number of arguments it places on the stack minus the number of arguments it takes off the stack. If we initialize in the reverse direction (from right to left as shown below) the cumulative **StackCount** never becomes positive until we reach the end at which point the overall sum still needs to be 1. One can also

```
int Program::Initialize (int depth, int max_args, int loc)
{
    if (depth == 0) Genome[loc].Initialize(0) ; // init to terminal
    else {
        int num_args = /* random number from 0 to max_args */ ;
        for (int i=0; i<num_args; i++) { // make the arguments
            loc = 1 + Initialize (depth-1, max_args, loc) ;
        }
        Genome[loc].Initialize(num_args) ;
    }
}
```

Figure 13.5
This method recursively initializes an instance of a **Program** class. It returns the last location in the array that it initialized, and inputs the depth of the tree to create, the maximum number of arguments that a node can have, and the location in the array at which to start.

```
Initialize (int TargetSize, int MinCount)
{
    int i, CurrentSize = 0;
    Node *NodePtr = Genome + MaxGenomeSize;

    while (1) {
        if ((++CurrentSize + abs(StackCount)) == TargetSize) ||
            (StackCount == MinCount))
            StackCount += GetRandomTerminal(--NodePtr);
        else if (StackCount == 0)
            StackCount += GetRandomFuntion(--NodePtr);
        else
            StackCount += GetRandomToken(--NodePtr);
        if (CurrentSize == TargetSize) break;
    }

    for (i=0; i<TargetSize; i++) //left shift the genome
    *NodePtr = *NodePtr++ + (MaxGenomeSize-TargetSize);
}
```

Figure 13.6
A method to initialize a genome using the constraint that the cumulative stack count must never become positive.
Note that it initializes the genome from right to left, then shifts it to start at the proper place.

add another constraint, **MinCount**, which allows us to loosely control the implied shape of
the linearized tree (see Figure 13.6).

13.3.2.2 Postfix Crossover

Our golden rule, that the **StackCount** sums to 1 over any legal postfix expression, can be
extended to subtrees also. That is, if one starts at any point X on a postfix expression and
sums leftwise until one reaches point Y where the **StackCount** first goes to 1, then points
X and Y will span the subtree whose root node is X. Note that terminals constitute a subtree of size one and are, on average, the most commonly exchanged subtrees.

Since it is a simple matter of determining where a subtree ends, implementing the traditional GP crossover as a subtree exchange operation is straightforward. But again, since
we are exploring linear approaches, it seems suitable to consider the implementation of a
GA style, 1-point crossover operation [Holland 1975], [Goldberg 1989]. The stack counting mechanism described in the previous section is used for such a routine. The basic rule
is that any 2 loci on the 2 parent genomes can serve as crossover points as long as the
ongoing **StackCount** just before those points is the same. This requirement forces the
syntax to be maintained. For example, consider the 2 parent genomes below, with the
ongoing **StackCount** shown in brackets:

```
P1 = { a[1], b[2], a[3], -[2], +[1], c[2], -[1] }
P2 = { d[1], e[2], +[1], b[2], +[1] }
```

We can exchange the segment {+, b, +} on P2 with the segment {a, -, +, c, -}, {+, c, -} or {-} on P1 since the ongoing `StackCount` (shown in the brackets) just before all of these segments are the same (2).

13.3.2.3   Postfix Mutation

Mutation can be performed using part of the initialization and crossover methods. We select a subtree to replace as we do in crossover, and then generate a subtree just like we do in initialization. Finally, we may need to shift part of the non mutating segment in order to make room for the mutating part.

13.3.3   The Flow Control Problem with Postfix

The major problem with the postfix ordering scheme is its inability to handle flow control. The obvious dilemma is that with postfix, the arguments to a function are always executed before the function itself making it impossible to avoid the execution of conditional subexpressions which we want to skip.

13.4   Mixfix

There is a way to avoid executing parts of a tree while satisfying the requirement that functions get and return values through the stack. We simply allow certain functions to perform arbitrary computations *between* evaluating arguments, and use the results of these computations to decide whether to evaluate the next subtree or skip it. Consider a hybrid ordering we call *mixfix* in which the arguments for flow control constructs are interspersed with code for the actual function. For example, the following coding:

```
( if Y then X - (a+b) ) + c
```

results in the following mixfix expression:

```
Y IF X a b + - ENDIF c +
```

Here we introduced a special marker token "**ENDIF**" which indicates where the "**IF**" function should skip to when its conditional (**Y**) is false. The delimiters are also needed in order to maintain syntax and avoid an expression where the **IF** construct is left without an **ENDIF**:

`X Y Z IF + - //the AritySum is OK, but no ENDIF`

This scheme adds complexity to the genetic operators since finding or creating subtrees for both the **IF** node and the **ENDIF** node must be accomplished in a different manner than the other tokens. In addition, nested conditional structures will require a counting operation so that a given conditional can find the **ENDIF** which belongs to it. This will certainly add execution overhead to the system. One way to minimize this overhead is to omit the **ENDIF** and use a *Jump-Offset* as part of the **IF** construct:

`X IF:5 X a b + - c + //the IF will skip 5 tokens if Y is FALSE`

Now, the **IF** function simply adds its Jump Offset to the current token index in order skip its "then" subtree as opposed to searching for an end marker. This scheme minimizes the execution overhead associated with the delimiter approach above. However, it adds even more complexity to the genetic operators since for expressions containing nested conditional structures, the Jump Offsets of the outer flow constructs must be adjusted when any of their inner subtrees are altered.

For flow constructs with more than one argument, we must distribute the individual sub-constructs as follows:

`Y IF:4 a b + ELSE:3 c d -`

In this case, if **Y** is true, the **IF** function immediately returns allowing the execution to continue at the **a** node. When the **ELSE** sub-construct is invoked it simply skips the **c d -** sub-expression. When **Y** is false, the **IF** function skips to the node immediately following the **ELSE** token (in this example the **c** node).

So now we have added even more complexity to the syntax preserving operations since the entire distributed **IF-THEN-ELSE** construct must be dealt with as a whole for various operations. For example, the crossover method obviously could not exchange the (**ELSE:3 c d -**) fragment by itself leaving the **IF** without an **ELSE**.

Although the stack based approaches just discussed have obvious drawbacks, they do offer some distinct advantages over the other implementations discussed throughout this chapter. The primary advantage of the stack based approaches are that they allows us to easily evaluate programs in parallel manner by maintaining a different stack for each one. This is very useful, for example, in studying emergent behavior, where one wishes to simultaneously evaluate a number of agents interacting in the same simulation. A second advantage is that syntactically unconstrained expressions are possible thus allowing for experimentation along those lines. Finally, by having an explicit stack, we can control what happens when a stack overflow occurs which is not possible for implementations which execute recursively using a system stack.

13.5   Prefix Ordering

We realized that the best way to naturally solve the flow control problem while still taking advantage of the memory efficiency of the linear implementation, was to simply use a prefix ordering on the genotype array. Our prefix ordering scheme has 3 potential advantages over postfix:

1. arguments must be explicitly evaluated by their parent which allows control constructs to be implemented in a natural manner.

2. the coding required to skip a subtree is quite simple and does not require the special mechanisms discussed previously in the mixfix section (like using jump-offsets or bracket tokens).

3. an explicit stack mechanism is not needed which reduces coding complexity and performance overhead.

   We execute through the linear chromosome array recursively as shown in Figure 13.7. For efficiency, a global variable keeps track of the current position within the chromosome array. To evaluate the next argument (which is done with a call to **EvalNextArg()**) we

```
Node *current_node ; // The current Node being evaluated.

class Individual {
   Node *Genome ;   // pointer to an array of nodes.
public:
   RETURN_TYPE Eval()
       { current_node = Genome; return EvalNextArg() ; }
}

inline EvalNextArg() { return(current_node++).Eval() ; }
inline SkipNextArg()
 { for(int count=0; count>-1; count+=(current_node++).ArityM1()); }

class Node {
   RETURN_TYPE (*EvalFnct)() ;
public:
   RETURN_TYPE Eval() { return (*EvalFnct)() ; }
}

int x = 1.234 ;
float add() { return EvalNextArg() + EvalNextArg(); }
float varX() { return x ; }
```

Figure 13.7
The prefix representation and evaluation code, with example code for the nodes. Note that each**Node** must defin
a method called **Eval()** which uses **EvalNextArg()** to get it's arguments.

simply increment the global pointer and call the evaluation function of the node. If this evaluation function needs arguments, it can call **EvalNextArg()** to get them; each call gets a different argument.

Note that the distributed evaluation approach makes aborting an individual programs in midstream difficult. Although this can be accomplished using the standard C library functions **setjump()** and **longjump()**, the evaluation schemes used in the postfix and mixfix implementations allow a program to be suspended or aborted at any node.

Our results in Section 13.8 show that the prefix approach is the most efficient. Also, from inspection of the above test programs, it should be apparent that the PJT scheme provides the cleanest implementation. A key point is to note that **EvalNextArg()** includes the important side effect operation of incrementing the node pointer. This can lead to problems for logical operations like **AND** which will skip evaluating (and therefore skipping) it's second argument if the first argument is false. Therefore, a safe implementation of **AND** would be:

```
int AND()
{
    if (!EvalNextArg()) { SkipNextArg(); return FALSE; }
    else return EvalNextArg() ;
}
```

13.5.1 Initialization, Crossover and Mutation with Prefix

The prefix ordering scheme allows syntax to be maintained in the same fundamental way as postfix does. That is, by taking the arity of each node minus 1, and summing from left to right, the overall sum must equal -1 in order for the prefix expression to be valid.

A key point with respect to initialization is that with prefix, we add terminals at the end as we scan from left to right which allows the left-shift operation coded in the postfix initialization routine to be bypassed. Otherwise, the initialization, crossover, and mutation implementations used for prefix are analogous to the postfix routines discussed in Section 13.3.2

13.5.2 Handling Program Flow with Prefix

In this section we discuss various ways to handle program flow in the prefix scheme. The fundamental idea is to use the **EvalNextArg()** routine to evaluate arguments and **Skip-NextArg()** to skip over an argument without evaluating it. These routines are shown in Figure 13.8. Note that, in practice, we would keep track of how many iterations of the **while** loop we have performed and abort when the number becomes too large. The purpose of the given code, however, is just to show how to perform iteration in the prefix implementation.

```
RETURN_TYPE if_then_else()
{
   if (EvalNextArg()) {
      const RETURN_TYPE result = EvalNextArg();
      SkipNextArg() ;
      return result ;
   } else {
      SkipNextArg() ;
      return EvalNextArg() ;
   }
}

RETURN_TYPE while ()
{
   Node *Start = current_node;

   while (EvalNextArg()) {
      EvalNextArg() ;
      current_node = Start ;
   }
   SkipNextArg() ; // skip the body of the loop
}
```

Figure 13.8
Implementing **if-then-else** and **while** in a prefix implementation.

13.6   The Node Representation

Up to this point, we have been concerned only with how to represent the *topology* of the tree, and have simply used a function pointer to represent the information needed to evaluate a node. In this section, we present our preferred approach for representing a node.

13.6.1   General Data Support

The idea can be seen to follow from a simple observation. If there are 256 different types of node (functions and terminals combined, each constant counting as a different type of node), then we only need one byte to represent the node. This byte could be used as an index into an array of information about that type, including a function pointer, arity, name, etc. The function pointed to we call the *handler*, the array we call a *jump table* (since it's primary purpose is hold the function pointer), and the entire entry for one type of node is called a *token*, represented by the **Token** class:

```
class Token { char *Name; char ArityMinus1; float (*Funct)(); /*...*/};
```

This strategy of simply providing memory chunks (tables) based only on the data-types needed for an application as opposed to requiring a special token for each and every variable, allows the data-declaration portion of a program to be more loosely defined up front and even provides a framework for allowing such declarations to evolve if needed. We call this property *general data support,* and argue that for non-toy problems, the evolution of the programs data-model in addition to its parse tree might be essential and that general data support, while not a solution to this problem, is a step in the right direction.

13.6.2   The Opcode Format

For an application using 2 floating-point variables and many pre-defined constants, where one byte will do, one might set up the jump-table as follows:

Functions[0-15], Variables[16-17], Constants[18-255]

Note that in this scheme, the pointers to the constant handler is repeated over it's range and examines the node value itself to determine which constant is needed. See Figure 13.9. A pointer to the constant handler repeatedly appears in the jump-table in entries 18 through 255. When called, it finds which constant (between 18 and 255) is needed by examining the current node and returns the proper entry from it's array.

Although 256 different types may do in some applications, many applications need more (e.g. when using random ephemeral constants or other run-time-generated terminals such as modules). From here on we will assume that there are more than 256 but less than 65,536 types of node, so that two bytes is both necessary and sufficient. We could simply have an array of size 65,536, but this is cumbersome. Firstly, this takes a lot of memory and filling all entries is cumbersome. Secondly, many of these entries will be similar (e.g. many will be distinct constants). To take advantage of this, we can use part of the two bytes to represent the general type of node (e.g. "+", "sin", "constant"), and the rest of it (if needed) to specify the exact node type (e.g. "+" and "sin" need no further explanation; "constant" needs to know the particular value). The portion for the general type we call the *function index,* and the portion for the specific type we call the *specific lookup table index,* or simply the *table index.* We have one function for each general type of node; the function is passed the two-byte opcode and can use the table index however it sees fit.

There are many possibilities for splitting the 16 bits available to us. For most applications, a maximum of 16 functions would suffice. Therefore we considered using 4 bits to represent the function index and the remaining 12 bits for the table indexing. Thus our **Node** structure could be coded using bit-fields:

```
class Node {
   unsigned functIndex:4; //function index
   unsigned tableIndex:12; //table index
   // ...
} ;
```

```
#define RETURN_TYPE float

class Token JumpTable[256] ; //maps node number to info about that node.

class Node *current_node ;   // The node being evaluated.

/* Function handlers */
RETURN_TYPE Add() { return EvalNextArg() + EvalNextArg() ; }
RETURN_TYPE Mult() { return EvalNextArg() * EvalNextArg() ; }
// other functions

/* Variable handlers */
RETURN_TYPE x, y ;
RETURN_TYPE X() { return x ; }
RETURN_TYPE Y() { return y ; }

/* Constant handler */
RETURN_TYPE ConstTable[255-18+1] ;
RETURN_TYPE Const() { return ConstTable[*current_node-18] ; }

RETURN_TYPE EvalNextArg() { return (*JumpTable[current_node++].funct)() ; }
class Node genome[] = { 0/*add*/, 1/*mult*/, 16/*x*/, 17/*y*/, 16/*x*/ };

void InitJumpTable()
{
    JumpTable[0].name = "add" ; JumpTable[0].ArityM1 = 1 ;
    JumpTable[0].funct = Add ;
    /* other functs and vars*/

    for (int i=18; i<=255; i++)
        JumpTable[i].funct = Const, JumpTable[i].ArityMinus1 = -1 ;
}
```

Figure 13.9
An example of the workings of the one byte opcode. To complete the code we also need to initialize the name
and value of each constant, and to initialize the other function entries in the jump table. Also, the genome showr
would be for testing purposes; the real genomes would genomes would be generated at run time.

However, with this setup, all references of the function index and table index require
both a bitwise-AND operation and a bit-SHIFT operation which adds considerable node
overhead to the interpreter. Therefore, there is really no choice but to place the 2 indexes
on byte boundaries:

```
class Node { unsigned char functIndex, tableIndex; };
```

The only possible drawback to this layout is that now we have 256 possible tables (this
is probably too many), each with room for 256 entries for variable/module/constant
lookup (this is probably too few especially for constants). The solution is to take advantage of the extra function indexes. So if you want to have, let's say, 1024 possible float

constants for a given symbolic regression application, you simply reserve 4 function indexes (4*256 = 1024) which point you to 4 different (1 line) float handling routines. For example, the third such routine might look like:

```
float Float3() {
    return FloatConstantTable[2][current_node->tableIndex] ;
};
```

13.6.3   The Jump Table Mechanism

Our jump table mechanism is simply the array of **token** objects. The primary benefit of such a jump table is that now we can select which function to execute with an array deref-erence as opposed to a **case** statement. Although compilers will often compile a **case** statement into such a jump table, it will still do bounds checking on the index, and so will be slower than the hand coded jump table. If **current_node** points to a **Node** object, and **Tokens** points to the array of **token** objects, then **EvalNextArg()** can be implemented with the statement:

```
return (Tokens[(current_node++)->functIndex].Funct)() ;
```

In other words, this highly efficient code fragment amounts to nothing more than incre-menting and de-referencing a pointer, referencing an array and making a function call.

13.7   The Prefix, Jump-Table (PJT) Approach

We think the best overall implementation of the genome interpreter uses these 4 key con-cepts: a prefix ordering scheme, general data support, a 2-byte node representation, and a jump-table mechanism. It is the cleanest and most modular approach, since it avoids the need for **case** statements to determine the type of a node. It is the most flexible since it naturally handles all basic flow control constructs. Finally, in addition to the previous attributes, it is the most efficient approach in terms of its node space/time overhead. As any designer can testify, we were quite lucky to have such a clear winner.

13.8   Results

One must consider the relative importance of memory vs. speed. Koza [1992] provides empirical findings which demonstrate that by *tripling* the population size and keeping the case count constant, the overall efficiency of the simulation is usually *doubled* (the num-ber of individuals needed to be processed is reduced by 1/2). Based on this empirical relationship, we can derive an approximation relating the efficiency (e) of one interpreter with respect to another in terms of their node memory size (m) and evaluation time (t):

$$\frac{e_2}{e_1} = \left(\frac{t_1}{t_2}\right)\left(\frac{m_1}{m_2}\right)^{\log_3 2} \approx \left(\frac{t_1}{t_2}\right)\left(\frac{m_1}{m_2}\right)^{0.631}, \tag{13.3}$$

where efficiency is the simulation time required to have a 99% probability of convergence. Therefore, memory overhead is not as devastating as time overhead in the node implementation. An interpreter which evaluates nodes 4 times slower than an alternative interpreter but uses a node representation 1/9 the size, will be equally as efficient. We use Equation (13.3) to provide a general approximation of performance between the 5 different interpreter variants discussed above.

We tried our linear implementations out on a number of simple problems suggested by Koza [1992]. The results confirmed that we were able to converge as expected and essentially match the results of the established tree-based representation. This follows logically since we have a functionally equivalent system. Table 13.1 shows the performance results of 5 approaches presented in this chapter and also summarizes the advantages and disadvantages of each. We ran each test with 5 different compilers on 4 different platforms. The data shown below represents an average over the multiple runs.

Table 13.1

Results and comparison of the different approaches discussed in this chapter. t_2/t_1 and e_2/e_1 are the run-time and effeiency, respectively, relative to the PJT approach, and m is the number of bytes each node occupies. "Virtual Function Tree" refers to coding the **Eval()** function as a vertual function, "If Statement Tree" refers to having a **Type** field parsed by a **case** statement, and "Function Pointer Tree" refers to storing a pointer to the evaluation function at each node. We assume **MAX_ARGS** is 2, and we disregard the memory for the **ArgValues** field.

Representation	t_2/t_1	m	e_2/e_1	Advantages	Disadvantages
Prefix/Jump Table	1.00	1	1.00	Most Efficient and Flexible Approach.	Can't do Parallel Evaluation.
Postfix/Mixfix	1.10	1	0.91	Memory Efficient, Good for Parallel Evaluation, Unconstrained Syntax.	Flow Control is Awkward.
Virtual Function Tree	0.92	4	0.45	Effective, Re-usable Code.	Size Overhead, Memory Manager Needed.
If Statement Tree	1.47	7	0.20	Conceptually Simple	Size Overhead, Slow, Memory Manager Needed.
Function Pointer Tree	0.81	6	0.40	Overall Fastest Approach, Conceptually Simple.	Size Overhead, Memory Manager Needed.

13.9   Advanced Topics (Looking for Roadblocks)

Before we can claim that the PJT interpreter represents a generally usable implementation, we should first overview as many programming constructs which might be used in a GP setting, and see if such constructs pose any significant problems to the approach. Note that our focus in this discussion is on the mechanics of the implementation as opposed to the actual usage of such constructs in real simulations.

13.9.1   Beyond Closure: Handling Multiple Data Types

It is certainly possible that a given application has function nodes which work with a variety of different data types simultaneously. Such applications need an interpreter which allows all of these functions to work together in a transparent and flexible manner. C++ allows objects to overload their associated casting operators and constructors allowing automatic conversion between source and destination data objects. So, we can select a generic data-type which all of the other object types will convert to and from (using the automatic send/receive methods above) in a transparent manner.

This approach, which is more extensive than "closure" as defined by Koza [1992], works well provided that all of the different data types are convertible to/from the generic type. Montana [1993] discusses applications where the given data types include user defined structures like vectors and matrices. For such applications, the simple automatic-closure mechanism might not suffice since converting a **vector** to an **int** for example, might not make sense for certain handlers. Montana gets around the problem by providing special genetic operators which only allow certain connections to be made in the node structure. Thus a handler which expects an input of data type **T** will only be connected to handlers which output type **T**.

An alternative, is to have the genetic operators make connections based on a connection-strength table. For example, if a crossover method was to perform a subtree exchange resulting in a **string** to **float** connection, it could reference this table to see how viable that connection is. If the strength is less than some arbitrary value, then the operator simply looks again for a subtree which is more appropriate.

For certain applications, it also might be useful to have the interpreter pass parameters by reference as opposed to their value. The PJT scheme makes this easy since a node function can return a 2-byte address defined as {**table:entry**} which points to its return argument. The advantages of reference passing are threefold. First, functions can now return multiple arguments via structure references. Secondly, we now avoid the problem of passing the actual data of large data-structures (like a matrix) through the system stack thus avoiding unnecessary overhead. Finally, the address as defined above, allows the function-token to easily determine the data-type of the return argument (at run time) since the table value corresponds to the type. This run-time type information can then be used by the handler to convert arguments in a specific manner or to provide various services based on the receive types (allows the handler to overload itself).

Handling a set of mixed data-types does not seem to provide any roadblocks to any given interpreter implementation specifically. The PJT design does seem to provide a natural way to incorporate reference passing due to its table-driven emphasis. Overall, we feel that a more unified approach is needed and can be established as more difficult problems are attempted.

13.9.2   Module Implementation

This section discusses two aspects of modularization with respect to the linear PJT implementation. First we discuss a basic encapsulation technique and then we look into how modules are actually executed. One possible implementation of the module structure is:

```
class Module {
    int ArityMinus1;
    char ReturnType;
    int Size;
    Node *Genome;
} ;
```

Note that a **Module** has characteristics in common to both a **Program** (**Size**, **Genome**) and **Token**s (**ArityMinus1**, **ReturnType**). It is very important in a C implementation that all 3 of these structures be set up so that the module parameters correctly *overlay* the parameters it has in common with the other 2 structures. This overlaying allows us to treat a **Module** as a **Program** (during evaluation) or a **Token** (for referencing token description information) without any conditional coding needed. In C++, however, the *shadowing* mechanism used with multiple inheritance provides a superior approach to overlaying structures:

```
class Module : public Program, Token { /*...*/ };
```

Now we can have a polymorphic program pointer such that references like **P->Genome** are allowable where **P** can be a pointer to a **Program** or a **Module**. Furthermore, we can say **T->ArityMinus1** where **T** allows us to access the arity information in a **Module** or **Token** without having to know if **T** actually points to a **Module** or a **Token** object.

13.9.2.1   Encapsulation

There are various approaches for defining or encapsulating modules. Koza [1990] uses an approach called Automatically Defined Functions (ADFs). His approach is LISP based and involves every program being syntactically constrained. Angeline and Pollack [1993] suggest a free form approach where you simply encapsulate a subtree or a segment of a subtree which has already evolved in your current population.

Our technique follows from Angeline and Pollack. We first randomly select a subtree which is of a *minimum* size and inspect the terminals of that sub-tree. The number of different variable-terminals determines the arity of the encapsulated subroutine. Finally, we re-define each such variable-terminal in the subtree so that it references a general **Arg-Stack** instead of a global **VariableTable**. This then makes all of the parameters for that subtree "local".

For an example, consider encapsulating the following expression:

Prefix: * + + X a X - Y b   **Infix: (X+a+X)*(Y-b)**

If we were to compress the segment {+ + X a X}, we would have the following module defined after compression:

```
Module (M):
 arity = 1;
 size = 4;
 genome = {add, add, arg:0, const:a, arg:0}
```

The original program would also be compressed or re-defined such that the new second and third nodes now correspond to a "module-call":

Prefix: * M X - Y b   **Infix: M(X)*(Y-b)**

So now when the host program is executed, it will call the module-handler routine which will in turn find module **M** in the module table, and execute it's internal genome sequence. When this happens, the variable **X** will now be processed as an argument. That is, the module handler will **PUSH** the value of **X** on the **ArgStack** before actually executing the module's genome. Since, all of the variables in the module reference the **Arg-Stack**, the phenotype of the host program remains constant even though the genotype of the program has indeed been altered. Another point is that although the subtree contains 3 terminals, one of the terminals is a constant and the other 2 terminals are the same. Thus the arity of the newly defined module is only 1.

The encapsulation routine can be implemented with the following steps:

1. get a free module entry from the module table. If a free entry is not available, then abort.

2. find a random subtree of a minimum size from the host's genome array. Here we can use the same routine used to obtain subtrees in the crossover implementation.

3. inspect all of the terminals in the subtree and determine which ones are constants and which ones represent unique variables. The unique variables are re-defined to reference the **ArgStack** and **ArgHandler** as opposed to being variables.

4. define the module by copying over the subtree to the internal genome array of the module. Continue defining the module by setting the module's size to be equal to the subtree size and set the module's arity parameter to be the number of unique variables found in the subtree.

5. re-define the host program to reflect the compression/mutation operation. The subtree segment is replace by the call segment and the host-program's size parameter is decreased by the difference between the subtree size and the number of nodes in the call (this equals 1 + module's arity).

13.9.2.2 Module Execution

Module implementation with respect to the actual execution of modules seems like it might be easier to accomplish in a postfix implementation as opposed to prefix. This is due to the fact that subroutines naturally fit into a stack scheme. For postfix, this means that when a subroutine is to be executed, its parameters will automatically be waiting on the stack. For prefix, this is not the case. Each function or module is responsible for obtaining its own arguments. Since these arguments are outside the module, the module handler needs to load the **ArgStack** with the needed parameters on behalf of the given module before that module is actually invoked. Since such a handler can obtain the needed arguments by simply doing an **EvalNextArg()**, it turns out that our prefix scheme allows for a straightforward implementation and does not require any additional overhead compared to postfix. The code is shown in Figure 13.10.

13.9.3 Handling Recursion

Recursion can be promoted at two levels in a GP application. The first approach is to allow the main program itself to be called recursively by an internal handler. By adding a **Recurse-Handler** to the function set, the program itself can essentially be "called". This **Recurse-Handler** would invoke the program in the same way that the platform invokes the program except that the program would be passed an argument from an internal node as opposed to a "case" value. For example, consider a simple symbolic regression problem of 2 independent variables. For such an application, each program is essentially a subroutine of arity 2. A **Recurse-Handler** could be implemented as a function-token, with an arity equal to the program's arity, allowing it to obtain the proper number of arguments before calling the host program. In the example code shown in Figure 13.11 we use the argument stack. Each time the host program is recursively called, its variable handlers will

```
inline unsigned char CurrentIndex() { return current_node->t ; }

void module_handler ()
{
    Program *PrgSave; Node *NodeSave; int i;

    // initialize the local arg stack
    for (i=0; i<ModuleTable[INDEX].Arity; i++)
    ArgStack.Push = EvalNextArg();

    // now save the global prg pointer and token index
    PrgSave = Prg; NodeSave = current_node;

    // now make our gloable refs point to the subroutine at hand
    Prg = ModuleTable[INDEX]; NodePtr = Prg->Genome;

    // now execute the module at hand
    EvalNextArg() ;

    // restore the global references
    Prg = PrgSave; NodePtr = NodeSave;
}
```

Figure 13.10
The module handler.This is called when **Eval**ing a module node.

```
RETURN_TYPE RecurseHandler()
{
    //first load up our argument stack with the calling parameters
    for (i=0; i<PROGRAM_ARITY; i++)
    ArgStack.Push (EvalNextArg()); //PUSH automatically

    //save our current location so that the host eventually
    //returns here
    RecurseStack.Push (current_node);

    //now restart the host program and "call" it
    current_node = Genome - 1; EvalNextArg();
}

EvalRecursiveProgram ()
{
    current_node = Genome -1; EvalNextArg(); //normal invocation
        //assist recursion:
    while (current_node = RecurseStack.Pop) EvalNextArg();
}
```

Figure 13.11
A handler for nodes which mean "call this program recursively", and a wrapper to make the first (non-recursive) call.

access values further down this stack. We also use a special recursion stack so we can return to this point when the program completes.

Now instead of invoking programs with the usual **EvalNextArg()** macro, we need to keep restarting the program as long as the recursion stack is non-empty at the popped **current_node** location.

The second recursion scheme involves the use of modules. The idea is to allow a program to contain recursive subroutines. This is accomplished by mutating an existing module **M** by finding an internal function node within **M** which has the same arity as **M** itself. We then replace this internal function node with a Module-Recurse-Handler node which will "call" **M** in a similar manner that the handler outlined above called the main program. To maintain syntax, this Module-Recurse-Handler would have to able to assume an arity which is equal to the arity of **M**. For either recursive approach, we feel that it would be advantageous to bind the recurse-handler to an if-handler in order to increase the chances that the recursion process completes.

13.9.4 Simulated Multi-Tasking

Many applications require multiple trees to be evaluated in parallel. That is, one often wants to optimize something which consists of many parts in which each sub-part interacts with the other sub-parts in a parallel manner (i.e. the emergent behavior of ant colonies). Fortunately, it is not impossible to achieve multi-tasking with prefix. The basic idea is to somewhat combine prefix with a stack implementation. In other words, the genome ordering is still a prefix ordering, but the parameters are passed by an explicit stack and the genome is no longer executed by recursion but by the use of an execution stack which allows us to mimic what the system stack did. See Figure 13.12

13.9.5 Using Tables to Evaluate Diversity

The Jump Table mechanism results in a fair number of tables (arrays) being used for constants, variable, modules etc. An added bonus of a table-driven approach is that such tables can provide an easy means to dynamically evaluate diversity in the current population. This is accomplished by adding a "**Used**" bit to each entry in any of the tables just mentioned. To measure the ongoing diversity of the constants, we add a line to the constant handler:

```
void ConstHandler ()
{
 constTable[INDEX].Used = 1;
 return (constTable[INDEX]);
}
```

```
class Task { PROGRAM &prg; int Index, Status; };

main ()
{
   Task Tasks[2] = {{Prg[0], 0, RUNNING, {Prg[1], 0, RUNNING}};
   MultiTask (Tasks, 2);
}

multi_task (TASK &Tasks, int TaskCount)
{
   int t=0; //the task index

   for (;;) { // Forever
      switch (Functions[Tasks[t].Index].Type) {
      case FUNCTION:
         ExeStack.push (Tasks[t].Index); //save index for later exec
         Tasks[t].Index++; //goto the next node for this task
         break;
      case TERMINAL:
         //execute the terminal node which will push a value on stack
         EVAL (Tasks[t].Index++);

   //check if stack count is equal to the arity of the last func
   //on the ExeStack which is still waiting to execute. If it is,
   //then go ahead and evaluate this funct we saved earlier
         if (ArgStack[t].Count == ArityOfLastFunction)
            EVAL (ExeStack[t].pop);

   //now do a task switch
         if (++t == TaskCount) t = 0;
         break;
      case END:
         tasks[t].status = COMPLETED;
         if (all_tasks_completed ()) return; //****done****
      }
   }
}
```

Figure 13.12
Simulated Multi-tasking using prefix.

Now if we clear all of the "**Used**" bits to zero at the start of each generation, the bits which are not set when the generation has completed, correspond to constants which are no longer in use. We tried this out for our simple regression application and were amazed at how quickly the total number of unused constants increased.

13.10   Conclusion and Future Directions

The lower-level implementation issues surrounding what we call the Genome Interpreter have been presented. As opposed to just discussing issues along a single implementation direction, we have considered 5 different variations and evaluated each of these interpreters based on their efficiency, flexibility, and ease of coding. Our results clearly showed that a linear, prefix-ordered, jump-table approach (PJT) provides the best overall framework for the actual implementation.

One method of evaluation that could be even faster than a genome interpreter is a genome *compiler*. Although our interpreter has little time overhead, the overhead can still be significant if the functions and/or terminals take only a few machine instructions to implement. For example, when performing constant and variable lookup, and addition and multiplication, the overhead of the interpreter will still swamp the time needed to actually evaluate the node. However, the only way to do away with the function call overhead in an interpreter is to write many versions of each function, namely, one which does a function call when it's argument is a complicated subtree, and others to do in-line variable or constant lookup.

This could be solved using a genome compiler, which, given a tree, outputs machine language that doesn't perform any function calls. If, as is often the case, the same tree will be evaluated many times, the tree will only need to be compiled once instead of interpreted many times. Additionally, since crossover combines large chunks of different trees, we may be able to save compile time by not recompiling the chunks, but reusing the compiled code of the parents. One drawback, however, is that flow control may skip large parts of the tree, so that only a small part of the tree is ever interpreted whereas all of it must be compiled. This could make compiling slower than interpreting. And all sorts of variations are possible, such as compiling modules but interpreting the rest. Overall, attempting a genome compiler is a promising direction for speeding up evaluation even further.

Acknowledgments

Thanks to Greg Schmidt of Allen Bradley Controls for feedback on various issues presented in this chapter. Also, thanks to Graham Spencer of Stanford University for his early contributions with respect to testing. We also must recognize the importance of various discussions which took place on the GP internet mailing-list (organized by James Rice of Stanford University) related to implementation issues. Finally, thanks to those who reviewed this chapter for their many helpful comments.

Bibliography

Angeline, P.J. and J. B. Pollack (1993) Evolutionary Module Acquisition, *Proceedings of the Second Conference on Evolutionary Programming.*

Eckel, Bruce (1990) *Using C++.* Osborn McGraw-Hill.

Goldberg, D. (1989) *Genetic Algorithms in Search, Optimization, and Machine Learning*, Reading MA: Addison Wesley.

Holland J. (1975) *Adaption in Natural and Artificial Systems,* Ann Arbor, MI: The University of Michigan Press.

Koza, J.R. (1992) *Genetic Programming.* The MIT Press.

Montana, D.J. (1993) Strongly Typed Genetic Programming, BBN Technical Report #7866, May 1993.

Tackett, W.A. (1993) Genetic Programming for Feature Discovery and Image Discrimination, *Proceedings of the Fifth International Conference on Genetic Algorithms.* Morgan Kaufmann..

Winfield, A. (1983) *The Complete Forth,* Wiley Press

14 A Compiling Genetic Programming System that Directly Manipulates the Machine Code

Peter Nordin

Most genetic programming approaches use a technique where a problem specific language is executed by an interpreter. The individual code segments in the population are decoded at run time by a virtual machine. The disadvantage of this paradigm is that interpreting the program involves a large overhead. Often the complete system and the genetic operators themselves are written in an interpreting language like LISP. This reduces performance in most hardware environments. We have evaluated the idea of using the lowest level native binary machine code as the "programs" in the population. There is no intermediate language or any interpreting steps. The genetic program that administers these machine code segments is written in the 'C'-language. The algorithm is of steady state type and uses a small tournament as the selection mechanism. This approach has enhanced performance by a magnitude of three compared to a conventional system in an interpreting language. The increased performance is tested on a problem of symbolic regression of a classifier function in machine code. We evolved a machine code program that classifies Swedish words into nouns and non-nouns by spelling only. We compare the compiling genetic programming system (CGPS) with a Neural Network performing the same task. In our example, the results show superior performance of the CGPS compared to the connectionist approach. While the classification and generalisation capabilities were equal, the training time was more than 200 times faster, the classification time 500 times faster and the memory requirements are at least 10 times lower with the CGPS, as compared with the Neural Network.

14.0 Prologue

Consider the task of multiplying the two leftmost integer numbers in figure 14.1. This is a complicated task that takes a minute to complete with pen and paper. Today's PCs can perform a huge amount of instructions like this. While an average reader spends, half an hour reading this report, today's (1993) fastest PC executes about 200 billion instructions of this type [BYTE 1993]. If these instructions were to be written in the same way as in figure 1, the list of executed instructions would be more than six million kilometres long. This is equivalent to 150 times around the earth or 20 times to the moon, where every inch of the way represents an instruction so complicated that it could take a minute to complete by hand.

```
1265080939  *  2239753764  +  1675893667  -  1345928399=
```

Figure 14.1
Arithmetic instructions

Ask a LISP-interpreter to reverse a list. Half a second later it will give the reversed list. Is this all we get for those millions of instructions?

This work arose from this frustration of not being able to utilise the underlying low level performance of any computer. The research is as much driven by the urge to find a way to exploit this underlying power of every computer as it is to find an effective genetic programming system.

14.0.1 Reading Guidance

This paper is structured as follows: It consists of four major parts:

First we describe a Genetic Programming system written in the C-language where the individuals in the population, the evolving programs are represented in pure binary machine code. We describe the technique used for implementing the genetic operators on the binary code as well as the overall genetic algorithm used, section 14.1 and 14.2.

We apply this compiling genetic programming system to regression of a classifying function, described in section 14.3.

This approach is applied to the task of classifying Swedish words into nouns and non-nouns and compares the classification performance of the same task performed using a connectionist approach. The task, the training set, the neural network and the result from the comparison are presented in section 14.4 and 14.5.

Finally, we summarise the conclusions and present ideas for future work.

14.1 Introduction

Genetic programming provides a powerful technique for evolving programs to solve a variety of different tasks. The set of practically solvable tasks is highly related to the efficiency of the algorithm and implementation. It is therefore important to minimise the overhead involved in executing a genetic programming system.

Most genetic programming approaches use a technique where a problem specific language is executed by an interpreter. The individual code segments in the population are decoded at run time by a virtual machine. The data structures in the programs often have the form of a tree. This solution gives good flexibility and the ability to customise the language depending on the constraints of the problem at hand. The disadvantage of this paradigm is that interpreting the program involves a large overhead. There is also a need to define more complicated genetic operators, which also decreases performance. The more complicated structures, using pointers, makes execution even less efficient. Often the complete system and the genetic operators themselves are written in an interpreting language like LISP [Koza 1992, Page 71]. This reduces performance in most hardware environments. Recently some systems have been presented written in compiler languages like 'C' or 'C++', parsing structures equivalent to the programs used in a LISP implementation [Keith 1993]. This gives increased performance while it preserves the ability to be flexible with the representation and selection of a problem specific function set. Still there is a need to interpret the programs in the population that involves overheads both in execution time and memory consumption.

We have evaluated the idea of using the lowest level binary machine code as the "programs" in the population. Every individual is a piece of machine code that is called and manipulated by the genetic operators. There is no intermediate language or interpreting part of the program. The machine code program segments are invoked with a standard C function call. The system performs repeated type cast between pointers to arrays for individual manipulation and pointers to functions for the execution of the programs and evaluation of the fitness of the individuals. Legal and valid C-functions are put together, at run time, directly in memory by the genetic algorithm. This is a "compiling" genetic programming approach (CGPS).

14.2 The Compiling Genetic Programming System (CGPS)

We call our approach "compiling" as the system generates binary code from the example set and there are no interpreting steps. The idea is to use the real machine instead of a virtual machine and the hypothesis is that the loss in flexibility will be well compensated for by increased efficiency.

14.2.1 The Hardware Environment

We have experimented with CGPS on a PC with an Intel processor and on a SUN SPARC architecture. This report is based on an implementation in the SPARC environment which generally has a more stable architecture for low level manipulation and program patching. We have used a SUN SPARCSTATION 1+ in the examples described below.

14.2.2 The Language for the Genetic Algorithm Implementation

A CGPS is most naturally implemented using an efficient compiling language with the ability to make manipulations on a low level. There is also a required freedom in the use of data types and the ability to cast between different types, especially pointers to arrays and pointers to functions. We use the C-language and the standard SUN operating system compiler (version 4.1.1).

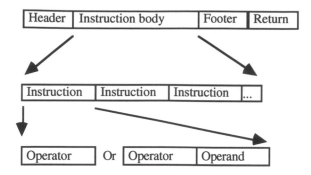

Figure 14.2
Generic machine code function structure

14.2.3 The Structure of a Machine Code Function Callable by a 'C'-function

The individuals in the population consist of machine code sequences resembling a standard C-function as stated above. Figure 14.2 above illustrates the structure of a function in machine code. This structure is fairly generic for different types of compilers and hardware architectures.

The function code consists of four major parts:

• **The Header** deals with the administration, necessary when entering a function. This normally means manipulation of the stack, for instance getting the arguments for the function from the stack. There could also be some processing to ensure consistency of processor registers. This section is often constant and can be added at the beginning of the initialisation of the individual machine code segments in the population. The mutation and crossover operators must be prevented from changing this field when they are applied to an individual during execution.

• **The Footer** is similar to the header but does the operations in the opposite order and 'cleans up' after the function call. The footer must also be protected from change by the genetic operators.

• **The return instruction** forces the system to leave the function and return program control to the calling procedure. If variable length programs are desired, then the return operator could be allowed to move within a range defining minimum and maximum program size.

• **The function body** consists of the actual program that evaluates the function.

The Instructions could in this context be divided into to classes: one class where a constant operand immediately follows the operator and another class where the operations are performed between registers or as unary functions and thus do not need an immediate operator constant. It is important to distinguish between these two types when applying the genetic operators to an instruction, see below.

14.2.4 The SPARC Architecture

Due to the specific characteristics of the SPARC processor, the compiler is, when certain criteria are fulfilled, able to optimise a function call in such a way that the header and footer part is not needed [sun 1988, 1990]. The processor has multiple internal register sets that work like a limited stack, and thus processing the stack in the header and footer becomes unnecessary. All examples in this report use this kind of simplified function call. The SPARC architecture is a 32 bit RISC architecture (Reduced Instruction Set) [sun 1988]. This means that instructions are simple, fast and normally limited to 32 bits in length. The return function is placed second to last in the function, not last as is most common. This enables the processor to fetch one more instruction to be performed during the jump back to the calling procedure. The fact that all instructions are 32 bits long means that you obviously can not fit both an op code and an operand of 32 bits within the 32 bits of instruction length. This implies that an immediate operand in SPARC only can have a smaller operand size, actually 13 bits, the remaining 19 bits are used by the operator, see figure 3 below. This is in spite of the fact that the ALU (Arithmetic and Logic Unit) normally operates with the precision of the full 32 bits. To perform these full 32 bit operations the operand must be taken from an other register or memory unit. This technique is the main disadvantage of this architecture used for the classification function regression described below. A CISC processor (Complex Instruction Set) like the Intel x86 in a PC solves this problem differently, by allowing instructions with variable sizes. This makes the architecture perform faster for our type of application. The SUN architecture however has other advantages like stability and function call simplicity. These features convinced us to choose the SUN architecture for our tests.

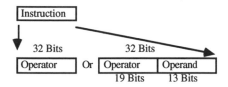

Figure 14.3
The structure of a simplified SUN SPARC instruction

Table 14.1
The function set. Selected Machine Code Instructions for SUN-4.

ADD (imm),ADD (reg2)	Add register one to an immediate operand or add register one to register two.
SUB (imm),SUB (reg2)	Subtract register one from an immediate operand or subtract register one from register two.
MUL (imm),MUL (reg2)	Multiply register one by an immediate operand or multiply register one by register two.
DIV (imm),DIV (reg2)	Divide register one by an immediate operand or divide register one by register two.
OR (imm),OR (reg2)	Logical OR of register one and and immediate operand or logical OR of register one and register two.
AND (imm),AND (reg2)	Logical AND of register one and an immediate operand or logical AND of register one and register two.
XOR (imm),XOR (reg2)	Logical EXCLUSIVE OR of register one and an immediate operand or logical EXCLUSIVE OR of register one and register two.
SETHI	Set the high bits of a register, used when an operand bigger than 13 bits needs to be loaded into a register
SRL	Logical shift right of register one to an immediate operand or logical shift right of register one a number of steps defined by register two.
SLL	Logical shift left of register one to an immediate operand or logical shift left of register one and register two.
XNOR	Logical EXCLUSIVE NOR of register one to an immediate operand or logical EXCLUSIVE NOR of register one and register two.

14.2.5 The Instruction Set

To limit the search space and to avoid complex control mechanisms and thus achieve maximum performance and get our CGPS of the ground, we have limited the set of machine instructions in a number of ways. We have used only two registers and only those machine code instructions of two addressing mode types. All of them operate internally in the processor. The first addressing mode takes one argument from memory immediately afterwards and performs an operation on this argument and a processor register and leaves the result in a register. The other takes the operand from the second register, performs the operation and leaves the result in the first register. With these kinds of instructions it is possible to reach the maximum processor throughput. An average individual program consisting for instance of a dozen instructions of this type can be executed in one fifth of a micro second on a 486 PC or a modern workstation, and the task that the genetic programming aims at, for instance the classification, is performed in the same amount of time. We have also constrained the problems to those that could be solved by a function taking an integer as argument and returning an integer. This however does not limit the problem space to mathematical problems as illustrated below. No control instructions like jump instructions are allowed which means that no loops can be formed. These limitations reduce the complexity and thus execution time of the individual programs. Table 14.1 lists the used operators used and summarises their usage. For further details see "Sun-4 Assembly Language Reference Manual",[sun 1990]. There is a total of 24 instructions.

Note that these functions can and are used to implement other common processor functions. Loading of operands into registers is performed by the OR instruction, clearing a register is performed by the AND instruction and the negation instruction can be performed by the XNOR instruction etc.

14.2.6 The Genetic Algorithm

The approach of machine code individuals could be applied to any type of genetic algorithm. We have selected some straight forward variants of the three basic genetic operators:

• **Selection**: Having tried more standard selection algorithms like proportional selection, we decided to chosen a simple form of steady state [Reynolds 1992] tournament selection. The procedure is simple and efficient and works as follows:

1 Randomly pick four individuals from the population.

2 Evaluate them in pairs, two at a time according to their fitness.

3 Let the two winners breed.

4 Replace the losers with the children of the two winners.

5 Repeat step 1-4 until the success predicate is true or the maximum number of tries is reached.

• The **mutation** operator has two parts one working on the operator and one on the operand. The one working on the operand randomly changes one bit of the operand if certain criteria are fulfilled. Before this can be done a check has to be made to see if this instruction has an operand. The other part working on the operator changes it to a member of the set of approved instructions to assure that there will be no jumps, illegal instructions, bus errors or loops etc. It also assures some arithmetic consistency, where for instance division by zero is prevented.

• The **crossover** is a standard uniform crossover. It works down to the bit level for operands, but preserves the integrity of instructions by preventing a crossover in the operand part of an instruction. A uniform crossover is always performed at instruction boundaries. If both of the instructions at crossover locus have operands then a secondary crossover locus is generated this time at the bit level of the operands, see figure 14.4. A second uniform crossover is performed on the operand. This technique emulates a crossover on one long continuos bit string. (Note that the operands are actually the thirteen last bits of the instruction as stated above, while figure 14.4 only illustrates three operand bits).

Figure 14.5 shows a flow chart of the system and gives further details of how the genetic operators are applied and in what sequence they are performed.

14.2.7 Comparison between CGPS and interpreting GP Systems.

Benchmark tests between the CGPS and an equivalent Genetic Programming system written in the interpreting language Prolog show that the compiling system executes at least 1000 times faster than the interpreted one. This ratio is the same for the individual classification program (with potential use in an application). The Prolog version used was Sictus Prolog for SUN-4.

Tests of the 'LITTLE LISP' system, in Common LISP [Koza 1992, Appendices B and C] show a similar performance advantage for the CGPS, more than 1000 times faster then the LISP system.

We are currently working on benchmark tests of GP systems written in the 'C'-language.

14.3 A Genetic Programming System for Heuristic Classification

We have applied our Compiling Genetic Programming System to a heuristic classification task. There are no details in the implementation that imply that this is a more suitable class of task than any other normally addressed by Genetic Programming. The reason for choosing this task is a combination of application interest and the ability to compare the system to a more established classification paradigm, Neural Networks. The goal is to evolve a classification function generating binary machine code and be able to classify objects from a domain learned by supervised learning. The result is intended to be a function that is able to generalise and classify examples that did not appear in the training set, but that presumably had something in common with the examples in the training set. The machine code functions, the individuals in the population take a 32 bit integer as input and always return a 32 bit output.

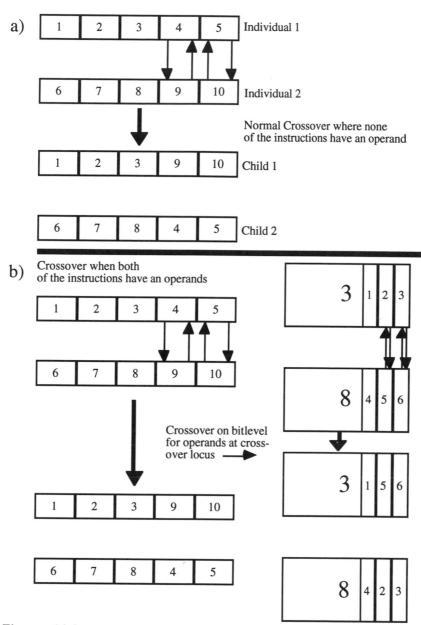

Figure 14.4
Two Types of Crossover

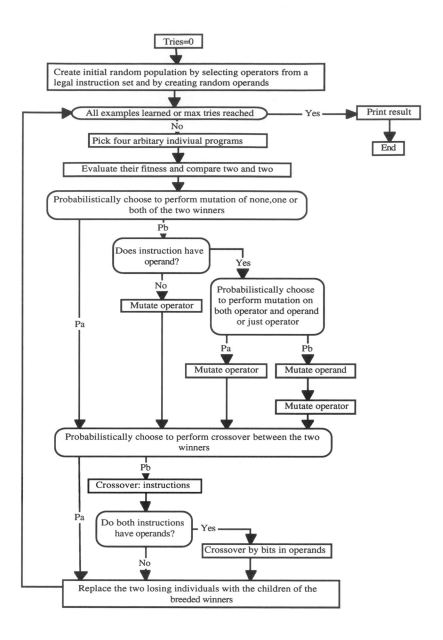

Figure   14.5
Flowchart of the CGPS

Table 14.2
Summary of the Noun Classification Problem

Objective:	Find a machine code program that realises a classification function able to classify Swedish words into nouns and non-nouns.
Terminal set:	Integers of max. 13 bits
Function set:	Machine code instructions for SPARC SUN-4. ADD, SUB, MUL, DIV, SLL, SRL, OR, AND, XOR, XNOR, SETHI.Between two registers and with an immediate 13 bit operand
Raw fitness:	The sum taken over 100 fitness cases of the absolute value of difference between the actual and the desired value of the classification function. Half of the fitness cases are nouns and half non-nouns, all coded as integers. If the integer represents a noun the desired function output is 0. If it is a non-noun the desired output is 65535 (2^16-1)
Standardised fitness:	Same as raw fitness for this problem.
Hits:	The number of correctly classified words. Where a word is considered to be correctly classified if it is smaller that 32768 for a noun and bigger for a non-noun.
Wrapper:	None. (Disassembler)
Parameters:	M=20-4000 (40000). Steady state. Number of individuals to be processed = 400000.
Success predicate:	A machine code program scores 0 as fitness value.

There is a training set divided into positive and negative examples. The goal of the function is: if it is called with a positive example as argument it then should return a value as low as possible, ideally zero. Similarly, when the function is called with a negative example, it should return the highest possible value. The output from the function is masked, i.e. it only returns the 16 least significant bits. In reality the highest output from the function individuals is $2^{16}-1$, i.e. 65535. This is thus the ideal output for a negative example. The fitness for a single positive example is simply the value returned, while the fitness for a negative example is 65535 subtracted from the value returned. The fitness function for an individual program is the sum of these single fitness case values over the complete training set.

This approach could be applied to a number of problem domains and we have chosen an initial evaluation of the task of machine parsing of natural language.

14.4   Comparison between the CGPS and a Neural Network

14.4.1   The Sample Problem

After testing our system with various simple standard problems, we decided that we wanted a problem that was more difficult and more interesting. We looked for a sample problem suitable for the comparison of CGPS and a connectionist paradigm. We wanted a problem with few known rules and with fuzzy boundaries, that could not be labelled as a trivial problem. We searched for a domain that had complex dependencies and were the optimum solution was unknown. However we wanted it to have some practical relevance and application. Finally we chose the task of classifying words into noun and non-noun using the spelling of the words only and not relating the words to any context or other information. It is not the intention here to make a complete investigation and evaluation of this problem space, neither to make a complete comparison between CGPS and Neural Networks. This test gives an illustration of the potential of a CGPS and its relation to a more established technique for heuristic classification. Table 14.2 summarises this problem according to conventions used in the book "Genetic Programming" [Koza 1992, Chapter 7].

14.4.1.1   The Training Set

The training set consists of 2100 Swedish words. These words were sequentially collected out of short newspaper articles. Very little processing was done after extracting them from these articles. Proper nouns were deleted but duplicates and homonyms were kept. The original order of the words was kept. This method is used to mimic the environment of a real application. From this large set of words, twenty-one training sets were extracted each consisting of 100 words, 50 nouns and 50 non-nouns. Every training set thus consisted of one negative training set and one positive training set of 50 words each. This ASCII information then had to be transformed and coded into numerical form to fit the requirements of the CGPS and the Neural Net. The coding was done in two different ways each reflecting the specific needs of the two paradigms.

14.4.1.2   Coding of Words for the Genetic Programming System.

As stated above the individual functions in the population of the genetic programming system only take a 32 bit integer as input. To squeeze the words into this compressed format we use a representation of 5 bits per letter, which gives a total of six letter maximum in every word to be used as input. If we have a word longer than six letters we use the last six letters only. We suspect that most information about a word's class is manifested in its last letters.

Table 14.3
Code Number for Letters in the Swedish Alphabet

Space=0	A=1	Å=2	O=3	Ä=4
Ö=5	E=6	I=7	U=8	Y=9
B=10	P=11	D=12	F=13	G=14
H=15	J=16	C=17	K=18	L=19
M=20	N=21	Q=22	R=23	S=24
T=25	V=26	X=27	Z=28	

For the coding of single letters into five bits we apply a straightforward method where every letter is given a number according to its place in a list of the twenty-eight letters in the Swedish alphabet. This list is composed to reflect some similarities between letters, vowels are for instance placed in the beginning of the list, see table 14.3.

These codes are then combined to 32 bits giving five bits to each letter, where the five least significant bits come from the last letter of the encoded word. More formally it can be represented as follows:

$$(14.1) \qquad\qquad C = \sum (32 \cdot 2^i \cdot l_i)$$

Here C represents the code number to be fed into the individual function in the population, and l_i is the number from table 14.3 where l_0 corresponds to the last letter of the word to be encoded. The most common Swedish word 'och' meaning 'and' will be coded to the number 3631 which thus frequently occurs in the non-noun training set.

14.4.1.3   Coding of words   for   the Neural Network

Coding the words for the Neural Network is somewhat less complex. The network has six input nodes one for each letter. The input to each node is computed by the formula:

$$(14.2) \qquad\qquad C_i = 0.033 \cdot l_i$$

Where C_i are the values fed into the input nodes. Note that the coding for the neural network is an advantageous since the coding divides the words by letter boundaries while the coding for the CGPS make no such distinction.

14.4.2   The Neural Network

We evaluated a few different types of Neural Networks and topologies before choosing the configuration and topology below. This combination gave a fair balance between learning capability, generalisation ability and efficiency.

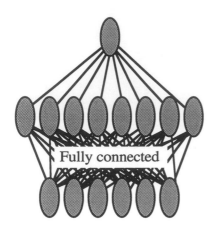

Figure 14.6
Topology of the Neural Network

The neural network is a standard three layer perceptron network. It has six input nodes corresponding to the letters in the words. It has eight hidden nodes fully connected with the input nodes and one output node, see figure 14.6.

The nodes have a sigmoid activation function and the weights are trained by the backpropagation delta rule with momentum. The table below shows a summary of the parameters used for the training algorithm.

Low End Initialisation Range	-0.1
High End Initialisation Range	0.1
Succes Criteria	0.3
Momentum	0.9
Weight Learning Rate	0.1
Bias Learning Rate	0.1

14.4.3 Training Method

The training and comparison was performed as follows: The twenty-one training sets were iterated through and used to train the CGPS and the Neural Network. The curves of the fitness function and the mean square error were plotted and the point where the derivative of the curves flattens out was measured. Figure 14.7 and 14.8 show two typical instances of these curves, one for the Neural Network and one for the Genetic programming system.

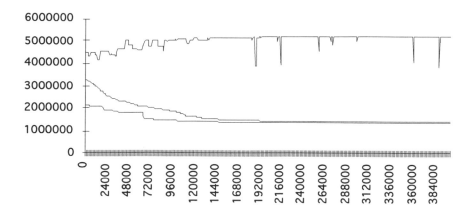

Figure 14.7
Training of CGPS with Population Size 4000 Individuals. Showing Number of Evaluated
Individuals and Their Fitness.

Figure 14.7 shows the development of the fitness value over a training run of 400000
individual evaluations. The figure shows three different curves. The lowest curve
illustrates the best individual in the population. The middle curve shows the average
fitness in the population and the curve at the top shows the worst performing individual
in the population. The population size was 50 in this run, see also Figure 14.9 and
14.10.

When the population has stabilised, the best individual is chosen and the number of
correct classifications are measured. The program is then tested on one of the other 20
training sets which it has not been trained on. This gives a value of generalisation
capability. The resulting Neural Network is also applied to an unknown validation set to
measure generalisation for this paradigm. A mean value over all twenty-one training sets
is calculated for classification in the current training set and in a fresh unknown
validation set.

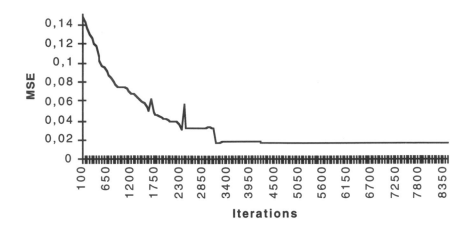

Figure 14.8
A Typical Training Output from the Neural Network Showing
Mean Square Error and Epochs.

14.4.4 Results of comparison

The results shows that both paradigms have about the same ability to learn the training set, an average of 89% of the training examples were learned by the Neural Network compared to 86% for the CGPS. When confronted with words that it has not been trained on, the Neural Network correctly classifies 69%, on average. The CGPS however classifies 72% correctly in the fresh validation set. These measurements are made with a population size of 4000 and an individual size of 12 instructions.

The average training time was 235 minutes for the neural network compared to approximately 1 minute for the CGPS. All times are clocked on a SUN SPARCSTATION 1+. A CGPS system clearly outperforms the network paradigm with regards to training time in this problem space. This should be considered in the light of the fact that the network has a more refined training algorithm while the implementation of the CGPS at this stage is to be thought of as a prototype system, where there still are many open questions regarding further performance improvements.

The training algortihm looks at 4000 examples to arrive at the final network and the CGPS processes 4000000 examples. In spite of this fact the genetic system is much faster. One possible enhancement for the future could be to look at different strategies how to present the training examples to the Genetic Programming system. In our work every one of the 100 fitness cases are iterated through in each individual evaluation. There are good reasons to believe that this number could be varied using just a few randomly picked samples from a large training set in every individual evaluation. Genetic Algorithms have often earlier been proved to perform well with an overall faster convergence time, in a noisy fitness environment as for instance in image processing [Fitzpatrick 1984].

The memory consumption of a CGPS individual consisting of 12 instructions and 32 bits per instruction is about 50 Bytes for a complete classifying program. The neural network needs 450 Bytes if it is written in assembler on a computer with a floating point processor. On a computer without a floating point processor a stand alone implementation of a neural network requires tens of thousands of bytes for linked libraries etc.

The classification time is less than one micro second with our hardware and the CGPS. The network could approach 500µs in assembler on a floating point processor. On a computer without a floating point co-processor it runs many times slower. The ratio is thus at least 500 times faster classification time. On a faster modern workstation, it is possible to run the CGPS individual 10 times faster, matching performance only otherwise achieved by hardware solutions.

Table 14.4 shows a summary of the comparison between the Compiling Genetic Programming System and the Neural Network.

14.4.5 Population Size and Efficiency

To investigate the impact of varying population and individual size we made a complete run over the twenty-one training sets varying the population size as follows: 10, 50, 100, 200, 300, 500, 1000, 1500, 2000, 3000 and 4000. We also made runs with a population size of 40000 on single training sets. The individual sizes used were 7, 12 and 28 instructions. Figure 14.7 shows a typical run with 4000 individuals in the population. Figure 14.9 and 14.10 show diagrams with 1000 and 50 as population size, note the different scales.

Table 14.4
Summary of comparison between a Neural Network and a CGPS.

	Neural Network	CGPS
Training time	235 min	1 min
Average percent of examples learned in training set	89%	86%
Average percent of examples correctly classified in unknown example set	69%	72%
Number of processed examples	4000	4000000
Storage consumption	450 ~20000 Bytes	50 Bytes
Classification time	500μs	1μs

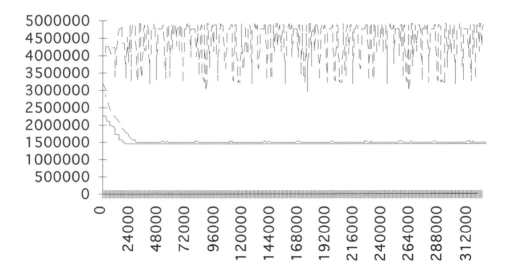

Figure 14.9
Training of CGPS, Population Size 1000 Individuals.

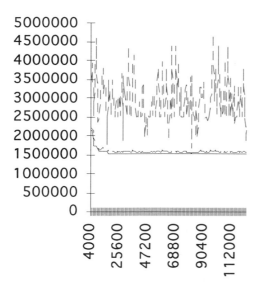

Figure   14.10
Training of CGPS, Population size 50 individuals.

The conclusion of this work is that a smaller population converges faster but with a higher fitness value, while a big population converges slower with a lower fitness value, of the best performing individuals.

14.5   Applicability

We have made some speculations about possible areas of application. The ability to evolve small subroutines in machine code could be of use where there is a need for fast execution and/or small size. Examples of environments with limited performance and memory capability could be handheld devices in consumer electronics running on one chip processors. The small and fast training algorithm gives the possibility of 'on-board' training in most environments.

On the other side of the spectrum there are applications with tight execution time limits for instance fast real-time system control. The ability to evolve fast but complex solutions in 'fuzzy' problem spaces makes pattern recognition a potential application area, as well.

14.6 Future Work

This report describes the initial results and tests with our CGPS. There are a number of questions we would like to investigate:

• How can the input capacity be increased?

• Which parameters affect generalisation capability?

• How will extended memory storage in the individual programs affect performance?

• Which real life applications are there?

To answer these questions we would like to study the following issues:

• The use of more processor registers in the code.

• Control structures like loops and simple recursion.

• Evolving programs that use RAM-memory programs.

• Programs with customised operators and functions in the form of fixed subroutines called at the machine code level.

• Investigation of automatic function definition and other methods to achieve automatic modularisation of programs.

• Evaluate program length constraints and the effect on generalisation capability [Li 1992].

• Different ways of increasing the input capabilities of the programs, for instance by using queue and stack primitives.

We would also like to try the technique on a real world problem. We are currently looking at a prediction problem in a real time petrochemical process environment, where tough execution time constraints are typical.

   We are working on benchmark tests against genetic programming systems written in 'C'-language.

14.7 Concluding Remarks

In this paper we have described a way to build a Compiling Genetic Programming System that evolves and manipulate the lowest of level binary machine code. We have shown that this CGPS is about 1000 times faster than a GP system written in an interpreting language. The CGPS can be used for heuristic classification, and competes with the more established paradigm Neural Networks. The learning and generalisation capabilities compare well with Neural Networks and the training time and classification time is more than two magnitudes faster with the CGPS, in our example. This performance combined with very low memory consumption, might make the system suitable for real time applications in low speed hardware or in applications where fast execution is required, like handheld devices, real time pattern recognition and process control.

Biblography

BYTE (1993) MAGAZINE, MAY, "INTEL LAUNCHES ROCKET IN A SOCKET",

Koza, John R. (1992) *Genetic Programming* MIT Press 1992

Keith,Mike J. (1993) and Martin C. Martin "Genetic Programming in C++: Implemtation and Design issues" chapter 13 of this book,

sun microsystems (1988) *A RISC Tutorial*, Part Number 800-1795-10

sun microsystems (1990) *Sun-4 Assembly Language Reference Manual*, Part Number 800-3806-10

Reynolds (1992) "An Evolved, Vision-Based Behavioral Model of Coordnated Group motion" in From Animals to Animats 2: *Proceedings of the second International Conference on Simulation of Adaptive Behavior*, J.A. Meyer, MA: MIT Press

Fitzpatrick,J.M. (1984) and Grefenstette, Van Gucht, "Image registration by genetic search", in *Proc. of IEEE Southeast Conf ,pp 460-464*

Li,Ming (1992),Paul Vitanui ,"Inductive Reasoning and Kolmorogov Complexity" in *Journal of computer and System Sciences*, pp343-384

III INNOVATIVE APPLICATIONS OF GENETIC PROGRAMMING

In this section, a wide range of applications of genetic programming are presented. Many of these contributions also include powerful and generally applicable techniques for improving the power of genetic programming. Chapter 15 leads off the section by evolving a program to enable a six-legged insect to walk in a simulated environment, using a minimum of *a priori* knowledge about the task of locomotion. In every trial of every experiment, genetic programming evolved programs capable of efficient walking, and often the solutions were better than that of hand-coded programs. Several enhancements to the genetic programming system are also presented with applicability to problems other than evolving walking behavior.

The *double auction* is the mechanism behind the minute-by-minute trading on many futures and commodity exchanges. Since 1990, double auction tournaments have been held by the Sante Fe Institute, where the competitors in the tournaments are strategies embodied in computer programs written by a variety of economists, computer scientist and mathematicians. In Chapter 16, genetic programming is used to evolve strategies superior to many hand-coded strategies from the Sante Fe Institute playoffs.

Chapter 17 looks first at an engineering design problem, that of the design of a filter for removal of noise from experimental data. A filter designed by genetic programming is compared to one designed by heuristic search and another designed by a classical genetic algorithm. The filter designed by genetic programming is shown to be superior. In addition, genetic programming is used to fit empirical equations to a chaotic time-series (the Mackey-Glass equation) and non-linear physiological data. The key role of the fitness function is demonstrated, along with several other conclusions relevant to non-linear time series experiments.

Planning is examined in Chapter 18, where genetic programming is used to create plans for tasks in a simulated world. An illustrative discussion of the creation of a fitness function is followed by a comparison of plans resulting from classical genetic programming and solutions to the same problem using automatically defined functions. An analysis of several successful individuals from the runs using automatically defined functions illustrates interesting sub-goal generation behavior.

Competitive fitness functions can generate superior performance for some problems, but in situations where the training cases for the fitness function are derived from real world data, actual evolution of the fitness cases is not possible. Chapter 19 presents a technique where competitive fitness functions can be used even with a fixed and non-evolvable set of training cases, illustrated by evolving decision trees for word sense disambiguation of natural language.

Chapter 20 discusses two experiments designed to test the applicability of genetic programming to the analysis and the unconstrained construction of pseudo-random number generators, or randomizers. The first experiment attempts to unravel the structure of a

number of generators recommended in the literature by guessing their output given previous output data. The second examines the possibility that co-evolving populations of programs in a competitive environment can produce randomizers which conform to specific
criteria without explicitly testing against those criteria as part of the evolutionary process.

Genetic programming is used to assign confidence measures for another machine learning paradigm, *memory based reasoning* (a k-nearest neighbor method) in Chapter 21.
News stories are given keywords using memory based reasoning, and genetic programming is used to evolve an algorithm designed to distinguish between news stories which
have correctly been assigned keywords, and those which have not and must be referred to
a human editor for classification. The algorithm evolved by genetic programming is superior to the best hand-coded algorithm for the same task.

Chapter 22 presents a system which evolves 3-D models over time, eventually producing novel models that are more desirable than initial models. Simulation results for 3-D jet
aircraft model evolution illustrate that this approach to model design and refinement is
feasible and effective. This approach is intended for use in automatic object recognition
systems, though the model fitness is currently determined by interactive human selection.

The automatic discovery of 2-dimensional features is the application examined in Chapter 23. It presents a hybrid system, with 2-dimensional hit-miss matrices evolved by a
genetic algorithm and these matrices then utilized by a program evolved with genetic programming. This system is applied to a subset of the problem of handwriting recognition,
specifically digit recognition, and individuals are evolved which can recognize very low
resolution digits. Possibilities for expansion into a full-size character recognition system
are discussed.

Chapter 24 presents a novel approach for the evolution of neural networks using genetic
programming. *Cellular encoding* is presented as a method for encoding families of similarly structured boolean neural networks, and is shown to be a particularly rich and effective representation mechanism for the evolution of such networks. This result is then
generalized in a discussion of the appropriate representation language for a variety of
tasks.

15 Automatic Generation of Programs for Crawling and Walking

Graham Spencer

This chapter demonstrates the use of genetic programming to automatically create programs capable of walking. Our goal was to generate programs that enabled a six-legged insect to walk in a simulated environment. Furthermore, we wished to accomplish this goal using a minimum of *a priori* knowledge about the task of locomotion. In every trial of every experiment, genetic programming evolved programs capable of efficient walking. Moreover, the performance of the evolved programs was at least as good and often better than that of hand-generated programs. One feature of our solution was an extensive use of side-effecting functions; we show that such functions simplify the genetic programming paradigm and enable a more powerful machine model. We also introduce a new genetic operator, constant perturbation, that allows the genetic programming system to fine-tune floating-point constants as part of the selection process.

15.1 Introduction

The problem of controlling a six-legged robot has been solved with many different approaches. Rodney Brooks and others have used a subsumption architecture in building a walking robot [Brooks 1989]. Randall Beer has demonstrated that a small neural net can be used to control locomotion in a simulated cockroach [Beer 1990]. We examined the same problem of six-legged walking; however, both the goals and the methods of our research were different. Our goal was to automatically generate programs that enabled a simulated six-legged insect to walk, and to accomplish this automatic generation with a minimum of assumptions about how walking should occur. Our method was genetic programming, or the controlled evolution of a population of programs.

Despite these differences in goals and methods, we attempted to follow the same physical model as Beer. This model describes a simulated two-dimensional insect acting under simplified laws of motion. Also, we judged the success of the generated programs using criteria similar to those used by Brooks: programs that caused the insect to travel farther were considered superior. We ran three experiments, with each experiment providing progressively less *a priori* information about how to walk. In every trial from all three experiments, a program evolved that was capable of sustained movement. Moreover, the performance achieved through genetic programming was at least as good and often better than that of hand-generated programs.

15.2 The problem

We have attempted to follow Beer's physical model as described in *Intelligence as Adaptive Behavior* [Beer 1990]. This model features an artificial insect with six legs,

Figure 15.1
Orthogonal and front views of the artificial insect.

each of which may be up or down. The insect can raise or lower its legs, and apply forces to them. When the leg is down, the forces on the legs translate into forces that move the body of the insect; when the leg is up, the forces on the legs swing the legs into different positions. The insect is unstable when its center of gravity is outside of the polygon formed by the feet that are down, and the insect falls when it has been unstable for a certain period of time.

We kept our physical model as close to Beer's as possible. However, because we provided less *a priori* knowledge about walking, many individuals in the population ignored one or more of their legs for the duration of the simulation, either leaving them up or dragging them. To make our simulation more realistic, we added a penalty for legs that dragged: each dragging leg reduced the speed of the insect by 25%. Also, we found it impossible to accurately model rigid legs in only two dimensions. To solve this problem, we chose to provide an imaginary joint in each leg that moved in a plane perpendicular to the plane of the simulation; this joint was never specifically controlled, but allowed the effective length of the leg in the plane of the simulation to vary within specific limits. Figure 15.1 shows an orthogonal view of the bug and a front view that demonstrates joint positions and the corresponding effective leg lengths.

15.3 The approach

We used genetic programming [Koza 1992] to generate programs to control the artificial insect. The programs were represented as expression trees. Each internal node of the tree was a function with 1, 2, or 3 arguments. Each leaf of the tree was a terminal that was either a floating-point constant or a variable representing a sensory input.

Table 15.1
Tableau for the problem.

Objective:	Evolve a program capable of causing a simulated insect to walk.
Terminal set:	Randomly generated floating-point constants. Depending on experiment, can also include **Oscillator** and **Get-Leg-Force-*n*** terminals.
Function set:	~, +, -, *, /, **min, max, fmod, if, Set-Leg-Force-*n*,** and **Set-Leg-State-*n***
Fitness cases:	Controlling the simulated insect with the evolved program.
Raw fitness:	The distance traveled by the insect under control of the program.
Standardized fitness:	Same as raw fitness.
Hits:	Not used.
Wrapper:	Not used.
Parameters:	Population size = 1000. Generations = 50, 65, or 100 depending on experiment. Selection by 6-way tournament.
Success predicate:	Whether the program was able to walk. However, once walking was achieved, programs were ranked by the distance they traveled.
Rules of construction:	None.

15.3.1 Functions and terminals

Most of the functions and terminals used in our experiments were conventional. We used the standard mathematical functions: ~ (unary negation), +, -, *, / (protected), **min**, **max**, and **fmod** (floating point modulo). The ternary **if** function returned its second argument if its first was positive; otherwise it returned its third argument. The unreturned argument was not evaluated.

In addition to floating-point constant terminals, two of our three experiments also used the **Oscillator** terminal. This terminal returned $10 \sin(t)$, where t was the number of elapsed time-steps since the beginning of the program's execution.

15.3.2 Side-effecting functions and simulated memory

One way in which our experiment differed from typical genetic programming applications was the heavy use of side-effecting functions. To control the state of the insect, we needed at least 12 outputs: 6 outputs to control the force applied to the legs, and 6 outputs to control whether the legs are raised or lowered. However, an expression tree used conventionally returns only one output: the value of its root node. One way to obtain 12 outputs is to simply assign one complete tree to each output. In other words, represent each program by a list of expression trees, with each tree computing one value in parallel with the other trees. This solution has two main limitations: it is complex because it requires us to maintain many trees per program, and it prevents the trees from sharing information.

The solution we chose instead was to use side-effecting functions. These are unary functions that evaluate their argument, set the force or position of a leg based on that argument, and then return the argument as their value. The argument 'passes through' the function unchanged, but the side effect is to move a leg. For each leg, there was a **Set-Leg-Force** function that set the force on the leg and a **Set-Leg-State** function that raised or lowered the leg depending on the sign of its argument.

Such side-effecting functions offer several advantages:

• Control for all six legs can be distributed throughout a single program; multiple branches are not needed.

• The order of leg movement can be set by the program. With multiple trees, the order in which legs are moved is fixed by the order in which the external genetic programming system evaluates the twelve leg branches.

• Intermediate results can be shared by several legs, because one leg can be set based on the force assigned to another leg. For example, (**Set-Leg-Force-1** (~ **Set-Leg-Force-0** (...))) assigns to leg 1 the negative of the force assigned to leg 0.

As a complement to the **Set-Leg** functions, in our later experiments we also provided **Get-Leg** terminals for each leg. These terminals returned the current force applied to the leg. Because we provided both reading and writing functions for the states of the legs, and because the states maintained their values through the course of the evaluation of the program, the states were equivalent to a conventional memory system (albeit a small one). The memory system analogy suggests an alternative to our abstract model of side-effecting functions: we can view the side-effecting functions as simply loads and stores operating on the memory of a small virtual machine. At each time-step elements of this memory are set by external sensors and read to control external devices (the legs). In fact, this is the way the programs were actually implemented in our system. We believe this model is significant, and we will discuss it further in section 15.3.5.

15.3.3 Constant perturbation

Another distinguishing feature of our experiment was the additional genetic operator we introduced. We used the standard operators of mutation (with a probability of 0.5%) and crossover (with a probability of 75%). However, we also used the *constant perturbation* operator with 25% probability. When this operator was applied to an expression tree, every floating-point constant in the tree was multiplied by a random number between 0.9 and 1.1; this had the effect of perturbing the constants by up to 10% of their original value. We feel that constant perturbation had two important positive effects on the efficiency of genetic programming in our experiment.

First, it allowed the genetic programming paradigm to perform local gradient searches as part of the reproduction mechanism. When invoked, the operator nudged individuals in a random direction through the space defined by the constant coefficients in the program. Such perturbation allowed the 'fine-tuning' of the coefficients in programs. Fine-tuning is difficult to accomplish through mutation alone, because a small deviation in a constant must be accomplished by replacing the constant with a similar one—an improbable event when mutation generates an arbitrary subtree. Furthermore, mutation tends to be a fairly destructive operator in that it can radically disrupt program structure; thus a high degree of fine-tuning by mutation would necessarily be accompanied by an even higher degree of 'coarse-tuning' as well. But constant perturbation has no effect on the overall structure of the program, and can therefore be applied frequently enough to explore the coefficient space.

Second, constant perturbation enriched the set of terminals by adding new constants that could be propagated through the population by crossover. One problem with using only the conventional operators is that they rely primarily on crossover to create new constants by combining existing constants with arithmetic operators. Mutation can also introduce new terminal constants; however, as we discussed in the preceding paragraph, mutation rates are necessarily low, so this effect is not significant. But since constant perturbation can be invoked frequently, it results in a rich range of coefficients that can be shared through crossover.

We believe this is an improvement over the constant creation used by Koza [Koza 1992], although the degree of improvement naturally varies with the problem domain. Earlier experiments with symbolic regression showed that constant perturbation can provide dramatic performance improvements for that particular application. Although we performed no quantitative tests to evaluate the usefulness of constant perturbation for the problem of locomotion, simple qualitative comparisons early in our research suggested that perturbation was beneficial.

15.3.4 Fitness evaluation

The fitness of each program was obtained by allowing the program to control the simulated insect for 500 time-steps. At each time-step, the program was evaluated, and the side-effects of the evaluation caused the legs to move. (Most expressions set each leg more than once during a time-step; only the last value set during the time-step affected the leg. See 15.3.5 for the effects of intermediate assignments.) The simulation was run until either the bug fell or all 500 time-steps had elapsed. The program's fitness was the distance (in units of simulation space) between its final and starting positions. Our method of fitness evaluation conveniently measured two important characteristics. First, the fitness evaluation gave programs a long time in which to walk; this allowed programs capable of walking to distinguish themselves from programs that fell immediately. Second, once most of the population was 'up and running,' the limitation of 500 time-steps favored those programs that could walk the fastest. Thus, our simple fitness calculation was appropriate for both the initial and final stages of each trial.

Early in our experimentation, we examined an alternate method of determining fitness: measuring the length of a path that an insect walked rather than the distance between start and end points. We rejected this method because it often resulted in degenerate solutions receiving high fitnesses. For example, many early individuals moved by dragging legs on one side of their body and moving only one leg on the other side; this resulted in a program that shuffled about in a circle. Such a program could accumulate a high path length, even though it did not exhibit desirable walking behavior.

15.3.5 Program structure

In section 15.3.2, we showed that the use of side-effecting functions was equivalent to a model based on a virtual machine with simulated memory. We mentioned the advantages of such a scheme in the context of our specific application. Although a full discussion of this idea is beyond the scope of this chapter, let us briefly list the general benefits of using this model:

• A single expression can generate many outputs. This eliminates the complexity of maintaining different trees for each desired parameter.

• Programs can evolve techniques that are found in every human programmer's tool-box—techniques such as temporary variables, intermediate results, and data reuse. Individuals in our experiments have exhibited all three of these devices.

• The genetic programming system gains a level of abstraction. The programs write to generic memory locations or device registers; in turn, the 'wrapper' function polls some memory locations and sets others according to the environment. This allows a cleaner division between the genetic-programming code and the problem-specific code. Successful programs will discover the semantics behind each memory location, just as they must discover the semantics behind each function in a more conventional model.

• It becomes easier to give programs scratch memory. Extra memory slots can be allocated that are neither read nor written by the wrapper function; these slots can be used by the programs for temporary results, etc. Even without such slots, memory locations are often used for temporary results before being overwritten by their final value; this happened frequently in our experiments (which did not allocate scratch memory for the programs).

A corollary effect of our use of side-effecting functions is that the return value of the expression is not used. This means that arithmetic functions early in the expression are relevant not because of the results they return but because of the ordering they impose on the evaluation of their arguments.

To give examples of memory use and ordering, we present a portion of a program that developed in generation 99 in our second set of trials and traveled a respectable distance of 1148.13 units in its allotted time. (Only a portion is shown because the program is too large to include in its entirety).

```
(+ (min (+ -1.59993 (* (* (~(fmod (~(Set-Leg-Force-2 (Set-Leg-State-0 (/ -
3.19355 Get-Leg-Force-2))) -2.29009) 10.5488) 4.98003)) (fmod (min 0.591271 (/
(/ (Set-Leg-Force-1 (~(- Get-Leg-Force-5 (if (min Get-Leg-Force-5 (~Get-Leg-
Force-2) Get-Leg-Force-4 Get-Leg-Force-2))) (Set-Leg-Force-4 (/ (if -42.0987
Get-Leg-Force-4 (if Get-Leg-Force-2 -23.4623 -45.7554)) -3.35099))) ...)
```

To gain an understanding of what this fragment is really doing, we can perform some simplifications. For example, we can fold constants and remove unused branches of **if** statements. Furthermore, we can show the sequence of evaluation of the **Set-Leg** functions in the form of a **(begin ...)** statement. Below is the same code rewritten in sequential form after simplification.

```
(begin
  (Set_Leg_State_0 (/ -3.19 (Get_Leg_Force_0)))
  (Set_Leg_Force_2 (Get_Leg_State_0))
  (if (> (min (Get_Leg_Force_5)  (- (Get_Leg_Force_2))) 0)
    (Set_Leg_Force_4 (- (Get_Leg_Force_4) (Get_Leg_Force_5)))
    (Set_Leg_Force_4 (- (Get_Leg_Force_2) (Get_Leg_Force_5))))
  (if (> (Get_Leg_Force_2) 0)
    (Set_Leg_Force_4 7.0)
    (Set_Leg_Force_4 13.66)) ...)
```

Note that we were able to eliminate the initial sequence (+ (min (+ -1.59993 (* (* (~(fmod ... because it was not an argument to any **Set** functions and thus could not affect the output of the program. Also note the way in which leg 2 is used as an intermediate result: it is set in line 3 and referenced in both line 4 and line 7. Finally, it is worth mentioning that while the nodes we removed have no effect on the expressed traits of the individual, these nodes (analogous to introns in DNA) are nonetheless useful during the evolutionary process because they provide fertile genetic material for use in future crossover operations.

15.3.6 Implementation

It has become increasingly common in genetic programming work to use C or C++ rather than Lisp. Our experiments were no exception; we made this choice for several reasons. First, the simulation portion of our experiment was computationally intensive, and we felt that C++ offered greater speed for such calculations. Second, C++ made it easier to implement simple networking features to allow parallelism across a cluster of workstations. Most importantly, not all of the platforms on which we wished to run our trials offered a dialect of Lisp, but all offered C++ compilers.

15.4 Results

We ran three different experiments (each composed of several trials), and in each experiment it became progressively more challenging for the insect to walk. We will describe the environments shortly, as well as the relative performance of the programs in each experiment. However, details of performance comparisons may obscure the most important result of the experiments, namely that programs capable of efficient, sustained locomotion evolved in all three experiments. This is especially significant considering that *none* of the initial, randomly generated individuals (28,000 in all—1000 individuals from each of 28 trials) were capable of walking. (Note that although no individuals from generation 0 were capable of walking, they often traveled short distances. This is because they fell immediately, but managed to push themselves forward as they fell.)

We began with unrealistic experimental conditions designed to make walking easy. In Experiment 1, we did not allow the `Get-Leg-State-n` terminals. Our intent was that providing only the `Oscillator` terminal as input to the program would encourage oscillatory motion in the legs, even though the programs would be less powerful since they lacked the ability to read from their general-purpose memory (as described in 15.3.5). The `Oscillator` also provided an analogy to the CPG (central pattern-generating) cells used in neural net models [Selverston, Rowat, and Boyle, 1993]. We further tried to facilitate oscillatory motion of the legs in Experiment 1 (and in Experiment 2) by reversing the specified force applied to the leg when the leg was up. Normally, a positive force applied to the leg would move the bug forward when the leg was down, and swing the leg towards the rear of the bug when the leg was up; this is the physically accurate

Table 15.2
Summary of experiments.

Experiment	Trials	Leg Reversal	Get-Leg	Oscillator	Description
1	8	Yes	No	Yes	Parameters chosen to make walking as easy as possible
2	8	Yes	Yes	No	Lack of oscillator terminal requires programs to evolve cyclic motion in the legs
3	12	No	Yes	Yes	Removal of leg reversal makes problem more realistic and more difficult

way to model the legs. However, reversing the force when the leg was up meant that a positive force would swing the leg forward. Thus, a program could successfully walk by applying constant positive force to the legs and raising and lowering them at the appropriate times. This made the task of walking drastically simpler; without the reversal, the program would have to vary the forces on the legs in coordination with raising and lowering them.

In Experiment 2, we maintained the leg reversal, but we removed the **Oscillator** terminal and added the **Get-Leg** terminals. In this experiment, we hoped to see gaits develop only from the feedback of the legs rather than from an external source of oscillation. Finally, in Experiment 3, we used both the **Oscillator** terminal and the **Get-Leg** terminals, but we did not use leg reversal. We hoped that the programs could learn to coordinate the necessary leg force oscillation with the raising and lowering of the legs. We summarize the experiments in Table 15.2.

To give a rough gauge of the performance of the solutions, we created the following program by hand:

```
(+ (Set-Leg-Force-0 (Set-Leg-State-0
      (Set-Leg-Force-2 (Set-Leg-State-2
        (Set-Leg-Force-4 (Set-Leg-State-4 Oscillator))))))
  (Set-Leg-Force-1 (Set-Leg-State-1
    (Set-Leg-Force-3 (Set-Leg-State-3
      (Set-Leg-Force-5 (Set-Leg-State-5 (~ Oscillator))))))))
```

This program traveled a distance of 851.79 units during the time allotted, and we will use it as a baseline for comparison in all three experiments. We should note, however, that all of the best programs from later generations achieved our goal of sustained locomotion for any distance. As discussed above, the distance traveled in the allotted time is a measure of how *efficiently* these programs caused the artificial insect to walk.

Because the hand-generated program uses the **Oscillator** terminal, it would not work in the conditions of Experiment 2; and because it assumes leg reversal, it would not work in the conditions of Experiment 3. However, we will continue to use it as a rough benchmark in the later experiments, since we could achieve similar performance in those environments with additional programming effort.

15.4.1 Experiment 1

Our first experiment used only the input from **Oscillator**, and reversed the forces on the legs when up. The initial random generation did very poorly in comparison to our hand-generated standard. We ran 8 trials of 1000 individuals each in this experiment, and none of the individuals in generation 0 were capable of walking. The average of the distances traveled by the best individuals from generation 0 in each of the 8 trials was only

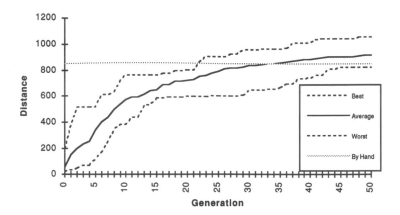

Figure 15.2
Results of Experiment 1, using leg reversal and the **Oscillator** terminal.

57.34. These individuals traveled as far as they did because they dragged four or five of their legs, and used one leg to push themselves until they finally fell. So despite the simplicity of our hand-generated program, this was not an easy problem for randomly created programs.

However, with successive generations, programs capable of continuous walking evolved. These programs had an obvious fitness advantage over programs that simply fell forward. By generation 22, an individual superior to our hand-generated program had emerged: one program traveled 865.58 units. The average distance traveled by the best individuals from each of the 8 trials was 728.78 units. By generation 33, the average of the 8 best-of-trial individuals had exceeded the distance traveled by the hand-constructed program, and by generation 50, the best individuals traveled an average of 915.49 units. The graph in Figure 15.2 summarizes the data about the best individuals from each of the 8 trials by showing the minimum, average, and maximum of these 8 individuals.

15.4.2 Experiment 2

In our second set of 8 trials, we examined the performance of genetic programming when the **Oscillator** terminal was removed. These trials required that the programs use feedback from the legs instead; this feedback was provided by the **Get-Leg** terminals. We expected this problem to be more difficult, and indeed it was.

The initial random generations without the oscillator performed about the same as the initial generations in Experiment 1. The average distance traveled by the best individuals from each initial generation was 35.13, and again, no individual was capable of sustained

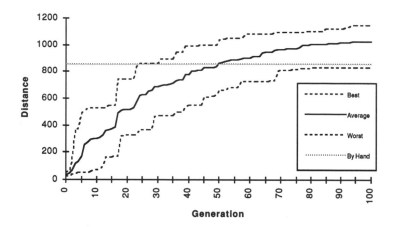

Figure 15.3
Results of Experiment 2, using leg reversal and the **Get-Leg** terminals.

walking. These trials developed more slowly than those in Experiment 1: not until generation 25 did an individual capable of beating our hand-constructed program emerge. By generation 100 (twice the total number of generations in Experiment 1), all but one of the trials had produced an individual better than our hand-generated program, and the best-of-trial average was 1020.99 units. Figure 15.3 summarizes the statistics of the best individuals from the 8 trials (note that the horizontal scale is different from Figure 15.2).

15.4.3 Experiment 3

In our last experiment, we removed our second simplification. In these 12 trials, we did *not* reverse the motion of the leg when the leg was up; this required the programs to coordinate swinging the legs with raising and lowering them. However, relying on the biological precedent that most organisms use both CPGs and sensory feedback [Beer and Gallagher, 1992], we used both the **Oscillator** and **Get-Leg** terminals in these runs.

These conditions proved to be the most difficult. The average distance traveled by all individuals from the first generation was only 0.000048. However, after 65 generations, the average of the best distances from each trial was 840.21. This distance is only slightly shorter than our baseline of 851.79. Figure 15.4 summarizes the results of this experiment.

The best individual in this experiment exhibited a sharp rise in fitness followed by a shallow plateau. In fact, the best individual actually performs better in this experiment

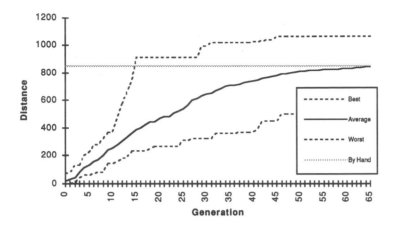

Figure 15.4
Results of Experiment 3, using the full set of terminals but without leg reversal.

than the best individuals from the first two experiments, despite the more difficult conditions. This is because the leg motions used for successful walking *without* leg reversal tend to be more powerful than those required for walking *with* leg reversal, even though the motions required without leg reversal are more complex. Since the necessary motions are more complex, these motions evolved more slowly in the environment of Experiment 3 (as reflected by the slow rise in average best-of-trial fitness); however, these motions conferred a greater fitness advantage than in previous experiments. So the best individual in Experiment 3 evolved the necessary motions early in its trial, and that caused its fitness to dramatically rise in the first few generations. The plateau occurred once the oscillatory motion had spread to most of the legs; the slower rate of increase that followed was due to refinement by crossover and constant perturbation.

15.5 Analysis of the results

Despite the increasing difficulty of the experiments, all three experiments evolved programs capable of walking. Furthermore, each trial followed a similar evolutionary path. Early individuals from each trial had no progressive leg motion; many fell immediately. Those individuals that did manage to achieve motion moved forward by pushing themselves with one foot until they became unbalanced. The chart in Figure 15.5 shows a gait generated by a typical program from generation 0. (Like Beer and many entomologists, we describe the gaits with charts of footfall. There is space in each box for 6 thick horizontal lines, and each of these lines corresponds to one leg of the insect. Where the line is black, the foot is down; where the line is gray, the foot is dragging; and where the line is white, the foot is up. The horizontal scale on our charts is time-steps, not distance.) In this early example, the insect is standing on three of its legs. Only two of the legs that are down are doing any real work; the third foot drags and thus contributes to balance but not to forward motion. The insect's actual path before it fell was semicircular.

However, improvement comes quickly: Figure 15.6 shows three programs (from generations 4, 10, and 18 respectively) that demonstrate the emergence of oscillatory motion in the legs. In Figure 15.6a, two of the program's legs have begun to exhibit oscillation. Although the dragging legs slow the insect down, they also prevent it from falling by helping it balance. This allows the insect to hobble forward, pushed by the two legs that move. The fact that dragging legs still contribute to balance means that oscillation can emerge in just one leg, and later spread to others. The program in Figure 15.6b, from generation 10, is a definite improvement—it moves four of its six legs (although one of

Figure 15.5
Gait of a program from generation 0.

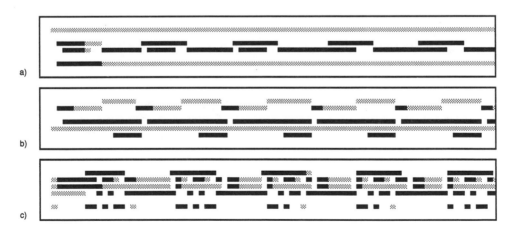

Figure 15.6
Gaits from generation 4 (a), generation 10 (b), and generation 18 (c).

the four only drags when it is down). Finally, the program in Figure 15.6c is moving five of its six legs, and all five of the moving legs exert forward force at some point in the gait.

Once oscillatory motion is prevalent in the population, refinement occurs as the sections of code responsible for oscillation spread to other legs through crossover. Successful individuals are likely to have several subprograms that generate oscillation, and crossover can copy these subprograms to other individuals. More legs become used for motion rather than simply for stability, and the use of the legs becomes more efficient.

Figure 15.7 shows three examples of highly successful strategies present in the final generations of three different trials. Note that there is very little dragging produced by these programs, and that most legs are used efficiently (although the second example completely ignores one of its legs). The diversity of locomotion strategies is especially interesting: Program 15.7a moves in surges by pumping its four outer legs at once; when it lifts these legs, it lowers its middle two legs for balance. Program 15.7b uses five of its six legs in slow, staggered strokes. And program 15.7c uses all its legs in a tightly controlled pattern of short hops. An important observation is that within each program, there are one or two dominant oscillation patterns that are exclusively responsible for the locomotion of the insect. This supports our conjecture that walking behavior evolves as sections of code responsible for oscillation propagate throughout the population. In each trial, a few useful sections of code (fingerprinted by their patterns of oscillation) confer significant fitness advantages to the programs in which they are embedded. This fitness

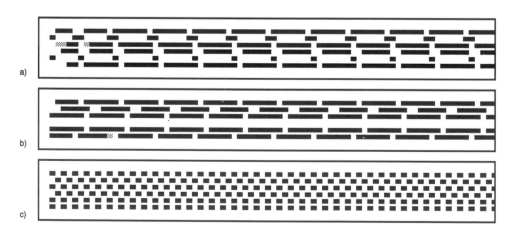

Figure 15.7
Gaits from the final generations of trial 1 (a), trial 2 (b), and trial 3 (c).

advantage in turn increases the propagation of the code fragments, because fit individuals are more likely to be chosen for reproduction and crossover. The gaits in Figure 15.7 demonstrate that by the final generations of a trial, the best individuals have several copies of these valuable oscillator fragments (with possible variations between the fragments due primarily to constant perturbation). Perhaps these code fragments are the genetic programming analogies to schema in genetic algorithms; more study is required to judge this conjecture.

A final observation is that the three evolved strategies shown in Figure 15.7 closely resemble gaits produced by a mechanical simulation of locomotion in the stick insect *Carausius morosus* [Beer, Chiel, Espenschied, Quinn 1993]. In that work, the authors used a hexapod robot to experiment with six-legged walking; by varying the speed of the controller, the authors generated a variety of gaits. The gaits in Figure 15.7 can be matched to gaits produced by different speeds of their controller; for example, 15.7c is similar to the tripod gait (maximum speed), while 15.7a and 15.7b resemble gaits produce by slightly slower controller speeds. It makes sense that our evolved gaits match the faster controller speeds, since speed was our fitness criterion.

15.6 Analysis of the method

15.6.1 Comparison with random search

One common criticism of genetic programming is that it is fundamentally no different from simple blind search. We feel such a claim is false for two reasons. First, out of the 28,000 randomly generated individuals, *not one individual* was capable of sustained locomotion. But in every trial we ran, after 28 generations (i.e., after the evaluation of an equivalent 28,000 individuals), many individuals were capable of continuous walking. Second, an analysis of the evolutionary history of individuals from each trial suggests that the crossover operation had a significant impact on the success of the experiments. It was crossover that allowed successful fragments of code to echo throughout the population and eventually become incorporated multiple times in the most successful individuals.

15.6.2 Comparison with other methods

In describing their work on using genetic algorithms to evolve neural nets capable of controlling locomotion, Beer and Gallagher cite a lack of *a priori* assumptions as an advantage to their approach [Beer and Gallagher 1992; Goldberg 1989]. Genetic programming is another approach that exemplifies this advantage. In fact, the genetic programming approach requires even less *a priori* knowledge than the method of Beer and

Gallagher. In order to keep the size of their neural nets manageable, Beer and Gallagher made a variety of assumptions about symmetries and other relationships between the legs of the insect. These assumptions drastically reduced the size of the problem space to be searched. In our experiments, we made no such assumptions: each leg could be moved independently of the other legs, and any cooperative or symmetrical motion had to be discovered by the evolutionary process. Indeed, we made only one *a priori* assumption about the behavior of our control programs, namely that it would be beneficial to reverse the response of the legs when they were raised. And as our third experiment demonstrated, we could discard even this assumption and still evolve programs capable of walking.

It could be argued that we are implicitly providing *a priori* information through our selection of operators because we provided only those operators necessary to solve the problem and no others. While it is true that we provided no other operators, our reason for this was that there were no other relevant operators that *could* be added to our set. Given the constraints of our physical system, every action that the insect was able to take could be accomplished by combinations of the primitive operators that we provided. In this respect, our lack of other operators made the problem *more* difficult because complex actions (such as repetitive motion) had to be assembled from sequences of more primitive ones. And while it is true that we could have included irrelevant operators, it is not likely that these would have affected our performance given that irrelevant sections of code in the form of unevaluated `if` branches were already common in all of our trials. (An example of this is `(if 13.4 (...) (...))`: the else clause of the `if` statement is never evaluated because the test is always true.)

There are other differences between the genetic programming approach and the approach of Beer and Gallagher. They note that because neural net models coincide so closely with biological systems, such models can give us insight into the operation of such biological systems. On the other hand, since we understand relatively little about the biological systems in question, the artificially generated neural nets are themselves difficult to understand. Genetic programming complements the neural net approach: since control is represented by a program, studies of successful strategies are unlikely to offer insight into the workings of biological systems. However, the programs generated are typically more easily understandable than neural nets because they can often be simplified and can always be translated into conventional programming languages.

15.6.3 Practical considerations

The success of genetic programming for a two-dimensional simulated insect begs two important questions: 1) Is the genetic programming approach scalable to three dimensions? 2) Is genetic programming at all applicable to a real robot?

15.6.3.1 Scalability

While walking in three dimensions is certainly more complex from a modeling point of view, the third dimension adds little complexity to the functions and terminals of our genetic programming system. A three-dimensional leg with two segments has three degrees of freedom: two where the leg meets the body, and one at the joint in the middle of the leg. This would require a total of 18 outputs from the genetic program—only 6 more than existed in our two-dimensional experiments. In other words, we would simply add six extra memory locations to our virtual machine model. Because of this manageable increase in problem complexity, we believe that genetic programming could be successful in three dimensions.

15.6.3.2 Real-Time control

In addressing the issue of real-time control, we must first determine how we wish to apply the genetic programming paradigm to the problem. It is clearly impractical to simply transplant our genetic programming system to a robot, since our system required thousands of trials to arrive at usable solutions—an overnight computing job on a workstation. However, one simple way to use genetic programming for robot control is to create solutions in simulation and then transfer only the best solution to the real robot. The performance of the real robot could be used to refine the physical simulation for future generations. With this approach powerful computers could generate solutions for simple robots. Such a method could be especially advantageous in situations where the robot is too simple to allow even basic learning strategies, much less a full-scale genetic programming environment.

The most significant disadvantage of such a 'transplant' solution is that it is not implicitly adaptive. The neural nets generated by Beer and Gallagher and the subsumption architecture created by Brooks are both effective at reacting to changes in their environment and even to changes in their own structure (loss of a leg, a damaged sensor, etc.). This is dramatically different from the conventional genetic programming approach that generates complete solutions in advance; these solutions are not adaptive and are therefore less robust. However, for many applications, this disadvantage may be outweighed by the benefits of having a complete (albeit brittle) solution in the form of a conventional program. Furthermore, it is possible that the robustness of a program could be increased by incorporating robustness into the fitness function. For example, we could base fitness on the distance traveled during several runs in which different sensors or legs were randomly damaged. This would select for those those programs that could cope with such hazards.

Ultimately, however, we would prefer a solution that can maintain the *static* flexibility of genetic programming (as provided by the ability to efficiently search a huge problem space) while gaining the *dynamic* flexibility of conventional adaptive systems (as measured by the ability to adjust to changing conditions in real-time). Clearly, to obtain such a solution, we must abandon the 'outer loop' of the genetic programming system which requires us to evaluate several generations of thousands of individuals in order to obtain an advancement in fitness (as manifested by each generation's best individual). Furthermore, we would like to abandon the simulation aspect altogether: rather than evaluating a program by its simulated performance, we would like a program's fitness to be based upon its real performance in an actual physical environment. One possible solution to this dilemma is to graft genetic programming onto a classifier system: this would allow complex programs to evolve while exploring a problem space whose parameters are no longer static, but rather change with the robot's environment. This is an area of study that we are currently pursuing.

15.7 Conclusion

At the beginning of this chapter, we stated that our goal was to automatically generate programs enabling a six-legged insect to walk. We achieved this goal, and demonstrated that even in the absence of *a priori* assumptions, genetic programming generates successful and efficient programs capable of making an artificial insect walk in our two-dimensional simulation. We believe that this approach would scale well in three dimensions. Compared with other work in this area, our approach has both advantages and disadvantages. The two primary advantages are that the genetic programming approach requires very little specific domain knowledge and that genetic programming produces solutions that resemble conventional computer programs. The primary disadvantage is that our current implementation of genetic programming is not adaptive.

Much work remains to be done in order to fully explore the application of genetic programming to the generation of control strategies. In particular, the territory of real-time genetic programming is largely uncharted.

Acknowledgments

I am grateful to John Koza for his invaluable advice in the conception and formulation of this project, as well as for his help in preparing this chapter. I am also grateful to John Nagle for providing code to help with the physical simulation. Finally, I thank the participants of the genetic programming list (`<genetic-programming-request@cs.stanford.edu>`) for the informative discussions that occur there.

Bibliography

Beer, R. D. (1990). *Intelligence as Adaptive Behavior: Experiments in Computational Neuroethology*. New York: Academic Press.

Beer, R. D., Chiel, H. J., Espenschied, K. S., and Quinn, R. D. (1993). Leg Coordination Mechanisms in the Stick Insect Applied to Hexapod Robot Locomotion in *Adaptive Behavior*, 4 (pp. 455-468). Cambridge: MIT Press.

Beer, R. D., and Gallagher, J. C. (1992). Evolving Dynamical Neural Networks for Adaptive Behavior in *Adaptive Behavior*, 1 (pp. 91-122). Cambridge: MIT Press.

Boyle, M. E. T., Rowat, P., and Selverston, A. I. (1993). Modeling a Reprogrammable Central Pattern Generating Network in *Biological Neural Networks in Invertebrate Neuroethology and Robotics*, Beer, R. D., McKenna, T., and Ritzmann, R. E. Eds. New York: Academic Press.

Brooks, R. A. (1989). A Robot That Walks: Emergent Behaviors from a Carefully Evolved Network in *Neural Computation* (pp. 253-262).

Goldberg, D. E. (1989). *Genetic Algorithms in Search, Optimization, and Machine Learning*. Reading: Addison-Wesley.

Koza, J. (1992) *Genetic Programming: On the Programming of Computers by Means of Natural Selection*. Cambridge: MIT Press.

16 Genetic Programming for the Acquisition of Double Auction Market Strategies

Martin Andrews and Richard Prager

The Double Auction (DA) is the mechanism behind the minute-by-minute trading on many futures and commodity exchanges. Since 1990, DA tournaments have been held by the Santa Fe Institute. The competitors in the tournaments are strategies embodied in computer programmes written by a variety of economists, computer scientists and mathematicians. This paper describes how Genetic Programming (GP) methods have been used to create strategies superior, in local DA playoffs, to many of the hand-coded strategies.

To isolate the contribution that the evolutionary process makes to the search for good strategies, we compare GP and Simulated Annealing optimisation of programmes. To reduce the cost of learning, we also investigate an approach that uses statistical measures to maintain a uniform population pressure.

16.1 Double Auction Markets and the Santa Fe Tournaments

Most of the world's financial markets operate by the organised meeting of buyers and sellers, with all participants declaring prices at which they wish to trade. The potential for computer optimised trading strategies is evident.

16.1.1 The Double Auction Mechanism

Suppose that the commodity in question is coffee. The market takes place in a room in which buyers and sellers can both shout out their trading prices. So as to prevent cliques of traders manipulating prices available to outsiders, these markets abide by rules designed to force the *free and equitable flow of information*. Under the AURORA rules (used on the Chicago Board Of Trade), each new bid must be made at a higher price than the last, and similarly each new offer must be lower than the last. Actual trades in coffee may only take place between the buyer and seller who made the latest bid and offer. Thus, the trading process involves three stages:

1. The bid and offer prices start some distance apart, with buyers and sellers unsure of where they will meet.

2. The bid and offer prices converge as new bids and offers are announced.

3. The buyer and seller who finally agree on a common price will now satisfy their requirements by trading contracts at that price. The remaining participants must restart the bargaining process.

The trading price of coffee fluctuates when one set of participants feels greater urgency to deal. For instance, if buyers are more eager to trade then the sellers, the buyers must accept trades at the sellers' descending series of offers more quickly, and the trading price rises.

16.1.2 Measuring Trading Efficiency

Given a collection of buyers and sellers, each of whom knows how much they value a quantity of coffee, it is possible to determine the Competitive Equilibrium (CE) price, at which they should ideally all be prepared to trade. Knowledge of a group of traders' internal prices enables their expected profit to be calculated (if all favourable trades were done at the CE), and thus their *Efficiency* (defined as the ratio of realised profit to expected profit). Traders should, in a market with full disclosure of information, all have *Efficiencies* of 100%.

Of course, it is in none of the participants interests to announce how much they *actually* value coffee, rather, they repeatedly test whether they can have a cheaper bid accepted (or a more expensive offer). Good traders can gauge what the equilibrium price may be, and fool others into trading away from it.

16.1.3 The Santa Fe Tournaments

Economists who examine double auctions can directly observe the behaviour of markets [Smith 1988 and 1992]. However, whilst this gives information about the overall market efficiency, it is difficult to examine the strategies that lead to the observed behaviour of individuals. One way around this problem is to define the strategies of the participants via computer programmes, and then hold simulated double auctions for them to take part in. The first DA tournaments were held by the Santa Fe Institute in 1990, and a share of $10,000 was offered to the writers of algorithms that could perform well in a double auction competition. The tournament attracted around 25 devious and well thought-out strategies.

The computer programmes submitted to the competition were entered into a long series of separate auction games with the other entries, and were finally ranked for their performance in a selection of different market environments (for example equal numbers of buyers and sellers; trading against a monopoly seller; etc.).

The behaviour of the *markets* was analysed from an ecological perspective [Rust et al 1992a], looking at the broad profitability of each strategy and the effect of poor strategies being displaced in the market by good ones.

The tournament was also studied from an auction strategy point of view [Rust et al 1992b]. This was possible because the programmes were open to inspection, and hence the individual strategies and relative performances were known. By taking apart a few of the winning strategies, it was possible to discover what they were doing to enable them to consistently outperform the market.

16.1.4 Structure of Local Double Auctions

For these Genetic Programming (GP) experiments, a double auction simulation was run locally. It should be possible to use the strategies developed by simulation directly, as the tournaments are continuing in the form of the Arizona Token Exchange (AZTE), which allows the marketplace to exist across the Internet.

In the Santa Fe marketplace, the commodity traded is notional tokens, and all prices are in integer units. Trading takes place in discrete time-steps rather than continuously for simplicity. At the start of every time-step, traders are told the current bid and offer; a history of previous trades; and other useful information, for example the identities of the last traders. They then return their own bid (or offer) values. The new holders of the current bid and offer are then given the option of doing a trade. The information that the traders possess enables them to keep within the AURORA rules, *and* avoid trading tokens at a loss. However, as a certain amount of profit is to be expected, merely not losing does not indicate competence.

For each auction game, each trader is informed whether it is to be a buyer or a seller, and the values it is to place on the tokens that it is to buy or sell. Typically, there are four sellers and four buyers in a game, and they all have/want four tokens. Of these tokens, perhaps two or three would be profitable if they were traded at the equilibrium price (the traders are only told their own token values, and the distribution from which the others' tokens have been drawn). Since each trader has a different profit potential in each game, the traders' performances can be measured consistently by computing their respective *Efficiencies*, as defined earlier. A typical double auction game trace is given in Figure 16.1.

16.1.5 The Local Experiments

In the local double auctions, a broad selection of strategies that had been ranked in the Santa Fe tournament were played against each other. The strategies were pitted in auctions with 4 buyers and 4 sellers. One of the sellers was a strategy defined by a GP; one of the buyers was Skeleton, a programme available to all entrants before the competition. The six other competing strategies were chosen at random from the set {Kaplan, Skeleton, Anon1, Anon2, Kindred, Leinweber, Gamer}. An auction game was played with four rounds, with the 4 groups of 4 tokens being rotated between the participants on the buy and sell sides, so that all participants played with each set of

Genetic Program :(Max cBid cBid)

```
Buyers 786 788 794 801 804 806 818 823 828 834 841 842 845 859 860 863
Sellers881 872 870 869 818 814 806 806 768 762 756 742 740 736 735 729
Equilibrium Range :{ 806, 814}
```

Players		Buyers				Sellers					
	Lin	Anon2	Skel	Kapln	GenP	Kind	Lin	Skel			
Tokens											
T1	845+	859+	863+	860+	736+	735+	740+	729+			
T2	834+	828+	841+	842+	756+	762+	768+	742+			
T3	818+	801-	823+	804-	806+	818-	814-	806+			
T4	786-	794-	806-	788-	869-	872-	870-	881-			
Start!											
1 BA		397	789*	1	923	741*	920		789	741	
BS									Buyer 3 : Seller 3 at 789		
2 BA	790*	397	786	1	911	790*	933		790	790	
BS									Buyer 1 : Seller 3 at 790		
3 BA	789*	397	787	1	930	815*	930		789	815	
BS											
4 BA	790		797	815*	789*	811		805	815	789	
BS									Buyer 4 : Seller 1 at 789		
5 BA	789*	397	786	1	930	815*	922		789	815	
BS											
6 BA	789		796	815*	789*	811		807	815	789	
BS									Buyer 4 : Seller 1 at 789		
7 BA	789	627	791*	1	928	815*	926		791	815	
BS											
8 BA	792	804*	798	1	814			806*	804	806	
BS									Buyer 2 : Seller 4 at 806		
9 BA	784	629	786*	1	928	815*	923		786	815	
BS											
10 BA	794	800*	795	1	814			804*	800	804	
BS											
11 BA	801	805*	801	1	803			802*	805	802	
BS									Buyer 2 : Seller 4 at 802		
12 BA	795*	631	787	1	922	815*	930		795	815	
BS											
13 BA	798		802*	1	814			811*	802	811	
BS									Buyer 3 : Seller 4 at 811		
14 BA	795*	632	788	1	909	815*	934		795	815	
BS											
15 BA	796		801*	1	814*				801	814	
BS									Buyer 3 : Seller 2 at 801		
16 BA	799*	633	790	1	911	815*	921		799	815	
BS											
17 BA	800		801*	1	814*				801	814	
BS											
18 BA	802		802	803*	813*				803	813	
BS											
19 BA	804		804*		812*				804	812	
BS											
20 BA	805		805*		811*				805	811	
BS									Buyer 3 : Seller 2 at 805		
21 BA	793*	633		1	912	815*	920		793	815	
BS											
22 BA	793			803*		*			803	815	
BS											
23 BA	804*					*			804	815	
BS											
24 BA	805*					*			805	815	
BS											
25 BA	806*					*			806	815	
BS											

Figure 16.1

A single DA game run.

Notes:

A '+' next to a token value in the first rows indicates that it may be traded profitably.

A '*' next to a bid or offer indicates that it is the Current Bid or Offer (i.e. that trader is the only one given the option of actually trading at this price). Actual trades are shown in the right-most column.

The current bids ascend until a trade takes place (AURORA).

All the local auctions were limited to 25 time-steps.

tokens. A number of such games were played, and the average *Efficiency* of each strategy was noted.

The GP trader paid no explicit attention to the identities of those with whom it was competing. This was prevented to put the GP traders on a more equal footing with the other competitors (which were written independently of each other).

16.2 Genetic Programming of Strategies

The learning of strategies for financial trading is problem of some complexity. Its main difficulties are :

- Time delay of reward information

- Noisy nature of the market participants' interactions

- Good strategies may work covertly

- Tactics may be very difficult to learn whilst trading against experts

These problems mean that straight Neural Network (NN) algorithms (either with a recurrent architecture or coupled with a delayed-reward mechanism such as Temporal Difference Learning [Sutton 1988]) must be trained for enormous lengths of time before they begin to learn the subtleties of the environment. Even then, short-horizon tactics may prevent the acquisition of long-term strategies because of local-minima.

With evolutionary methods, on the other hand, no individual is faced with the task of learning a good strategy; rather, the *process* learns by the accumulation of combinations of good tactics.

The learning of trading strategies was previously tried via the GA control of neural network weights. This met with limited success, however, as the training times were excessively long, and it is unclear whether the network even learnt the rudiments of how to trade [Rust et al 1992b; Andrews 1992]. The GP paradigm has natural appeal for researchers engaged in strategy optimisation:

- GP allows a completely natural expression of a problem to be used. There need be no translation into an abstract intermediate representation.

- GP seems to offer a way of discovering effective strategies for playing complex games [Koza 1992, *PacMan*].

- Recently, a GP approach was very successful in learning strategies for playing a simple game with complex dynamics [Knobeln Contest 1993].

Evolutionary methods certainly reduce the problem of learning with limited feedback (as no learning is done by any individual). However, all learning algorithms must suffer from the noisiness of the trading environment, as this is really a defining characteristic of markets - if patterns were obvious, people would already trade them out of existence.

One additional difficulty that the double auction trading environment presents is that some of the hand-coded trading algorithms developed for the Santa Fe competition are of the covert variety. For instance, the Kaplan strategy waits in the background until the other participants have almost negotiated a trade (the bid/ask spread is small), and then jumps in and steals the deal if it is profitable (see buyer 4 in Figure 16.1). This behaviour means that neophyte traders, who negotiate badly, cannot ever take advantage of Kaplan's strategy, as it outmanoeuvres them, by not even taking part in the negotiations. An analogy with learning tennis from MacEnroe [Andreoni & Miller 1990, p20] seems particularly apt - it is far easier to learn in environments that are only slightly challenging.

16.2.1 The GP Environment

The GP was constructed to evaluate a function of what the trader's next offer should be (the trader was only cast in the seller's role - the buyer's role is fairly symmetric). This set-up was used (rather than a programme consisting of side-effects) as it was the natural way to use the information presented by the simulation to all the other computer strategies.

Table 16.1

Objective:	Traders that are successful against a random selection of strategies hand coded for double auction market trading.
Terminal set:	`HighestToken, LowestToken, NextToken, CurrentBid, CurrentOffer, LastPrice, TimeLeft` (rescaled),.
Function set:	`+, -, If>, Max, 30%, Abs.`
Fitness cases:	Auctions with the GP player as one of 4 sellers trading against 4 buyers (one of which was defined to be Skeleton, so that relative *Efficiency* could be assessed). All other players chosen at random from a selection of successful Santa Fe competitors.
Fitness:	The ratio of *Efficiency* of the GP seller to that of the Skeleton buyer, for a number of sets of four auctions (with token values being rotated within each set).
Parameters:	Population size : 300
Termination criterion:	Final generation reached.
Success measure:	Performance over 100% implies a likely good level of success in the DA tournaments.

The performance criterion used for fitness was the ratio of the *Efficiency* of the GP seller to that of a Skeleton buyer (known to be a good average). This fitness measure was chosen as it measures the abilities of the GP relative to the other traders in the market.

The GP environment itself was written in 'C' for speed, portability and ease of interface to the existing DA simulation code. The time taken for a typical run was around 3 hours on a Sun SparcStation 10.

16.2.2 The Programming Constructs

The function and terminal sets are described in Table 16.1. The values of all the terminals are calculated from the information given to each participant by the Santa Fe simulation, and are the ones commonly used by the other (human-coded) programmes.

All the terminal nodes (environment sensors) were defined so that they returned 'pricelike' values. This was necessary in the case of `TimeLeft`, for instance, as this was naturally measured as small integer values, and these could not contribute usefully to the return of good offers. The value of `TimeLeft` was rescaled to lie between `LowToken` and `HighToken`, and to increase between these two endpoints. Thus, the GP could easily find out how late in the game it was, simple by comparing this to many other sensor value combinations. The reason that the scaling problem was approached in this way, rather than allowing the GP to discover a scaling itself, is that the tournament rules themselves allowed for large variation in the game parameters, though in fact relatively little variability was tested. It was thought best to provide the GP with a set of automatically robust sensors.

In the same vein, it was felt that providing the GP with all the arithmetic operators was unnecessary - so a predefined percentage operator was used. In this case **(30% A)** returns **A*.3**. Examination of the programmes generated shows that the **30%** operator was also used to synthesise other percentages.

16.2.3 GP Selection Parameters

The populations of individuals were created a whole generation at a time. No mutation operator was used, and side experiments seemed to show that there was no real gain from allowing the possibility of straight reproduction of parents.

The GP crossover parents were chosen by a uniformly weighted tournament selection with 2 candidates each. As noted in [Koza 1992], tournament selection with two individuals is equivalent to simple linear rank based selection. If, in addition, a probabilistic choice is made between the tournament winner and the other individual, any linear rank distribution can be created. With three individuals, similarly, any quadratic function of rank can be created.

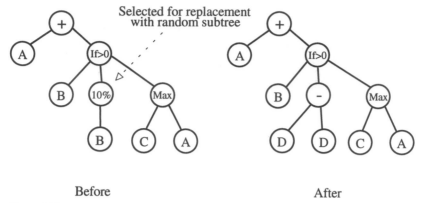

Figure 16.2
The possible action of a single mutation by the SA algorithm.

Tournament selection is particularly important for large population sizes, as it does not require the sorting of the entire population, nor does it require all fitness values to be available before selection can take place. This becomes important for a population distributed amongst multiple processors.

16.2.4 A Comparable Simulated Annealing Environment

Whilst the GP paradigm searches many problem spaces effectively, and largely beats pure random tree generation; how much useful information is generated by the genetic process? It seemed reasonable to test GP search against that of a leading global optimisation algorithm: Simulated Annealing (SA) [Ingber 1993].

The SA algorithm used takes the current tree and mutates it by substituting a random tree for one of the sub-trees (selected uniformly at random from all nodes), as illustrated in Figure 16.2. The size of the substituting subtree is limited to being of depth(3), but is otherwise generated in the natural way. Let the current fitness be called f_c and the mutated tree's fitness be called f_m. Then, at the next time step, the mutated tree replaces the current tree if and only if :

either $F_m > F_c$

or $U < \exp \dfrac{-(F_c - F_m)}{Temperature}$ where U is a Uniform [0,1] R.V.

The *temperature* parameter was reduced by 0.03% after each new acceptance. This rate was chosen so that after 100 generations, SA would still allow a reasonable downward step in fitness, to enable it to backtrack.

16.3 Is Genetic Search Useful ?

Initially, a comparison between the GP and SA optimisation methods was made by looking at the performance of the best-of-generation individual. It was found that the best individual's fitness produced by the SA was very comparable to that produced by GP. Upon closer examination, however, it was determined that the SA performance was gained by the generation and repeated testing of extremely risky strategies. The SA process would remember the individual who succeeded in the occasional fluke set of auctions.

The results in Figure 16.3 are of the distribution of performance for every member of the population, averaged over a period of 10 generations, with time (generation number) moving towards the reader. The difference in optimisation strategy between the SA and evolutionary techniques is significant. Because the two optimisation methods are set to optimise the same thing, one might expect both to produce similar high-flying individuals. However, as Figure 16.4 shows, the best of the final generation individuals, when repeatedly tested, have achieved their high fitnesses by different approaches. Both of these individuals has had its measured fitness drawn from a distribution of possible scores (since they have both ranked best, the best of these scores both exceeded 500%). However, the distributions that these scores have been drawn from differ. The GP individual is plainly less risky and more consistent. That is to say, its best-of-generation ranking has been obtained largely because of its high average fitness measurement, rather than repeated sampling of a rather skewed distribution. This demonstrates the common sense point that GP retains more information about the fitness landscape than the one-at-a-time method of simulated annealing.

Of course, it is important to note that a performance of 100% implies that the GP produced traders are on a par with the hand-coded strategies produced by economists and mathematicians. That this can be achieved in the mean is encouraging, and that the best-of-generation individual can exceed this over 65% of the time (median score 127%) is remarkable.

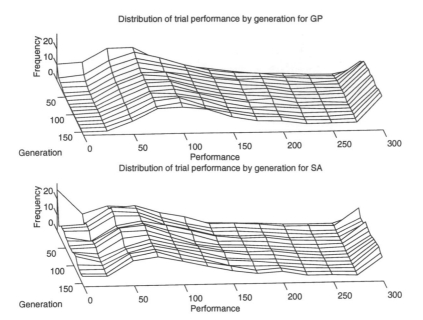

Figure 16.3
Graphical view of optimisation strategy for GP and simulated annealing.

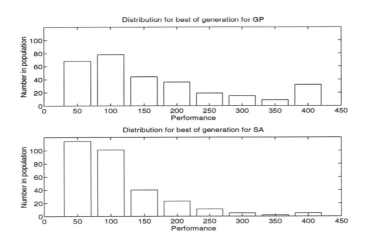

Figure 16.4
Distribution of best of final generation performance.

16.4 Economising on Fitness Evaluations

The costly step in the evolution of DA strategies is the evaluation of a statistically accurate performance value, to be used as the fitness. Reduction in standard deviation of this measure takes place at the rate of (Sqrt NumberOfAuctionsPlayed). Thus, if very accurate fitness measurements are required to make the evolution progress, a great number of costly experiments must be done. A reduction in this cost was sought.

The results earlier were gained with the same fitness measure for every generation. However, an analogy with monkeys and typewriters is relevant. If our fitness function is the creation of Hamlet, then an accurate assessment of how Hamlet-like the monkeys' output is would be unnecessary in the early stages. It would be more appropriate to check whether any keys are being pressed, followed by checking for any words at all. Thus time-consuming checking would be avoided whilst the monkeys evolved simple sentences.

Many fitness measures can be calculated more accurately by spending more time calculating them, for instance numerical integration of functions, classification of sets, simulation of behaviour, etc. However, the evolutionary process is only dependent on the ordering of the finesses, so numerical accuracy is not paramount, as long as one can be confident about the ordering. Alternatively, one could look at ordering errors as solely causing a reduction in population pressure.

Given the estimates of performance of two individuals, and the variance of those estimates, one can calculate the probability that the two individuals are, in fact, correctly ordered (by testing the hypothesis that one mean is higher than the other. See, for instance, [Press et al 1988]). This probability can be used in several ways :

• If a constant population pressure is desired, then the variance of the estimates can be manipulated (for instance, by taking more samples), so as to produce a constant probability of correct ordering (i.e. constant population pressure).

• The probability may be noted, so that the experiment can be terminated when the existing measurement of fitness is not sufficiently accurate to drive evolution forward (i.e. experimental error swamps the underlying differences in fitness).

Experiments were carried out with the number of games per fitness case being controlled by a calculation of the number required for constant population pressure. Here, the two individuals chosen as reference points were those at the 50% and 75% levels within the population (i.e. median and upper-quartile individuals). The pressure was held constant by choosing the appropriate number of games to play to keep these individuals correctly ordered with 95% confidence.

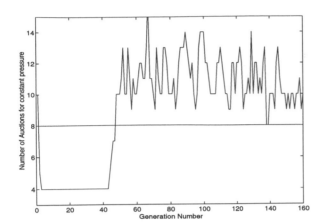

Figure 16.5
The 'ideal' number of games schedule for a typical DA run.

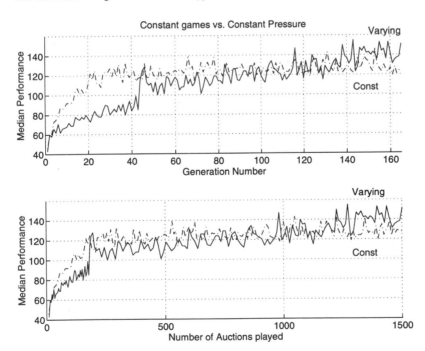

Figure 16.6
Median performances with the constant and variable schedules for game playing.

As can be seen from Figure 16.5, initially the number of games required for constant pressure was low (4 games was a set minimum). This saves time when it is easy to distinguish the reasonable player from the utterly hopeless. Later on, the number of games suggested rises above the number played by the constant games run (which was set at 8 games per individual). This improves the chances that the evolution is making use of actual differences in performance, rather than noise.

The median performance is shown in Figure 16.6 for both a constant-games and a constant-pressure run, in both per-generation and per-game time (which is far closer to wall-clock time). The advantage of being able to vary the number of games dynamically can be seen in both cases.

16.5 Further Work

One interesting direction to take GP double auction learning is to allow the GP strategies to compete against each other in the same auction. One of the problems with the DA game is the lack of an obvious dominant strategy. For instance, in the game of Backgammon [Tesauro and Sejnowski 1989], the randomness of the dice seems to ensure that a player's optimal move is more-or-less independent of his opponent's style. The correct strategy for the DA depends entirely on the opponents, however, so that the evolutionary arms race that occurs between GP buyers and sellers does not necessarily encourage a good strategy for playing previously unseen opponents.

A further avenue for experimentation is to use the GP methodology to trade on the financial markets from a slightly more distant perspective. For instance, rather than learning to handle the double auction game, use GP for generation of buy and sell signals for commodities by detection of trend signals given by the mid-price process. Strategies for this task would be of more direct relevance to traders off the open-outcry floor.

Acknowledgements

This work was supported financially by the generosity of the Science and Engineering Research Council. As yet, no backing has been sought from financial institutions. Parties interested in specialist knowledge about this and other applications are welcome to contact the authors directly.

Many thanks are due to the organisers of the AZTE, for providing an excellent framework of research with their tournament, manuals, programmes and enthusiasm.

Bibliography

Andreoni, J. & Miller, J. (1990) "Auctions with Adaptive Artificially Intelligent Agents", Santa Fe Technical Report : SFI 90-01-004

Andrews, M. (1992) "The Impact of Reward Strategy on Automaton Learning", Cambridge University Engineering Department Technical Report.

Dawkins, Richard (1982 "The Extended Phenotype", W.H. Freeman & Company, Oxford.

Knobeln Contest (1993) Rules and results available via ftp from : Sanfrancisco.ira.uka.de [129.13.13.110] in directory /pub/knobeln.

Ingbcr, L. (1993) "Simulated Annealing: Practice versus Theory", submitted to *Statistics and Computing*.

Press, W.H., Flannery, B.P., Teukolsky, S.A., & Vetterling, W.T. (1988). "Numerical Recipes in C", Cambridge University Press, Cambridge.

Rust, J., Palmer, R., & Miller, J. (1992a) "Behaviour of Trading Automata in a Computerized Double Auction Market", in *The Double Auction Market : Institutions, Theories, and Evidence*, D. Friedman and J. Rust (eds.), Addison Wesley.

Rust, J., Miller, J., & Palmer, R. (1992b) "Characterizing Effective Trading Strategies : Insights from a Computerized Double Auction Market", *Journal of Economic Dynamics and Control*.

Smith, Vernon L., Suchanek, G., & Williams, A. (1988) "Bubbles, Crashes and Endogenous Expectations in Experimental Spot Asset Markets", Econometrica, Vol 56, No. 5, p1119-51.

Smith, Vernon L., & Williams, A.W. (1992) "Experimental Market Economics", *Scientific American*, December 1992, p72-77.

Sutton, R.S. (1988) "Learning to Predict by the Methods of Temporal Differences", *Machine Learning 3*, p9-34.

Tesauro, G. and Sejnowski (1989) "A parallel network that learns to play backgammon", *Artificial Intelligence*, 39, p357-390.

17 Two Scientific Applications of Genetic Programming: Stack Filters and Non-Linear Equation Fitting to Chaotic Data

Howard Oakley

Optimization within nearly infinite search space is a common problem in applied science, for which two examples illustrate the application of genetic programming. Three different techniques were used to develop filters for the removal of noise from experimental data. Heuristic search was used to develop a median filter, a classical genetic algorithm optimized a 7-tap moving average (FIR) filter, and genetic programming was used to optimize a stack filter. The latter had the highest fitness, and was computationally more efficient than the best median filter, which was in turn superior in fitness to the best moving average filter. Genetic programming was also used to fit empirical equations to a chaotic time-series (the Mackey-Glass equation) and non-linear physiological data. Initial results confirm the key role of the fitness measure in such work; oscillatory series are readily fitted with linear functions unless the computation of fitness includes an appropriate measure such as incremental comparison of Fourier power series. The use of Lyapunov exponents and dimension estimation is suggested in more sophisticated compound fitness measures. Genetic programming may prove to be useful in both forecasting and structural studies of non-linear systems, at both local and global levels.

17.1 Introduction

Experimental science abounds with problems which seek optima from near-infinite search spaces. Remarkably, Koza's monograph [Koza 1992] introducing the technique of genetic programming includes a large number of examples which illustrate the potential of genetic programming in such searches. The work reported here was undertaken to meet a need for improved investigative tools, rather than as a study of genetic programming *per se*, although it may usefully extend the repertoire of examples.

17.2 Development of Stack Filters

17.2.1 Background

The selection of a filter to remove noise from a signal is widely recognized as being empirical at best (see for example [DeFatta 1988]). Not only are there many different types of filter, but each type can be configured with a near-infinite range of different coefficients (e.g. tap weights). A new instrumentation system employing laser Doppler rheometry was being prepared for use in human experiments (in which it was to measure blood flow in the skin) when it was decided to investigate the most appropriate final stage output filter for the rheometer. Typically, these devices are used with 'RC single pole' (exponential) filters with time constants of between 0.1 and 5.0 s, making each output reading the result of the equation

$$x_t = \sum_{n=-\infty}^{t} \frac{e^{-n/t_c}}{\sum e^{-n_t/t_c}} x_n$$

where x_t is the filter output at time t, t_c is the time constant of the filter, and x_n is the unfiltered value at time n.

Three other filter types were deemed suitable for comparison. The simplest was a 7-tap finite impulse response (FIR) filter, in which output readings resulted from

$$x_t = \sum_{n=t-3}^{t+3} w_i x_n$$

where w_i is the ith weight such that the sum of the seven weights is 1.0. Median filters were also used, in which output readings resulted from

$$x_t = median(x_{t-n}, ..., x_{t+n})$$

where the window size is 2n + 1 and the median is the (n+1)th value in rank.

Stack filters are a recent generalization of median filters, particularly suited to implementation in hardware [Wendt 1986]. Again operating over a window of finite width, each raw measurement within the window is first decomposed into a column of binary digits of height equal to the maximum input value. Thus, for a maximum input value of 255, the number 128 would be represented as a column of 127 0s above 128 1s:

0	255	top
...	...	
0	129	
1	128	
1	127	
...	...	
1	1	bottom

The columns for each window are then juxtaposed to produce a Boolean matrix thus:

0	0	0	0	1	0	0	255
...	...	...	...	...	...	...	
0	0	0	0	1	0	1	204
...	...	...	...	...	...	...	
0	0	0	0	1	1	1	195
...	...	...	...	...	...	...	
0	0	0	1	1	1	1	163
...	...	...	...	...	...	...	
0	1	0	1	1	1	1	126
...	...	...	...	...	...	...	
1	1	0	1	1	1	1	94
...	...	...	...	...	...	...	
1	1	1	1	1	1	1	66
...	...	...	...	...	...	...	
1	1	1	1	1	1	1	1
94	126	66	163	255	195	204	1

Boolean and other mathematical operations can then be applied across the rows (between the columns) of the matrix, to produce a Boolean or binary digital column of the same height, which can then be summed (the reverse of the original threshold decomposition) to return a single number. In the case of a median filter, the row operator can be simple summation of the binary digits followed by testing against the median threshold of $n+1$ for a window width of $2n+1$.

17.2.2 Methods

A typical unfiltered dataset, containing 600 time-ordered blood flow measurements, was obtained under standard experimental conditions from a volunteer, informed and consenting human subject. This dataset was then cleaned up by eye, to produce an idealistically 'clean' series which fitted the experimenter's view as to what the data should have looked like (there being no other way of achieving a 'gold standard' output data series). Rather than use the original data to represent a 'noisy' equivalent, a mixture of uniform (rectangular) and Gaussian distributed noise, some of which was proportionate to the signal, was then developed, tested to ensure statistical similarity with the residuals between the clean and original data, and added to produce a synthetically 'dirty' dataset.

A global fitness measure, independent of any fitness formulae used locally by particular optimization techniques, was used to compare the efficacy of different filters, according to:

$$raw \ fitness = \frac{1}{1 + \sqrt{\sum (x_f - x_c)^2 / n}}$$

where x_f is the filtered and x_c the clean value for a given position in the series of length n. This fitness measure tends to 0.0 for the least fit, and to 1.0 for the perfectly fit filter results.

Heuristic search encompassed two 7-tap FIR filters (with even and bell-shaped weights), RC single-pole filters with time constants from 0.0375 to 3.0 s, and median filters with window widths 3, 5, 7 and 9.

Table 17.1
Tableau for filter optimization. % is the protected divide operation [Koza 1992].

Objective:	Find best 7-tap FIR filter	Find best stack filter
Terminal set:	Window of 7 data points, $\Re$	Window of 7 data points, threshold decomposed
Function set:	+, −, %, *	logical NOT, AND, OR
Fitness cases:	600 'clean' points	600 'clean' points
Raw fitness:	Sum over 600 fitness cases of squared error between filtered and 'clean' points	Sum over 600 fitness cases of squared error between filtered and 'clean' points
Standardized fitness:	Same as raw fitness	Same as raw fitness
Hits:	Filtered and 'clean' points are within 0.01 of each other	Filtered and 'clean' points are within 0.01 of each other
Wrapper:	None	None
Parameters:	M = 500, 1000, 2000. G = 51, 101	M = 500, 1000, 2000. G = 51, 101
Success predicate:	None	None
Max. depth of new individuals:	6	6
Max depth of new subtrees for mutants:	4	4
Max depth of individuals after crossover:	17	17
Fitness-proportionate reproduction fraction:	0.1	0.1
Crossover at any point fraction:	0.2	0.2
Crossover at function points fraction:	0.7	0.7
Selection method:	fitness-proportionate	fitness-proportionate
Generation method:	ramped half-and-half	ramped half-and-half

A genetic algorithm search was performed using 7-tap FIR filters with weights between 0 and 255 (integer). Spears' GAL program [Spears 1991], which uses Baker's SUS selection method, was run with a string of length 56 bits, encoding 7 x 8 bit words to represent the tap weights in plain and grayscale encoding. Production runs employed populations of size 5000, mutation rate 0.001, and crossover rate 0.6, terminating when best fitness had long stabilized. Test runs with different mutation and crossover rates were also performed.

Genetic programming was performed using the Simple Lisp implementation [Koza 1992] with local performance enhancements. Two groups of runs were undertaken, the first to optimize 7-tap FIR filters, and the second to optimize stack filters of window width 7. Initial exploratory series employing small population sizes and generation numbers were followed by production series which took several days or weeks to complete 2 to 10 runs. The tableau is shown in Table 17.1.

17.2.3 Results

Ranked fittest filters are summarized in Table 17.2, and a selection illustrated in Figure 17.1.

Table 17.2
Comparison of Fitness Values for Best Filters Tested. Fitness is expressed on the scale 0.0 (unfit) to 1.0 (perfectly fit). GA is the generic algorithm, GP is genetic programming.

Rank	Class	Filter	Optimization	Fitness
1	Stack	Mod. median 3 window 7	GP	0.0816
2	Stack	Median 5	heuristic	0.0787
3	Stack	Median 3	heuristic	0.0776
4	FIR	Modified 0, 60, 133 ...	seeded GP	0.0695
5	FIR	0, 60, 133, 250, 143, 61, 6	GA	0.0685
6	FIR	0, 10, 20, 40, 20, 10, 0	heuristic	0.0685
7	FIR	1, 64, 131, 236, 148, 64, 3	grayscale GA	0.0684
8	FIR	6, 61, 143, 250, 133, 60, 0	GA + heuristic	0.0682
9	Stack	Median 7	heuristic	0.0673
10	Stack	Median 9	heuristic	0.0572
11	FIR	10, 10, 10, 10, 10, 10, 10	heuristic	0.0521
12	FIR	(+ (% (– (+ X5 0.944 ...	GP	0.0489
13	RC	0.0375 s	heuristic	0.0459
14	none	no filtering	heuristic	0.0437
15	RC	0.1 s	heuristic	0.0266

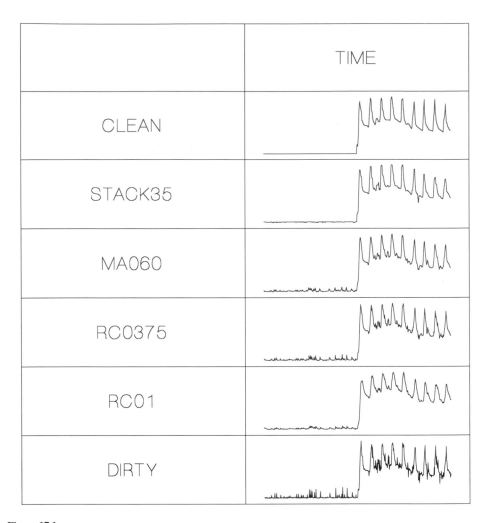

Figure 17.1
A selection of filters used, together with the 'clean' and 'dirty' datasets. Each is plotted as blood flow in arbitrary units against the index of the data point. STACK35 has been filtered with the fittest stack filter, MA060 with the fittest FIR filter, RC0375 with the RC filter with time constant 0.0375 s, and RC01 with the 0.1 s RC filter.

The fittest filter discovered resulted from the optimization of stack filters by means of genetic programming, which had a root mean squared error of 11.2 (less than 6% of the mean data value). The form of this filter, as a simplified Common Lisp S-expression, was:

```
(OR (AND Y2 Y3) (AND Y2 Y4) (AND Y3 Y4) (AND Y0 Y3 Y6))
```

where the window extends from Y0 to Y6 to derive the filtered value at Y3. This is a window width 3 median filter (any 2 of [Y2, Y3, Y4]) with an additional logical term to encompass all 3 of [Y0, Y3, Y6]. Application of this filter to other test datasets appears to produce excellent results (e.g. Figure 17.2). Genetic algorithm runs stabilized by the time that 300 000 individuals had been evaluated, yielding a fittest filter which had a root mean squared error of 13.6. Changing the mutation rate between 0.00001 and 0.1 had very little effect on the median best fitness achieved (ranging from 0.0664 to 0.0656), whilst changing the crossover rate between 0.0 and 0.8 had more effect (median best fitnesses from 0.0635 at a rate of 0.0 to 0.0662 at 0.8). Using grayscale rather than plain encoding resulted in a slightly lower fitness.

17.2.4 Discussion

Of all the filter techniques developed, the stack filter optimized using genetic programming was the fittest, and it was also computationally simpler than its nearest two competitors (both median filters resulting from heuristic search). Whilst the derivation of filters by genetic programming required more computer processing time than by simple application of the genetic algorithm, the latter required more programming, and would have required a much more substantial investment of time had it been decided to apply it to stack filters, for which such a simple encoding scheme would have been impossible.

Several previous studies have used evolutionary and other techniques to optimize stack filters. Chu [Chu 1989] employed the genetic algorithm with considerable success, whilst others [Ansari 1992] have tried simple neural networks which appeared to be as efficacious. Unfortunately, the data used in those studies is different from that used here, of necessity, and neither paper described its best-performing filter. Detailed comparison with those studies is thus not currently possible.

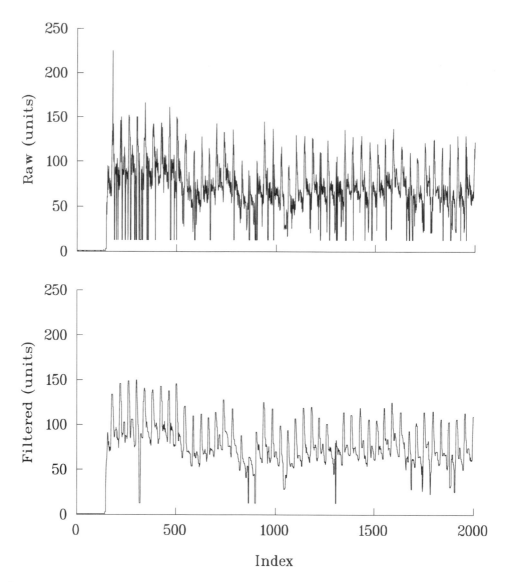

Figure 17.2
A typical experimental dataset, before (upper) and after (lower) filtering with the best filter found with genetic programming.

Although no particular effort was taken to make the techniques used efficient, the fittest filters were drawn from very small samples of a very large search space. Even given the narrow restrictions imposed in the genetic algorithm example, there are approximately 7.2×10^{16} different 7-tap FIR moving average filters with tap weights in the integral range 0–255. An exhaustive search of these would clearly be impractical – even assuming a computer capable of assessing 1000 every second, this would take approximately 2.3 million years. This contrasts with the several days taken to arrive at the fittest 7-tap FIR moving average filters by genetic algorithm and genetic programming, although there is of course no guarantee that these are optimal. Search spaces for stack filters are even larger, of course.

The genetic algorithm approach performed well, and demonstrated that crossover rather than mutation was key to its performance. However, median filters selected heuristically outperformed the best FIR filter, which is in agreement with much of the literature [Gallagher 1981].

Genetic programming proved capable of out-performing all other techniques, although no effort was devoted to refining the approach used for efficiency. The fitness measure used was also conventional, in using the root mean squared error, although absolute error was an alternative. In the circumstance described, there are theoretical preferences for the former technique [Alexander 1986], and the desire to fit high peaks well is also supported.

The filter developed by genetic programming is now in daily use, and future rheometers are expected to incorporate it. Interestingly, the fittest stack filter is closely related to the second and third fittest heuristically-found median filters, suggesting that it may have good general performance when applied to other types of data. Further investigation is required.

17.3 Fitting of Non-Linear Equations to Chaotic Data

17.3.1 Background

The investigation of non-linear and potentially chaotic systems has been studied extensively, but remains fraught with difficulty. In the first instance, discrimination of chaos can be approached conservatively (e.g. [Ruelle 1990]), or more liberally (e.g. [Farmer 1987]). Particularly when considering biological systems, the requirements of tests for chaos – including very long series of noise-free data – are usually very difficult to achieve, although many such systems would appear to be likely to be chaotic [Glass 1992].

One diagnostic criterion of chaos is that of unpredictability, which is often formalized in the Lyapunov exponents. In spite of this, considerable efforts have been devoted to the prediction of chaotic series, predominantly in the short term [Casdagli 1989] or locally within phase space [Meyer 1992]. Casdagli [Casdagli 1989] has succinctly cast the problem of prediction thus:

"The standard problem in dynamical systems is, given a nonlinear map, describe the asymptotic behavior of iterates. The inverse problem is, given a sequence of iterates, construct a nonlinear map that gives rise to them. This map would then be a candidate for a predictive model."

Koza's monograph [Koza 1992] includes two demonstrations of simple non-linear systems, one of which is also chaotic, which have been fitted with predictive equations by means of genetic programming. Although the chaotic example is the logistic map, one of the simplest instances of chaos, his limited exploration of short term prediction is promising, and consistent with Casdagli's suggestion that genetic programming may be useful in forecasting chaotic series [Casdagli 1992].

In the course of collecting data from human subjects using the laser Doppler rheometer, it became apparent that at least some time series data for peripheral blood flow was non-linear, and appeared to be chaotic. Accordingly, the work reported here is part of a study eventually intended to develop quantitative techniques for the investigation of non-linearities in the control of peripheral blood flow in the human. The first step in this was to examine their potential and problems using a standard and similar system which is known to be chaotic, after which noisy theoretical and experimental data could be studied.

The chaotic system chosen as being similar in visual appearance to that of the physiological data was the Mackey-Glass equation [Mackey 1977]:

$$\frac{dx_t}{dt} = \frac{bx_{t-\Delta}}{1 + (x_{t-\Delta})^c} - ax_t$$

where a is usually taken as 0.1, b as 0.2, c as 10.0, and Δ as 30.0. Expressed as a flow, this is not only non-trivial to approximate, but is not explicitly soluble as an S-expression. For the purposes of this study, it was therefore used as a map:

$$x_{t+1} = x_t + \frac{bx_{t-\Delta}}{1 + (x_{t-\Delta})^c} - ax_t$$

which retains its chaotic characteristics (see Figure 17.3) whilst being expressible as an S-expression.

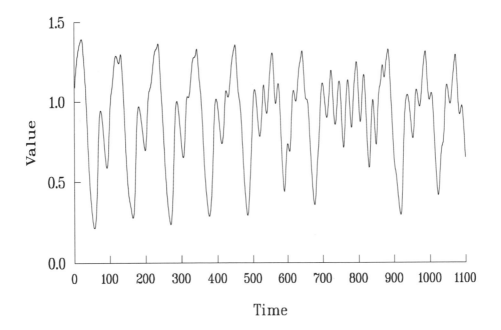

Figure 17.3
Members of the Mackey-Glass map following discarding of the initial 1000 in the series, as used in the experiments.

Rather than use very long training series, which might be achievable in many physical experiments, it was decided to restrict the training data close to that just sufficient to predict further members of the series, which if Δ is 30, would be at least 31 points. This is considerably less than most other studies of prediction of chaotic time series [Casdagli 1992], but is of similar order to that available from many physiological series. It is also of particular interest, as it is at the limit of the $D \leq 2\log_{10}N$ rule (D being the correlation dimension, and N the number of observations in the time series) expounded by Ruelle [Ruelle 1990] and derived from the Grassberger-Procaccia algorithm [Grassberger 1983]. Given that the correlation dimension of the Mackey-Glass series is 3 [Casdagli 1989], the expected minimum series from which useful information could be extracted is 32.

17.3.2 Methods

A large number of runs (several hundred in all) were undertaken using genetic programming, in each case to attempt to predict 20 to 1065 time steps into the future, given the same 35 sequential exemplars in the Mackey-Glass mapping. The exemplars were generated using double-precision floating point calculation, seeding the series with 35 random numbers between 0.5 and 1.5. The first 1000 in the generated series were discarded, and only data following those were used experimentally; these are illustrated in Figure 17.3.

Table 17.3
Typical tableau for fitting chaotic series. % is the protected divide operation [Koza 1992]. EXP10 is exponentiation to the power of 10 (x^{10}) protected from numeric overflow such that any operand of absolute value greater than 2.0 returns 2^{10}.

Objective:	Predict next n in Mackey-Glass mapping series
Terminal set:	Time-embedded data series from t = 1, 2, 3, 4, 5, 6, 11, 16, 21, 31, $\Re$
Function set:	+, −, %, *, SIN COS, EXP10
Fitness cases:	Actual members of the Mackey-Glass mapping (20 − 1065)
Raw fitness:	Sum over the fitness cases of squared error between predicted and actual points
Standardized fitness:	Same as raw fitness
Hits:	Filtered and 'clean' points are within 0.001 of each other
Wrapper:	None
Parameters:	M = 50–5000. G = 51, 101
Success predicate:	None
Max. depth of new individuals:	6
Max depth of new subtrees for mutants:	4
Max depth of individuals after crossover:	17
Fitness-proportionate reproduction fraction:	0.1
Crossover at any point fraction:	0.2
Crossover at function points fraction:	0.7
Selection method:	fitness-proportionate
Generation method:	ramped half-and-half

Genetic programming was performed using the Simple Lisp implementation [Koza 1992] with local performance enhancements. A typical tableau is shown in Table 17.3.

Time delay embedding was effectively achieved by the inclusion in the terminal set of previous values in time, with an upper limit to dimensionality of 10. Unusually, because previous values were deliberately not interspersed equally in time, it was possible for S-expressions to effectively incorporate multiple dimensional embedding, with unequal time delays.

Many runs were also augmented with fitness functions which included comparisons between Fourier power spectra of the predicted and real data series. These were used as the sole fitness measure, and in combination with the squared error measure given in Table 17.3. They were computed as the squared error between estimates of power across predicted and actual spectra following Fourier analysis. When used in combination with the untransformed squared error, the latter was estimated first, and only if it exceeded a given value (40.0, half the fitness achieved by the best linear extrapolation) was the power spectral fitness measure used by adding both power spectral and untransformed summed squared errors together.

17.3.3 Results

A common tendency was for the fittest S-expression to generate a constant, extrapolating along a straight line through the seed data, e.g. with S-expressions of the form

```
(+  -0.084783420006  (%  Y31  Y31))
```

Note that Y31 corresponds to a time delay of 30. This was most frequent when long forecasts were attempted using only the simple squared error fitness measure, and encouraged experiment with fitness measures incorporating power spectral analysis. When such compound fitness measures were employed, it was more common for fitter individuals to exhibit oscillatory behavior early in the period of prediction, damping down to a linear prediction after 50 to 100 time steps, as illustrated in Figure 17.4.

S-expressions shown are:

```
• GFFT1, long-term fit with FFT fitness (COS (* (- Y31
0.6168952889819135) (- (- Y31 Y1) -1.0653563570535542)))
• GFFT2, long-term fit with FFT fitness (% (EXP10 Y2) Y31)
• GST1, long-term fit without FFT fitness (COS (SQRT% (SQRT% (- Y31
Y1))))
• GST2, long-term fit without FFT fitness (COS (- Y31 (COS Y31)))
• L100, short-term fit over 100 points (SIN (* (* (COS Y31)
          (+ (SIN (* (COS Y31)
                 (+ (COS Y31)
                  (* (SIN (SIN (COS Y31)))
                   (SIN (* (COS Y31) (+ (+ (COS Y31) Y21)
Y21)))))))))
```

```
         Y21))
     (+ (* (COS Y31) (COS Y31)) Y21)))
• L40, short-term fit over 40 points (COS (- (* (COS Y4)
         (+ (* (COS (COS (- (* Y31
                              (+ (* (COS (COS (COS (+ (COS Y2)
Y31)))) Y31)
                                       Y31))
                               (COS (COS Y2)))))
             Y31)
           Y31))
        (COS (COS (COS (COS (COS (* (+ (* (COS Y4) Y31) Y31)
Y31)))))))))
• MG, original Mackey-Glass mapping (+ Y1 (- (% (* 0.2 Y31) (+ 1.0
(EXP10 Y31))) (* 0.1 Y1)))
```

A series of runs examining the effect of changing the function and terminal sets showed that even when they were reduced to exactly those occurring in the Mackey-Glass equation, the fittest S-expressions were no different in performance or structure from those in which much wider function and control sets were used.

The relationship between forecasting period and fitness was investigated by varying the number of fitness cases between 20 and 1065, between different series of runs. Normalized root mean squared errors [Farmer 1987] of results are shown in Figure 17.5, which demonstrates that forecast accuracy falls with increasing length of forecast (increasing number of fitness cases).

Given the *a priori* knowledge of the Mackey-Glass equation, a number of interesting S-expressions emerged from experimental runs. Surprisingly, the terminal set member at a time delay of 30 (corresponding to $x_{t-\Delta}$ in the Mackey-Glass mapping) was featured in the majority of fittest S-expressions, although it did not appear in the form $1/(x_{t-31})^{10}$. Of 22 runs to perform long-term prediction, this delay appeared in 20 of the fittest S-expressions (χ^2 test significant at $p \ll 0.001$), and of 22 runs to perform short-term prediction, the same delay appeared in 17 of the fittest S-expressions (χ^2 test significant at $p \ll 0.001$). In order to investigate whether this might be a consequence of the order of items within the terminal set, a limited number of runs were performed with the order reversed; no overall difference was apparent in the occurrence of items in the fittest S-expressions.

Most of these runs were performed to predict future values from one exemplar set alone, although a small series of runs was undertaken in which multiple predictions were made across several sets in the 1100 point series. Because the number of such runs is small, the fitnesses are likely to be poorer than those of production runs; however, they were reasonably close to those obtained from the production runs: forecasting N of 30 gave a best normalized root mean squared error of 0.1596, and N of 100 gave a best normalized root mean squared error of 0.9136.

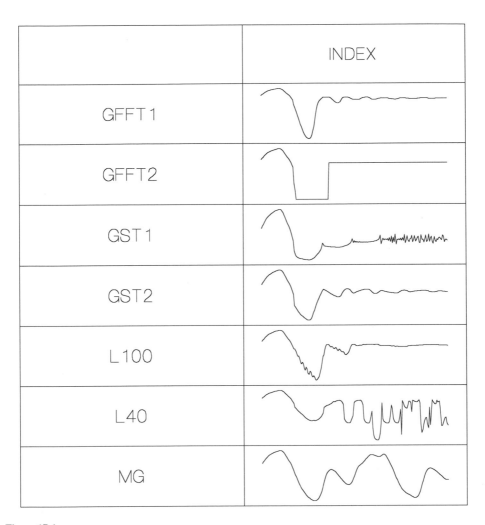

Figure 17.4
Plots of a selection of fitter S-expressions, and the original Mackey-Glass series. In each case, the first 35 points are seeded, and thus are common to the original Mackey-Glass mapping. A total of 200 points are shown for each example. S-expressions responsible for each are given in the text.

17.3.4   Discussion

17.3.4.1 Deception and Fitness

At first sight, although these results appear to confirm that genetic programming is of value in short-term prediction of chaotic series, it may appear disappointing that no S-expression came close to that for the Mackey-Glass mapping which generated the data. However, the series used is an effectively deceptive dataset for genetic programming, because of its extensive false optima, achieved by simply fitting an S-expression for a straight line through the data points. Indeed, some preliminary experiments with Koza's sine wave example [Koza 1992, p. 238 ff.] suggest that this may happen with any stationary series which exhibits periodicity, if the dataset being fitted contains several cycles, and there are relatively few points per cycle.

It is perhaps worth considering, for such a simple example as a sine wave, that the conventional fitness function based on error is prone to errors in phase, which is very unlikely to evolve as a single dimension of variation, although ADF may aid this. For example, an S-expression which generates a sine wave which is exactly out of phase with the fitness cases, although it is structurally very close to the correct S-expression, has a much poorer error-based fitness than a linear fit, which is structurally much further from the correct S-expression. The Mackey-Glass mapping presents an even tougher problem for genetic programming to solve, as it is only pseudo-periodic, and thus requires a complicated S-expression which is exactly 'in phase' with the fitness cases, before any longer term prediction achieves better fitness than that of a straight line. This very local true optimum is thus set in an extensive range of false optima, and the function is deceptive.

17.3.4.2 Fitness Measures

One approach to try to overcome this problem is to manipulate the fitness measure so as to make it less sensitive to the false optima, and easier to find the isolated true optimum. In these initial experiments, the power spectrum was seen as one approach to achieving this. However, the results make it clear that this too suffered from the same problem of deception.

One strategy which may be successful is what may be viewed as 'steered' genetic programming, in which the fitness function is composed of a series of sub-objectives, using the same 'divide and conquer' approach often adopted in developing computer programs. Thus, an initial objective in the fitness measure might be to achieve a particular peak in the power spectrum. Once this has been reached, then the fitness measure might be changed, to focus on another peak, and so on until an acceptable power

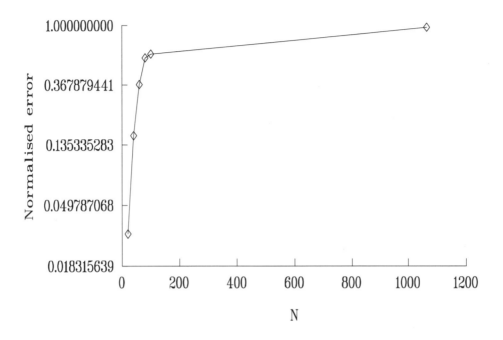

Figure 17.5
A semilogarithmic plot of the normalized root mean squared error of the fittest forecasting attempt, against the length of forecast attempted, N.

spectrum has been generated by the fittest S-expression in the population. Following that, there might be provision for fine tuning, based on the traditional measure of squared error. ADF (cf. Chapter 6) could also have a useful role.

Whether this particular approach is ideal or not, it seems likely that something of this kind is needed to fit such obdurate problems without having to run massive populations for many generations. It might be usefully coupled with segregation of the population into demes (each of which could use different fitness measures), and hierarchical automatic function definition [Koza 1992]. Although computationally intensive, the use of Lyapunov exponents and dimensions as part of the estimation of fitness may also prove worthwhile. Noting the effect of different prediction methods (Figure 17.4), a different approach might be to steer the fitness function across progressively longer periods of prediction.

One potential criticism of steering the fitness function is that it makes the search biased from the outset; however, this is true to some degree of all non-random search methods. Furthermore, the only relevance of bias in practical use is that it must not increase any tendency to deception. Given the results quoted here, it could be argued that an appropriate steering technique reduces bias.

17.3.4.3 Effectiveness of Prediction

Given the very short 'training' data series, both short- and long-term predictions appear surprisingly good. For example, comparison with data estimated from Casdagli's iterative radial basis function predictions [Casdagli 1989] – the best-performing technique reported in his study, and still [Stokbro 1992] one of the best techniques of all – are shown in Table 17.4.

One of the problems of most physiological time series is their brevity, relative to those derived from purely physical systems. For instance, the rheometer measurement system typically records blood flow at 40 Hz from each of two sites, but this is reduced to approximately 1 Hz when analyzed on a (heart) beat-by-beat basis. Periods of up to 10 min are usually collected, providing series of 100–600 data points. These would be insufficient for neural network techniques [Mead 1992, Stokbro 1992], particularly as few experiments are repeated on the same individual. However, this quantity of data could be expected to be sufficient to permit the study of series with correlation dimensions of 4–5 (according to the $2\log_{10}N$ rule). This would appear feasible using genetic programming, provided that the effect of noise is not too disruptive.

Table 17.4
Comparison of Normalized Root Mean Squared Errors with those of Casdagli. Errors stated are normalized root mean squared values, ranging from 0.0 (no error) to 1.0 (the error resulting from a linear fit through the data mean). Values attributed to Casdagli are estimated from iterative radial basis function plots in Figure 4 (c) of [Casdagli 1989].

Forecast N	Casdagli	This study, single	This study, multiple
20	0.0631	0.0311	
30	0.1585		0.1596
40	0.316	0.158	
60	0.631	0.371	
100	0.990	0.6170	0.9136

17.3.4.4 Structural Insights

A number of characteristics of the fittest S-expressions found may provide structural insights into a non-linear system, which could in turn help in the construction of a model. In the first instance, the form of the relationship in Figure 17.5 will be influenced by the presence or absence of chaos, and gradients reflect the resulting unpredictability of the system. Casdagli [Casdagli 1989] has pointed out that there are two likely underlying equations which express scaling laws, both of which are of exponential form. The two prevalent gradients in the original data are, using natural logarithms, 0.08121 (initial steep segment, N = 20 to 40) and 4.655×10^{-4} (later flatter segment, N = 100 to 1065). These compare with a computed metric entropy (equal to the sum of positive Lyapunov exponents) of approximately 0.01 for the Mackey-Glass flow [Meyer 1992].

Another clue to structure may lie in the remarkable preponderance of S-expressions incorporating a time delay Δ of 30. Although this could just be fortuitous, in the absence of *a priori* knowledge of the structure of a non-linear system, it could be a useful starting point for model construction. Altenberg describes this time delayed variable as having "high constructional fitness", cf. Chapter 3. Few other techniques appear capable of this, particularly as they may hide the empirical relationships discovered in the data series. In contrast, genetic programming encapsulates them in S-expressions which are rich ground for further examination.

17.4 Conclusions and Prospects

Genetic programming has optimized a stack filter which performs better on noisy physiological data than a range of other filters, optimized heuristically and by genetic algorithm. The filter developed may prove to be more generally robust, and the technique certainly appears effective.

Investigation of a well-studied chaotic system by means of prediction using genetic programming has resulted in short- and long-term forecasts which have lower errors than many previously published methods, even when 'trained' on a dataset of nearly minimal length (only slightly above the number predicted by the $2\log_{10}N$ rule). This suggests that, applied carefully, this technique may be useful in the forecasting of chaotic series.

Furthermore, the S-expressions evolved appear capable of providing structural information about the non-linear system which can be useful to its investigation and modeling. In particular, the graph of normalized root mean squared prediction error against the length of forecast, and examination of the composition of fitter S-expressions, may be useful.

This study has however highlighted the need for careful construction of fitness functions, and a role for 'steering' them to 'cope with' deception. Further work needs to be undertaken to examine this, and the 'divide and conquer' approach to selection, lest prodigious memory and processor speed be mandatory requirements for success using genetic programming.

These two examples of the application of genetic programming illustrate that it is a technique which can already solve problems involving optimization in near-infinite search spaces, and that it promises even greater utility in the future.

Acknowledgments

Significant assistance was provided by Dr Bart Bartholomew (formerly of NCSC), Dr Bill Spears (of NCARAI, NRL), Professor John Koza and James Rice (both of Stanford University), and Kim Barrett (of Apple Computer, Inc.), among others.

Bibliography

Alexander, S. T. (1986) *Adaptive Signal Processing. Theory and Applications.* New York, NY: Springer-Verlag.

Ansari, N. (1992), Y. Huang and J-H. Lin, Adaptive stack filtering by LMS and Perceptron learning. In *Dynamic, Genetic, and Chaotic Programming*, B. Soucek, Ed. New York, NY: Wiley Interscience.

Casdagli, M. (1989) Nonlinear prediction of chaotic time series. *Physica D*. 35:335–356.

Casdagli, M. (1992), D. des Jardins, S. Eubank, J. D. Farmer, J. Gibson, and J. Theiler, Nonlinear modeling of chaotic time series: theory and applications. In *Applied Chaos*, J. H. Kim and J. Stringer, Eds. New York, NY: Wiley-Interscience.

Chu, C-H. H. (1989) A genetic algorithm approach to the configuration of stack filters. In *Proceedings of the Third International Conference on Genetic Algorithms*, J. D. Schaffer, Ed. San Mateo, CA: Morgan Kaufmann.

DeFatta, D. J. (1988), J. G. Lucas and W. S. Hodgkiss, *Digital Signal Processing*. New York, NY: John Wiley & Sons.

Farmer, J. D. (1987) and J. J. Sidorowich, Predicting chaotic time series. *Phys. Rev. Lett.* 59:845–848.

Gallagher, N. C. (1981) and G. L. Wise, A theoretical analysis of the properties of median filters. *IEEE Trans. Acoust. Speech Sig. Process.* ASSP-29:1136–1141.

Glass, L. (1992) and D. Kaplan, Time series analysis of complex dynamics in physiology and medicine. *Proceedings of NATO Workshop on Comparative Time Series Analysis* held at the Santa Fe Institute, 14–17 May 1992.

Grassberger, P. (1983) and I. Procaccia, Measuring the strangeness of strange attractors. *Physica* 9D:189–208.

Koza, J. R. (1992) *Genetic Programming. On the Programming of Computers by Means of Natural Selection.* Cambridge, MA: MIT Press.

Mackey, M. C. (1977) and L. Glass, Oscillation and chaos in physiological control systems. *Science* 197:287–289.

Mead, W. C. (1992), R. D. Jones, Y. C. Lee, C. W. Barnes, G. W. Flake, L. A. Lee and M. K. O'Rourke, Prediction of chaotic time series using CNLS-net – example: the Mackey-Glass equation. In *Nonlinear Modeling and Forecasting*, M. Casdagli and S. Eubank, Eds. Redwood City, CA: Addison-Wesley.

Meyer, T. P. (1992) and N. H. Packard, Local forecasting of high-dimensional chaotic dynamics. In *Nonlinear Modeling and Forecasting*, M. Casdagli and S. Eubank, Eds. Redwood City, CA: Addison-Wesley.

Ruelle, D. (1990) Deterministic chaos: the science and the fiction. *Proc Roy Soc Lond A* 427:241–248.

Spears, W. M. (1991) *GAL*. Common Lisp source code published electronically at ftp.aic.nrl.navy.mil.

Stokbro, L. (1992) and D. K. Umberger, Forecasting with weighted maps. In *Nonlinear Modeling and Forecasting*, M. Casdagli and S. Eubank, Eds. Redwood City, CA: Addison-Wesley.

Wendt, P. D. (1986), E. J. Coyle and N. C. Gallagher, Stack filters. *IEEE Trans. Acoust. Speech Sig. Process.* ASSP-34:898–911.

18 The Automatic Generation of Plans for a Mobile Robot via Genetic Programming with Automatically Defined Functions

Simon G. Handley

Planning is the creation of programs to control an agent, such as a robot. Traditionally, planners have maintained a logical model of the agent's world and planned by reasoning about what plans do to that world. In this chapter I describe a new planner, the Genetic Planner, that uses artificial selection, sexual mixing (recombination) and fitness proportionate reproduction to breed computer programs (i.e., to plan). This planner uses a simulation of the world to execute candidate computer programs (i.e., candidate plans). I first describe this planner and then I show it at work on a simple problem—a robot on a 2-D grid. Also, Koza's Automatically Defined Functions (ADFs) are used and the results compared with the non-ADF genetic programming system.

18.1   Introduction

Planning is the creation of computer programs (i.e., plans) that will be executed in the future and that will purposively control an agent, such as a robot.

Traditionally, planners have reasoned about the worlds they are planning for. This started with the recognition [Green 1969] that planning is reducible to automatic theorem proving in the situation calculus [McCarthy and Hayes 1969] and was made operational by the "STRIPS assumption" [Fikes and Nilsson 1972]. Non-linear planners (planners that explicitly search a space of incomplete plans) such as NOAH [Sacerdoti 1977] and TWEAK [Chapman 1987] also reason about their environment.

The Genetic Planner does not reason about the world it is planning to act in. Rather, it has a procedural model of the world and it simply *runs* candidate plans to see how well they work.

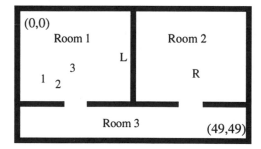

Figure 18.1
The simulated world: a 49×49 grid containing three rooms, a robot ('R'), three blocks ('1,' '2' and '3') and a light switch ('L'). The robot, the blocks, and the light switch each occupy one cell.

Table 18.1
The Tableau for the Motion Planning problem.

Objective:	Find a plan that achieves a given goal.
Terminal set without automatic function definition:	(left), (right), (move), (climb), (light).
Function set without automatic function definition:	prog2, prog3, prog4.
Fitness cases:	One: the goal to be achieved.
Raw fitness:	A linear combination of the cost of the plan and the fitness of the goal expression.
Standardized fitness:	same as raw fitness (but normalized to the range [0..1]).
Hits:	Number of predicates in the goal expression that are 100% satisfied.
Wrapper:	None.
Parameters:	M = 2,000; G = 51.
Success predicate:	A plan successfully achieves the goal and incurs the minimum cost. (This minimum cost is given by the user for each goal.)
Overall program structure with automatic function definition:	One result-producing branch; one function-defining branch. The defined function takes no arguments.
Terminal set for the result-producing branch:	(left), (right), (move), (climb), (light).
Function set for the result-producing branch:	prog2, prog3, prog4.
Terminal set for the function-defining branch:	(left), (right), (move), (climb), (light).
Function set for the function-defining branch:	prog2, prog3, prog4.

The Genetic Planner uses artificial selection, sexual mixing (recombination) and fitness proportionate reproduction to breed computer programs that will control the agent. The Genetic Planner doesn't need a logical representation of the world or the agent; rather, it uses a simulation (i.e., a procedural representation) of the agent and the world to execute candidate computer programs (i.e., candidate plans).

18.2   The Genetic Planner

In this section I'll describe the Genetic Planner. In the next section I'll encode an example problem for the Genetic Planner. This section will be rather brief as I'll rely on the next section to make things clearer.

For the Genetic Planner to produce plans to control an agent in some world, it needs the following description of that world:

- a set of procedurally-defined actions (or operators) that the agent can take,

- a function that returns the fitness (with respect to some goal) of each state of the world, and

- a simulation of the world.

The Genetic Planner uses the procedurally defined operators to execute candidate plans in the simulation. The fitness function evaluates the state of the world after each candidate plan has been simulated. The fitness of a plan is just the fitness of the state of the world after that plan has finished executing (in the simulation).

It is often convenient to represent the state of the world as a set of ground predicate calculus terms. Given such a representation of the world it is also convenient to represent goals as predicate calculus expressions. I have extended the Genetic Planner to facilitate such representations. In this case, the user must provide the following description of the world:

- a set of procedurally-defined operators (as before),

- a set of predicates that together completely describe the world (such as on, clear and handempty in the blocks world),

- a fitness function for each of the just-mentioned predicates (these take as arguments a predicate and a state of the world and return a number that is the fitness of that predicate in that state),

- a ground goal expression with no quantifiers (i.e., a conjunction, disjunction or negation of either a goal expression or one of the above predicates), and

- a simulation of the world (as before).

For the remainder of this chapter, the term "Genetic Planner" will refer to this extended version.

Notice the lack of a fitness function that acts on states in the extended version of the Genetic Planner. The fitness functions that the user provides in the extended version act on predicates, not on expressions. For example, in the blocks world a conceivable goal expression is 'on(a,b) $\land$ on(b,c)'. The user-supplied fitness functions will compute the fitnesses of 'on(a,b)' and 'on(b,c)' but don't give the Genetic Planner any idea how these two fitnesses should be combined. To make up for this, I have used the following scheme (where $0 \leq \text{fitness}(\cdot) \leq 1$ and a fitness of 0 is best):

$$\text{fitness}(a \lor b) = \text{fitness}(a) \cdot \text{fitness}(b),$$
$$\text{fitness}(a \land b) = \text{fitness}(a) + \text{fitness}(b) - \text{fitness}(a) \cdot \text{fitness}(b), \text{ and}$$

$$\text{fitness}(\neg a) \quad = \quad 1 - \text{fitness}(a).$$

This conversion from expressions to fitnesses is adapted from fuzzy logic [Zadeh 1975, 1983].

Minimum and maximum could also be used for conjunction and disjunction but they're not as good as the above definitions because they discard valuable information. For example, consider the expression a ∧ b. If fitness(a) = 0.5 and fitness(b) = 0.3 then 0.5·0.3 = 0.15 and min(0.5,0.3) = 0.3. If, however, fitness(a) = 0.6 and fitness(b) = 0.3 then 0.6·0.3 = 0.18 and min(0.6,0.3) = 0.3. The fitness of a ∧ b improves when the fitness of the conjunction is the product of the fitness of its components but is unchanged when it is the maximum of its components. Two programs, one of which is slightly better in some way (as reflected in the slightly higher fitness of 'a'), will have the same fitness by the max/min definition of fitness: this is the sense in which information is discarded.

If the plans produced by the Genetic Planner will control a real agent in a real world then the efficiency of the plans becomes important. The fitness functions described above will result in plans that achieve the required goal but in extraordinarily inventive and bizarre ways—there is no selective pressure to produce minimal plans. Such selective pressure can be introduced by modifying fitness to be a linear combination of fitness (as defined above) and the cost incurred during the execution of the plan,

"real" fitness = "naive" fitness $\times (1-p)$ + cost-normalized-to-the-range-[0,1] $\times p$,

where p is the proportion of "real" fitness that comes from cost (recall that fitness is in the range [0,1]).

18.3   An Example World: A Robot on a 2-D Grid.

In this section I'll show how to formulate an example world—a robot on a 2-D grid—so that the Genetic Planner can create plans that achieve goals in this world.

The problem is as follows: the world is finite, two-dimensional and divided into a square grid of 49×49 tiles. This world contains: a robot (that can move North, East, South or West), three boxes, a light switch and various walls. Each grid cell contains either the robot, the light switch (only one is allowed per world and its location is fixed), a box, or a part of a wall, a robot and a box (the robot can climb onto the boxes) or nothing. Figure 18.1 shows this world.

This problem is a simplification of the problem on which Fikes and Nilsson demonstrated STRIPS in their seminal paper [Fikes and Nilsson 1971] (the simplification is that their world had 4 rooms). I chose this problem to demonstrate the Genetic Planner on because it has no known bias for or against this approach to planning.

I'll now show each of the pieces of information that the Genetic Planner requires.

18.3.1   A set of procedurally-defined operators

The actions that the robot can take in this world are:

- `left` (the robot turns 90 degrees anti-clockwise),

- `right` (the robot turns 90 degrees clockwise),

- `move` (the robot moves one square in the direction it is currently facing, as long as there is no wall cell in the way and no blocks are pushed into wall cells; the robot can push multiple adjacent blocks simultaneously if there is an empty cell in which to push the furthermost block),

- `climb` (the robot climbs onto a box if there is one in an adjacent cell, otherwise it just moves), and

- `light` (toggles the light switch on/off if the robot is standing on a box in a cell adjacent to the light switch, otherwise the robot just moves).

The Genetic Planner takes these operators, adds three sequencing operators (`prog2`, `prog3` and `prog4`), and produces programs that operate the robot. Some examples: (`prog2` (`move`) (`move`)) moves the robot two squares in the current direction, (`prog3` (`move`) (`move`) (`prog2` (`right`) (`move`))) does a Knight's move (two squares forward, one to the right).

18.3.2   A set of predicates that describe the world

The predicates that describe the world are:

- 'at(*object*,*x*,*y*),' which is true if the *object* is at the location (x, y) in the current state, where $x, y \in [0...48]$. The *object* is one of 'robot,' 'light-switch,' 'box-1,' 'box-2' and 'box-3.'

- 'next-to(*object₁*,*object₂*),' which is true if the two objects are in adjacent grid cells in the current state.

- 'light-switch(x),' which is true if the status of the light switch in the current state is x.

18.3.3   Fitness functions for each of the predicates.

The raw fitness of the predicate 'at(*object*,*x*,*y*)' is defined as the manhattan distance between *object* and the cell (x, y). (The manhattan distance between points (x_1,y_1) and (x_2,y_2) is $|x_1-x_2| + |y_1-y_2|$.) The fitness of 'next-to(*object₁*, *object₂*)' is similar: it is just the

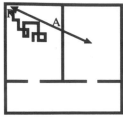
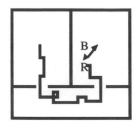

Figure 18.2
Two programs that are trying to direct the robot from the center of room 1 to the center of room 2. The arrows indicate the error of each plan.

distance between the two objects. The third predicate, 'light-switch(x),' lacks a spatial component so its fitness function is two-valued: 0 if the predicate is true and 1 otherwise.

Recall that the Genetic Planner combines the fitness of individual predicates to produce the fitness of an entire expression. So, for example,

$$\text{fitness(next-to(x,y)} \vee \text{next-to(y,z)))} = \text{fitness(next-to(x,y))} \cdot \text{fitness(next-to(y,z))}$$
$$= \text{distance-between(x,y)} \cdot \text{distance-between(y,z)}.$$

The Genetic Planner will use these fitness functions to discriminate between imperfect programs. Consider, for example, the goal of moving the robot from the center of room 1 (of figure 18.1) to the center of room 2. Consider the two programs shown in figure 18.2: one that heads for the top left of room 1 and one that gets almost to the center of room 2.

The plan on the left of figure 18.2 leaves the robot further from the center of room 2 (the room on the right) than does the plan on the right (i.e., distance A > distance B). Without a distance-based measure of fitness, all the Genetic Planner can say about these two plans is that they both don't make it to the goal cell. Adding the notion of fitness allows the Genetic Planner to recognize and nurture plans that do some things right, but don't necessarily do everything right.

18.3.4 A ground goal expression

The Genetic Planner accepts goal expressions that are conjunctions, disjunctions and negations of at($\cdot$), next-to($\cdot$) and light-switch($\cdot$). Some examples: 'next-to(box-1,box-2) $\wedge$ next-to(box-2,box-3)' is true of any state in which all three boxes are adjacent to each other, 'light-switch(on) $\wedge$ next-to(robot,box-1) $\wedge$ $\neg$(next-to(robot,box-2) $\vee$ next-to(robot,box-3))' is true of states in which the light switch is turned on and the robot is next to box-1 but not next to either of boxes 2 and 3.

18.3.5 A simulation of the world

Finally, a simulation is needed in which the operators—left, right, move, climb and light—can be defined. This is straightforward and the operator definitions are what one would expect.

The tableau in table 18.1 summarizes this problem.

18.4 A Demonstration of The Genetic Planner

To show that the Genetic Planner actually works, I've used it to plan for the three goals that Fikes and Nilsson (1972) demonstrated STRIPS on. These goals are (in both English and predicate calculus):

1. Turning on the light switch: light-switch(on),
2. Pushing the three boxes together: next-to(box-1,box-2) ∧ next-to(box-2,box-3), and
3. Going to a location in another room: at(robot, x, y).

The Genetic Planner could not find a plan to achieve the first goal, light-switch(on). Because the Genetic Planner doesn't have a logical representation of the world, it doesn't realize that light-switch(on) = ∃Box . on(robot,Box) ∧ next-to(Box,light-switch) ∧ light-switch(on) (where on(a,b) has the obvious semantics). If the Genetic Planner could perform this deduction then it would have a spatial predicate—next-to(Box,light-switch)—that converts into multiple degrees of fitness. Since the Genetic Planner can't reason about the world, it can only randomly generate plans until one happens to push a box up to the light switch, climb on the box and then turn on the light switch.

The second goal—to push the three boxes together—is much better suited to the Genetic Planner than the first goal. I did 65 runs with a population size of 2,000 and each run lasted for a maximum of 51 generations (1 initial random generation and 50 subsequent evolved generations). For all runs reported here (and in the next section) 20% of fitness was the cost of the plan.

To evaluate the difficulty of these goals I'll use the following notation introduced by Koza [Koza 1992]. $P(M,i)$ is the probability of a single run of at most i generations (each consisting of M individuals) finishing successfully. This is approximated by doing a large number of runs. The number of runs required to produce a successful individual with probability z is defined in terms of $P(M,i)$,

$$R(z) = \left\lceil \frac{\log(1-z)}{\log(1-P(M,i))} \right\rceil.$$

The number of individuals that must be processed to find a successful individual with probability z is $I(M,i,z) = R(z) \cdot M \cdot i$. The minimum of the $I(Mi,iz)$ curve is a measure of the difficulty of the problem.

For the problem of pushing the boxes together, the minimum of the $I(M,i,z)$ curve is 52,000 and is attained at generation 12. That is, 2 runs, each having a population size of 2,000 and each lasting for at most 13 generations (12 evolving generations + generation 0), will process 52,000 individuals (= 2 runs × 2,000 individuals × 13 generations) and there is a 99% chance that one or more of those 52,000 individuals will be a solution to the problem.

The third goal—to go to another room—is also easily solved by the Genetic Planner. I did 63 runs, with each run having, as before, a population size of 2,000 and a maximum of 51 generations. The minimum of the $I(M,i,z)$ curve is 616,000 and is attained at generation 13. That is, if we do 22 runs, each having a population size of 2,000 and each lasting for at most 14 generations then 616,000 individuals will be processed and there is a 99% chance that one or more of those 616,000 individuals will be a solution to the problem.

18.5   Automatically Defined Functions

Genetic programming with Automatically Defined Functions (ADFs) refers to the ability to evolve both a program and a collection of subroutines that are called by the program [Koza 1992, 1994]. For example, if we are trying to evolve a program that computes $x^4 + 1$ then a possible program is

```
(progn (defun adf0 (arg0)
          (* arg0 arg0))
       (values (+ (* (adf0 x) (adf0 x)) 1))).
```

The above s-expression is evaluated in two steps: First, the function adf0 is defined to take one argument, arg0, and to return the square of that argument, (* arg0 arg0). Second, the expression (+ (* (adf0 x) (adf0 x)) 1)) is evaluated with the above definition of adf0, thus returning (+ (* (* x x) (* x x)) 1) = x^4+1.

Crossovers are restricted to transfer subtrees only between one or other of the above underlined regions. If the function or terminal sets of the first underlined region (the *function-defining branch*) and the second underlined region (the *result-producing branch*) are different then crossovers must be restricted to exchange subtrees only between the same branches of different individuals. (For example, the terminal sets of the two branches in the above example are different: the function-defining branch contains `arg0` and the result-producing branch contains `x`.)

When using genetic programming with ADFs there is another step to take to prepare a problem for solution by genetic programming: deciding how many defined functions to have, how many arguments each defined function should take and deciding what calls what.

For this problem, I decided that each individual would contain a single function-defining branch, that the defined function would take no arguments and that only the result-producing branch can call the defined function (i.e., `adf0` is in the result-producing branch's function set but isn't in the function-defining branch's function set).

I did 40 runs to solve the second goal (pushing the boxes together) with ADFs. The minimum of the $I(M,i,z)$ curve is 76,000 and is attained at generation 18.

I also did 82 runs of the third goal (moving to the other room) with ADFs. The minimum of the $I(M,i,z)$ curve is 96,000 and is attained at generation 11.

Table 18.2 summarizes the above results. The following statistics are also useful: The *structural complexity ratio* is the difference in size between successful non-ADF and ADF programs. Similarly, the *efficiency ratio* compares the difference between the minimum of the $I(M,i,z)$ curves.

	Structural Complexity Ratio	Efficiency Ratio
Boxes together	1.2	0.7
Other room	1.0	6.4

For a given problem, a structural-complexity ratio greater than 1 means that ADF programs are smaller than non-ADF programs. The efficiency ratio is a measure of the relative ease of solution for a problem with ADFs compared to without ADFs; an efficiency ratio greater than 1 means that fewer individuals need to be processed to find a solution with ADFs than without them.

For the goal of pushing the boxes together, the structural-complexity ratio is 1.2 which means that the ADF programs are smaller than the non-ADF programs. The efficiency ratio is 0.7 which means that finding a solution is harder with ADFs than without them. This would seem to be because this problem is too easy to justify the extra overheads of evolving the subroutines and integrating them together in the result-producing branch. (Note, however, that the efficiency ratio for this goal is not very meaningful. This is because of the quantization in the definition of $R(z)$ and the fact that all but one run succeeded (see Table 18.2). For example, if the one unsuccessful run had actually succeeded before generation 20 then the efficiency ratio would be 1.2. Many more runs need to be done to determine a statistically valid value for this ratio.)

For the goal of moving to the other room, the structural-complexity ratio is 1.0 which means that the ADF programs are the same size as the non-ADF programs. The efficiency ratio is 6.4 which means that finding a solution is easier with ADFs than without ADFs. This would seem to be because the genetic programming system is able to hierarchically decompose the problem and to recombine the sub-solutions (i.e., the defined function) in such a way as to avoid repeatedly re-evolving a solution to the sub-problem represented by the defined function.

18.6   Analysis of Some Best-Of-Run ADFs

In this section I'll analyze some of the ADFs that the genetic programming system created.

Figure 18.4 shows the path followed by a robot under the control of a best-of-run plan that solves the second goal (pushing the boxes together). The individual was

```
(progn (defun rft0-0 ()
          (prog2 (prog2 (prog3 (right) (light) (light)) (prog4 (left)
            (move) (climb) (move))) (prog2 (prog4 (right) (left)
            (left) (left)) (prog3 (left) (left) (left)))))
       (values (prog4 (prog2 (prog4 (rft0-0) (rft0-0) (climb)
          (climb)) (prog4 (climb) (move) (move) (move))) (prog2
          (prog3 (rft0-0) (rft0-0) (climb)) (prog4 (rft0-0) (move)
          (light) (move))) (prog4 (prog3 (light) (move) (move))
          (prog2 (rft0-0) (light)) (prog3 (prog3 (move) (right)
          (left)) (right) (light)) (prog2 (right) (left))) (prog2
          (left) (prog3 (climb) (move) (move)))))))
```

Table 18.2
Summary of ADF and non-ADF runs for two of the three goals. The "#runs" column is the total number of runs done, "Unsucc." is the number of those runs that were unsuccessful, "Min I(M,i,z)" is the minimum of the I(M,i,z) curve, and "Size" is the average size of successful individuals.

	Blocks Together				Other Room			
	#runs	Unsucc.	Min I(M,i,z)	Size	#runs	Unsucc.	Min I(M,i,z)	Size
non-adf	65	3	52,000	149.9	63	41	616,000	123.2
adf	40	1	76,000	122.2	82	11	96,000	122.8

and its defined function can be summarized as RM^2LMCML where L,R = turn left/right, M = move/light and C = climb. This ADF moves the robot in an L-shaped trajectory and leaves the robot pointing 90° anticlockwise of its original direction (see figure 18.3(a)).

Note that the climb operation doesn't change the robot's direction. When climbing off a box, the robot first tries to move into the square that's adjacent in the North direction; if that square contains a box or a wall then it tries the adjacent squares in the South, East and West directions, in that order. If all four adjacent squares are full then the climb operation fails and the robot stays put.

Note also that although the defined functions don't have explicit arguments, they are implicitly parameterized by the location and direction of the robot when the defined function is called. For example, in Figure 18.5 the two calls to the defined function do quite different things.

Figure 18.5 shows another best-of-run individual that solves the same goal (pushing the boxes together). This individual is

```
(progn (defun rft0-0 ()
          (prog2 (prog4 (prog3 (light) (right) (climb)) (light)
          (prog4 (light) (left) (right) (climb)) (prog3 (left)
          (prog4 (light) (left) (right) (climb)) (light))) (prog4
```

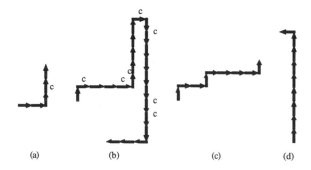

(a)   (b)   (c)   (d)

Figure 18.3

The defined functions: they correspond to (a) figure 18.4 (RM^2LMCML); (b) figure 18.5 ($MRCM^2CLMCM^3RCRMCM^4C^2MRM^3$)—Note that the robot is left pointing 90° anticlockwise relative to its initial heading.; (c) figure 18.6 (MRM^2LMRM^4LM); and (d) figure 18.7 (M^8LM). Arcs annotated with a "c" are climbs—it is important to distinguish between climbs and moves for the pushing-boxes goal but not for the moving-to-the-other-room goal.

```
      (prog2 (light) (light)) (prog4 (right) (climb) (right)
      (light)) (prog2 (climb) (move)) (prog3 (move) (right)
      (prog3 (prog2 (prog2 (left) (light)) (prog4 (light)
      (left) (right) (climb))) (prog2 (climb) (light)) (prog2
      (prog3 (right) (light) (light)) (light))))))))
  (values (prog2 (prog2 (prog4 (light) (climb) (rft0-0) (left))
      (prog2 (left) (light))) (prog4 (prog3 (rft0-0) (prog4
      (prog4 (move) (left) (right) (right)) (left) (move) (move))
      (right)) (prog3 (prog2 (left) (right)) (climb) (left))
      (prog2 (move) (right)) (prog4 (right) (light) (move)
      (right))))))
```

and its defined function can be written as

$$MRCM^2CLMCM^3RCRMCM^4C^2MRM^3$$

and its trajectory is shown in figure 18.3(b).

Figure 18.6 shows a best-of-run individual that solves the third goal (moving to the other room). This individual is

```
(progn (defun rft0-0 ()
      (prog2 (prog2 (prog3 (prog3 (climb) (right) (light))
        (light) (left)) (prog3 (climb) (right) (light))) (prog2
        (prog4 (climb) (climb) (climb) (left)) (light)))
      (values (prog3 (prog4 (right) (prog4 (climb) (climb) (rft0-0)
        (prog4 (light) (prog3 (climb) (prog2 (move) (climb))) (prog2
        (rft0-0) (move))) (climb) (climb))) (prog2 (rft0-0)
        (light)) (climb)) (prog3 (prog3 (left) (left) (left))
        (prog4 (rft0-0) (climb) (left) (light)) (prog4 (light)
        (prog3 (light) (rft0-0) (climb)) (climb) (climb))) (prog4
        (prog2 (left) (rft0-0)) (prog3 (climb) (move) (left))
        (prog3 (rft0-0) (right) (climb)) (prog4 (rft0-0) (move)
        (left) (climb))))))
```

and its ADF can be written as MRM^2LMRM^4LM; it moves the robot right six squares and straight ahead three squares by a zig-zag movement (see figure 18.3(c)).

Figure 18.7 shows another best-of-run individual that solves the same goal (moving to the other room). This individual is

```
(progn (defun rft0-0 ()
      (prog2 (prog4 (prog3 (light) (climb) (left)) (prog3 (left)
        (right) (right))(prog2 (left) (left)) (left)) (prog4
```

```
        (prog4 (left) (light) (light) (move)) (prog2 (light)
          (move)) (prog2 (climb) (left)) (climb)))
   (values (prog2 (prog4 (left) (prog3 (left) (light) (light))
       (rft0-0) (prog3 (climb) (move) (move))) (prog4 (prog4
       (climb) (rft0-0) (light) (prog3 (light) (move) (move)))
       (prog3 (right) (climb) (light)) (prog4 (rft0-0) (light)
       (prog4 (rft0-0) (left) (left) (right)) (left)) (left))))).
```

This defined function is equivalent to M^8LM and is shown in figure 18.3(d).

18.7   Discussion

The Genetic Planner does not reason about the world it is planning to act in. Rather, it has a procedural model of the world and it simply *runs* candidate plans to see how well they work. Many traditional "problems"—such as the Sussman anomaly and dynamic

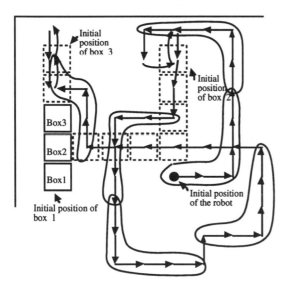

Figure 18.4
The trajectory of the robot following a best-of-run individual. The following conventions are used in this (and the following) figures: the circled moves are those generated by the defined function; straight arrows are successful moves (and unsuccessful climbs and lights); curved arrows are successful climbs; dashed squares indicate the initial and intermediate positions of the boxes, solid squares indicate the final positions of the boxes. This figure (and figure 18.5) is a close-up of the top-left 10x10 portion of room 1.

worlds—simply disappear.

The Genetic Planner doesn't care if its goal is conjunctive or not—it can equally well plan to achieve goals that contain disjunctions and negations. The Sussman anomaly (which consists of goals that contain a conjunction the conjuncts of which cannot be solved independent of the others) is not anomalous for the Genetic Planner since it implicitly plans for all components of the goal expression simultaneously.

Planners that reason about the world they're planning in have difficulty with dynamic worlds because when the world changes they have to examine and possibly discard all deductions that were based on the previous state of the world. The Genetic Planner, however, is not thrown totally off track if the world the plans are running in changes between generations (or even between individuals in a "steady state" [Syswerda 1991] system). Koza's discussion of "catastrophic damage" [Koza 1992] shows empirically that genetic programming can recover from surprisingly severe changes in the fitness function.

The Genetic Planner trivially handles actions that occur over a finite time (i.e., "temporal world models"). The Genetic Planner as described in this chapter will construct plans that contain such actions: only the simulation must be extended. Remember that the Genetic Planner won't try to reason about such actions so complicated time-period algebras are not required.

Planners that reason about the world typically reason for an extended time and then come back with an answer. This is not very useful if there is a time constraint such as "go

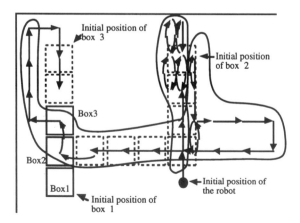

Figure 18.5
The trajectory of another best-of-run individual.

away and come back with the best plan you can find in t seconds." Algorithms that output a stream of increasingly accurate answers are called *anytime algorithms* [Dean and Boddy 1988]. The Genetic Planner is such an algorithm: it can be thought of as a coroutine that outputs a stream of monotonically increasingly fit plans—at the end of each generation it can output the best individual seen so far.

The Genetic Planner has two primary drawbacks. First, it needs feedback from the world about the relative success of unsuccessful plans. The Genetic Planner cannot efficiently plan in worlds in which reasoning is required to recognize the relative worth of unsuccessful plans—such as a plan for the turning-on-the-light-switch problem that pushes a box up next to the light switch but doesn't climb up and turn the light switch on. The Genetic Planner is ideally suited to *motion planning* (planning trajectories in two- or three-dimensional space) because the spatial components in the goal predicates convert trivially into multiple shades or degrees of fitness.

The second drawback with the Genetic Planner is speed. Evolving large populations of plans is time- and space-consuming. This, however, is also true of any planner that is planning for a non-trivial world—since planning is intractable [Chapman 1987].

18.8 Future Work

The plans created by the Genetic Planner are currently very simple. I'd like to add sensors, conditionals and iteration to the actions that the robot can perform. "Complex"

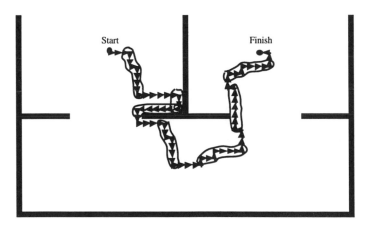

Figure 18.6
A best-of-run individual that solves the move-to-the-other-room goal.

actions such as iteration are easily added to the Genetic Planner because it doesn't need to reason about them—it just has to execute them in its simulator.

If the robot is capable of sensing the world then the programs being evolved could be considered as mappings from worlds to actions, rather than as a sequence of actions. That is, instead of calling each program once, the simulator would call each program multiple times: once for each move (this is similar to how Koza solved the "artificial ant" problem [Koza 1992]). This complicates the simulator (it now has to work out when to stop simulating programs) but reduces memory usage (the programs currently evolved by the Genetic Planner get very large).

It would also be interesting to try varying the influence of cost in the fitness function. A simple and—I suspect—profitable scheme would be linearly to ramp the percentage of fitness that is cost from 0% in generation 0 up to, say, 25% in the final generation. This would enable the Genetic Planner to clamp down on unruly plans later in the planning process while not unduly stifling innovation in the early generations.

18.9   Conclusions

This chapter has introduced a new planner, the Genetic Planner, that plans by using genetic methods to evolve programs (i.e., plans). I encoded a simple world for the Genetic Planner—a robot on a 2-D grid—and three simple goals were planned for by it. The Genetic Planner easily created plans to achieve two of the goals while the third was troublesome. This third goal—turning on the light switch—was an example of a goal for which the Genetic Planner is not suited—the "Discussion" section explains why this is so.

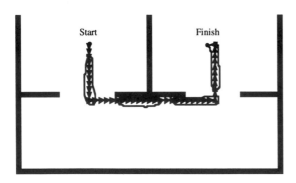

Figure 18.7
The trajectory of another best-of-run plan for the moving-to-the-other-room goal.

The Genetic Planner easily created plans that achieved the other two goals because these goals contained a spatial component which the Genetic Planner used as a fitness measure. I have also shown that using Automatically Defined Functions can facilitate the evolutionary process, if the problem is hard enough.

Acknowledgments

I'd like to thank John Koza and James Rice for their help with genetic programming; the reviewers for their helpful comments; and the Knowledge Systems Laboratory, Stanford University, for the use of their computers.

References

Chapman, D. (1987) Planning for conjunctive goals. *Artificial Intelligence*, 32, pages 333–77.

Dean, T. and M. Boddy (1988) An analysis of time-dependent planning. Proceedings of the Seventh National Conference on Artificial Intelligence, pages 49–54, Menlo Park, CA: AAAI Press.

Fikes, R. E. and N. J. Nilsson (1971) STRIPS: A New Approach to the Application of Theorem Proving to Problem Solving. *Artificial Intelligence*, 2, pages 189–208.

Green, C. (1969) Application of theorem proving to problem solving. Proceedings of the First International Joint Conference on Artificial Intelligence, pages 219–39, Los Altos, CA: Morgan Kaufmann. (Also in *Readings in Artificial Intelligence*, B. L. Webber and N. J. Nilsson (eds.), pages 202–22, Los Altos, CA: Morgan Kaufmann, 1981.)

Koza, J. R. (1994) *Genetic Programming II*. In press. Cambridge, MA: The MIT Press.

McCarthy, J. and P. Hayes (1969) Some philosophical problems from the standpoint of Artificial Intelligence. *Machine Intelligence*, 4, pages 463–502, B. Meltzer and D. Michie (eds.), Edinburgh University Press.

Sacerdoti, E. D. (1977) *A Structure for Plans and Behavior*. New York: Elsevier.

Syswerda, G. (1991) A study of reproduction in generational and steady-state genetic algorithms. In *Foundations of Genetic Algorithms*, pages 94–101, G. J. E. Rawlins (Ed), San Mateo, CA: Morgan Kaufmann.

Zadeh, L. A. (1975) Fuzzy Logic and Approximate Reasoning. *Synthese*, 30, pages 407–28.

Zadeh, L. A. (1983) Commonsense Knowledge Representation Based on Fuzzy Logic. *IEEE Computation*, 16(10), pages 61–6.

19   Competitively Evolving Decision Trees Against Fixed Training Cases for Natural Language Processing

Eric V. Siegel

Competitive fitness functions can generate performance superior to absolute fitness functions [Angeline and Pollack 1993], [Hillis 1992]. This chapter describes a method by which competition can be implemented when training over a fixed (static) set of examples. Since new training cases cannot be generated by mutation or crossover, the probabilistic frequencies by which individual training cases are selected competitively adapt. We evolve decision trees for the problem of word sense disambiguation. The decision trees contain embedded bit strings; bit string crossover is intermingled with subtree-swapping. To approach the problem of overlearning, we have implemented a fitness penalty function specialized for decision trees which is dependent on the partition of the set of training cases implied by a decision tree.

19.1   Introduction

Competitive fitness functions can generate performance superior to absolute fitness functions [Angeline and Pollack 1993], [Hillis 1992]. A competitive fitness function is computed through some type of interaction between co-adapting individuals. For example, [Angeline and Pollack 1993] evolves Tic Tac Toe players whose fitness measurements are computed by having population members participate in a Tic Tac Toe tournament. Hillis [1992] evolves sorting networks in competition with a separate, evolving population of lists of numbers to be sorted. A sorting network's fitness depends on how well it sorts lists of numbers from the competing population, and a list of number's fitness is dependent on how poorly it is sorted by sorting networks. Competitive fitness functions guide the adaptive process since weaknesses of adapting individuals are discovered and therefore accented by competing adapting individuals.

Hillis' sorting networks are evaluated against a dynamic population of training cases whose adaptation involves crossover and mutation. However, many induction tasks involve a fixed, static "population" of training examples which cannot participate in these creative evolutionary operations. Such situations arise when the training data have been empirically collected, as in symbolic regression. In this chapter, we present a method by which the training set can competitively adapt without the generation of new training examples.

Decision trees [Quinlan 1986] are appropriate for many pattern classification problems. Koza [1991] has demonstrated that the genetic programming paradigm is capable of inducing decision tree structures. In this chapter, we evolve decision trees for a real world problem with noisy, empirical data.

In [Tackett 1993], genetic programming is used for a two class classification problem. It is compared to a binary tree classifier, which statistically induces trees similar to the decision trees presented in this chapter.

Table 19.1
Tableau for the competitive evolution of decision trees for word sense disambiguation

Objective:	Evolve a decision tree which classifies occurrences of *discourse cue words* as to their usage.
Terminal set:	The two classes of this classification problem, specifically, *discourse* and *sentencial*.
Function set:	Each internal node is like a `switch` statement in C. A series of comparisons is made, and one downward arc is selected. See section 19.4 for detail.
Fitness cases:	513 examples of *discourse cue words*, their immediate context as used in spoken English, and their meaning as used in that context.
Raw fitness:	A decision tree is evaluated over a competitively selected distribution of training cases (513 cases), and raw fitness is the number of training cases correctly classified. See section 19.6 for details on competition against fixed training cases. See section 19.7 for the description of a fitness penalty which averts overlearning.
Standardized fitness:	513 minus raw fitness.
Parameters:	Number of generations = 500, population size = 900.
Termination predicate:	Reach final generation of run.
Identification of best:	Every 20 generation, the population of decision trees is tested for absolute fitness (as evaluated uniformly across the 513 training cases).

Sections 19.2 and 19.3 describe the problem domain and the set of training examples. Sections 19.4 and 19.5 explain how decision trees work and how crossover is implemented for the decision tree representation. Section 19.6 describes how fixed training data can participate in competitive adaptation. Section 19.7 describes a method to avert overlearning when inducing decision trees over a limited training set. Finally, section 19.8 draws conclusions.

19.2   The Domain: Word Sense Disambiguation

In natural language, many words have multiple *senses* (meanings). For example, *anyway* can mean "in any case", as in, "I did it *anyway*." (This is considered the *sentencial* meaning of *anyway*.) It can also mean, "Let's return to a previous topic", as in "*Anyway*, what were you saying before?" (This is considered the *discourse* meaning of *anyway*.) The sense of an ambiguous word such as *anyway* is dependent on the context in which it is used. A major thrust of the natural language understanding field is deriving mechanisms to resolve such ambiguities, a process called *disambiguation*.

The approach to word sense disambiguation taken here is to evolve decision trees which attempt to establish word sense by looking only at *immediate* context, that is, the *tokens*, (words and punctuations marks) residing within a small distance of the word to be

Table 19.2
Example discourse cue words – those which occur most frequently in this study. A total of 35 cue words occur. The number of times each discourse cue word occurs as a training case is listed.

Cue word	Instances	Cue word	Instances	Cue word	Instances	Cue word	Intances
and	405	*like*	73	*say*	36	*actually*	33
now	76	*but*	69	*well*	35	*see*	29
so	75	*or*	63	*look*	35	*first*	25

disambiguated. Specifically, a decision tree can examine the tokens residing immediately to the left of the word, and those up to four positions to the right (positions -1 through 4).

The class of words being disambiguated are *discourse cue words* (e.g. *anyway*). Table 19.2 contains example discourse cue words. A discourse cue word is used by a speaker to convey intentions with respect to the "flow" of a discourse. Cue words often indicate how a sentence or clause relates to the current topic of conversation, e.g. digression, conclusion, etc. Each discourse cue word also has at least one alternative sense as a verb, adverb or connective, its *sentencial* sense. Therefore, any instance of such a word must be disambiguated as to whether it is being used in a *discourse* sense or a *sentencial* sense.

Hirschberg and Litman [1993] explore several methods for cue word sense disambiguation, including the examination of intonational features. They also measure the ability to perform this task by looking only at punctuation marks immediately before and after the cue word, suggesting the strategy embodied by the decision tree in Figure 19.1.[1] This decision tree correctly classifies 79.16% of the training cases used in these experiments. If the word is the first in a sentence, i.e. following a period, it is classified as *discourse*.

The next section introduces the set of training examples we have for the problem of word sense disambiguation, and formalizes word sense disambiguation as a classification problem.

19.3 The Training Cases

For these experiments, we have access to the transcript of spoken English used by Hirschberg and Litman [1993]. In this transcript, each discourse cue word has been manually marked by a linguist as to whether its sense is *discourse* or *sentencial*. The training examples are therefore empirical data – measurements of human perception. The transcript provides 1,027 examples.

---

[1] This decision tree is extrapolated from Table 11 in [Hirschberg and Litman 1993]. Their study used the same transcript as this one, and primarily used the same training cases.

Table 19.3
Example training cases. Each training case has a value for each of 6 attributes, and a class. Attribute **0** is the cue word to be disambiguated.

-1	0	1	2	3	4	**Class:**
work	*and*	*we*	*are*	*really*	*pleased*	`discourse`
.	*But*	*we*	*stop*	*there*	*because*	`discourse`
.	*Now*	*that*	*doesn't*	*mean*	*we*	`discourse`
very	*well*	*founded*	*principle*	*principled*	*in*	`sentencial`
to	*look*	*more*	*like*	*sentences*	.	`sentencial`
,	*and*	*that's*	*on*	*the*	*second*	`sentencial`
description	*ok*	.	*Is*	*a*	*surgeon*	`discourse`

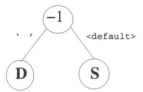

Figure 19.1
This small, manually created decision tree correctly classifies 79.16% of the training examples. At leaves, "D" stands for *discourse* and "S" for *sentencial*.

Table 19.3 contains sample training data. Each training example has 6 *attributes*: positions -1 through 4. This includes position zero, the discourse cue word to be disambiguated. For any given training example, each attribute has a corresponding *value*, that is, a token. Each training example also has a *class*, that is, the word sense. Note that we have formalized word sense disambiguation as a two class classification problem.

Decision trees operate on one training example at a time, attempting to derive the correct classification. We evolutionarily induce decision trees which correctly classify a high percentage of the training cases. In order to ascertain the generalization performance of these trees, induction takes place over one half of the training cases, and evolved trees are tested over the remaining cases, the *test cases*. The division into training and test sets is made randomly upon each run.

The next section describes the decision tree mechanism, and how it is used for word sense disambiguation.

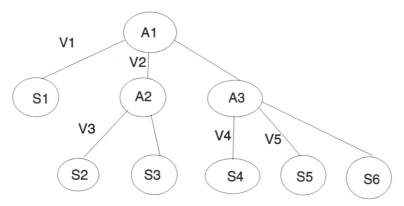

Figure 19.2
Formal representation of a decision tree. Internal nodes are labeled with attribute sets, arcs are labeled with value sets, and leaves are labeled with a class. Rightmost arcs are "default paths".

19.4 How Decision Trees Work

Figure 19.2 shows the formal representation of a decision tree, and Figures 19.3 and 19.4 show example decision trees generated by evolution.[2] Each internal node of a tree is labeled with a set of attributes (token positions), each arc is labeled with a set of values (tokens), and each leaf is labeled with a class name (word sense).

The internal nodes are treated like a `switch` statement in C. A series of comparisons is made, and one downward arc is selected. For a given training example, the tree is traversed deterministically from root to leaf, thus classifying the example, by the following recursive process:

At the current internal node, the set of values from the training example which correspond to the node's attributes is identified, and the first arc with an intersecting value set, going from left to right, is selected.

The rightmost arc under each internal node is a "default" arc which has no explicit value set. This arc is traversed if none of its sister arcs has an intersecting value set.

For example, to classify the first training example from Table 19.3 with the decision tree in Figure 19.3, the tree traversal starts at the root node. The right arc is traversed, since position -1 has neither a period nor *and*. Then the leftmost arc is traversed, since position **0**

---

[2]These trees have been automatically edited to remove most redundant and useless data for the purpose of inspection. The editing process preserves semantics and is *not* part of the evolutionary process.

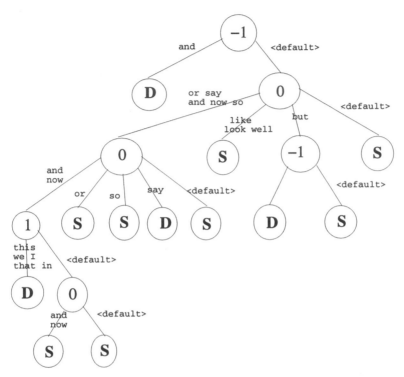

Figure 19.3
Example decision tree, induced by evolution. At leaves, "D" stands for *discourse* and "S" for *sentencial*. This tree scored 81.09% over the training set and 83.27% over the test set. The original (unedited) tree has 37 nodes.

has value *and*. Then two more leftmost arcs are traversed, leading to a leaf which classifies the training example as *discourse*, the correct classification.

The superset of values which can be members of a decision tree's value sets is the compilation of all values which occur in the training examples. A value frequency threshold, 15, has been selected manually; only values occurring frequently enough are considered for explicit inclusion in decision trees. The resulting superset of values is of size approximately 26.[3] Since some values cannot appear explicitly in value sets, the default arcs are necessary.

If a node contains attribute set {**0**}, its arcs may only contain discourse cue words. Therefore, a separate superset of values is used for the downward arcs leading from a node with attribute set {**0**} – The set of all discourse cue words. A frequency threshold of 4

[3] Since the partition of train/test sets is randomly selected at the beginning of a run, the frequency count of the tokens varies, so the number of tokens with frequency above the frequency threshold varies.

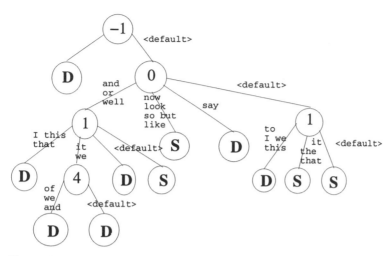

Figure 19.4
Example decision tree, induced by evolution. At leaves, "D" stands for *discourse* and "S" for *sentencial*. This tree scored 83.24% over the training set and 81.52% over the test set. The original (unedited) tree has 42 nodes.

has been selected for the superset of discourse cue words, resulting in a superset of size approximately 20. The attribute sets are therefore restricted to either being {**0**} or a subset of {**-1, 1, 2, 3, 4**}.

Note that although the attribute sets of the decision trees in Figures 19.3 and 19.4 are all one-member sets, the representation allows for more than one attribute to appear at each internal node. Also note that a decision tree does not necessarily have to look at the word it is disambiguating; there are some generalizations that hold for all discourse cue words.

19.5   Crossover Operations on Decision Trees

The crossover mechanism is designed to allow for any decision tree architecture to emerge. There are two representational issues which simple subtree swapping does not address. First, there are a variable number of daughters per node. Second, there are attribute and value sets at the internal nodes and arcs, respectively.

The attribute sets at internal nodes and value sets on tree arcs embedded in decision trees are represented as bit strings. In order to represent a set as a bit string, each member of the superset is assigned a location on the bit string. A bit string contains member x of the superset if and only if x's bit location has a 1. Therefore, attribute set bit strings are of length 6, and value set bit strings are of length approximately 20 or 26.

One-point bit string crossover is intermingled with the subtree-swapping crossover of genetic programming.[4] Once two trees have been selected for crossover, a random node from each is selected. If neither node is a leaf, and their attribute sets are *compatible*, i.e. are both $\{0\}$ or neither contain 0, then with 66% probability bit string crossover will take place. In all other cases, the subtrees which have the chosen nodes as roots are swapped.

When bit string crossover is selected, crossover takes place between the nodes' attribute sets. Then, a random downward arc is selected from each node, *arc1* and *arc2*. The value sets corresponding to these arcs are crossed over. Then, the set of sister arcs to the right of *arc1*, as well as the subtrees they lead to, are exchanged with the set of arcs to the right of *arc2*, as well as the subtrees they lead to. Note that this operation alters the number of daughters per node.

19.6   How Fixed Training Data Participate in Competitive Adaptation

In order to measure the fitness of a decision tree, a subset of training cases is selected, and raw fitness is the number of these training cases correctly classified by the decision tree. The canonical method for induction over a training set is to select training cases with uniform distribution. This section describes a method by which training cases are selected competitively.

Fixed training cases cannot participate in creative evolutionary operations such as crossover and mutation. Therefore, it does not make sense to use fitness-based selection for selecting reproduction participants. Instead, fitness-based selection of individuals is used to select training cases when computing the fitness of a decision tree.

In a competitive environment, weaknesses are sought out by competitors. In our implementation of competition, the training cases which tend to be incorrectly classified by decision trees become more fit, and therefore selected more frequently during fitness measurements.

Competition is implemented as follows:

• Each training case has a fitness measure which is initialized to zero before the first generation of decision trees are evaluated. This fitness measure continuously adjusts. It is never again initialized.

---

[4]David Andre's chapter in this book presents work in which subtree-swapping is intermingled with bitmap crossover.

Table 19.4
Performance improvement generated by a competitive fitness function in terms of performance over the training set. The last row shows the average performance of the individual scoring highest over the training cases during 500 generations of evolution. For the other rows, the best individual of the given generation is selected. Runs without competition compute decision tree fitness across 513 random training cases.

	Without competition			With competition		
Generation	Number of runs	Average score over training cases	Standard deviation	Number of runs	Average score over training cases	Standard deviation
100	58	80.85%	1.24	46	81.96%	1.35
200	52	81.22%	1.34	44	82.81%	1.35
300	44	81.49%	1.53	44	83.34%	1.31
400	42	81.65%	1.60	43	83.83%	1.26
500	42	81.79%	1.59	42	84.07%	1.29
Overall best	42	81.95%	1.55	42	84.09%	1.31

- Each time a decision tree is tested on a training case, the training case's fitness is incremented if the tree makes the incorrect prediction, and is decremented if the tree makes the correct prediction.

- When calculating the fitness of a decision tree, 2-member tournament selection[5] over the set of training cases is repeatedly used to select 513 (non-unique) cases, so the decision tree's raw score is between 0 and 513. Note that the same training case may be used more than one time during a fitness measure. Also note that the fitnesses of training cases change *during* the fitness calculation of one decision tree.

Competitive fitness measurements are *relative*, that is, they are computed across a non-uniform distribution of training cases, and measure fitness with respect to the current fitnesses of the training cases. Therefore, in order to ascertain how much is being learned on an *absolute* scale, it is necessary to periodically compute the absolute fitness measurements (that is, uniformly across the entire training set) of the population of decision trees. The measurements of absolute fitness are used only to keep track of the best decision tree created so far; it is *never* used by the evolutionary process. Absolute fitness is computed every 20 generations.

Since 2-member tournament selection is used to select training cases, the distribution of selected cases is not as skewed as it could be for fitness proportional selection, and decision trees are given less of an opportunity to "forget" what they've already learned. Further

---

[5]*Tournament selection* is accomplished by selecting two or more individuals at random, and keeping only the one with highest fitness.

Table 19.5
Performance improvement generated by a competitive fitness function in terms of score over the test set. See caption of Figure 19.4 for more information.

| | Without competition | | | | With competition | | |
Generation	Number of runs	Average score over test cases	Standard deviation		Number of runs	Average score over test cases	Standard deviation
100	58	78.51%	1.33		46	78.49%	1.18
200	52	78.59%	1.34		44	78.80%	1.62
300	44	78.72%	1.35		44	79.02%	1.49
400	42	78.54%	1.45		43	79.11%	1.68
500	42	78.64%	1.52		42	78.92%	1.71
Overall best	42	78.45%	1.44		42	79.12%	1.72

experiments would be necessary to determine the effect of other selection procedures. It is possible that cyclic behavior could emerge, during which absolute fitness stops increasing.

Table 19.4 shows the improvements gained from a competitive fitness function in terms of the score attained over the training cases, and Table 19.5 shows the scores over the test cases for the same batch of runs. Section 19.8.2 discusses these results.

The decision trees in Figures 19.3 and 19.4 were evolved with competition.

19.7   Averting Overlearning with Decision Trees: Fitness Penalty

When evolving decision trees, there is an intrinsic tendency towards generalized learning. This is because subtrees with smaller depth have the survival advantage that they have a greater probability of remaining intact, and since shorter subtrees have fewer choice points they have less of an opportunity to over-tune to the training data than subtrees with greater depth. Also, the frequency threshold imposed on members of the value sets will tend to avert overlearning. In spite of these factors, average performance of an evolved decision tree over the test cases does not out-perform the decision tree in Figure 19.1. We therefore have implemented a fitness penalty to avert overlearning.

A decision tree can be viewed as the compilation of many *rules*. Any traversal of the tree from root to leaf in which one attribute and one value is selected at each choice point (i.e. node) is a rule of the form:

if (*attribute1* = *value1*) and (*attribute2* = *value2*) and (*attribute3* = *value3*) . . .
then *classification* = *class1*

When a decision tree is used to classify a set of training cases, each training case will be classified by one and only one of these rules. Therefore, the rules indicate a partition of the set; each rule corresponds to one partition. When training over a small set of examples, the same examples must be used repeatedly for fitness measure. Therefore, it is possible for decision tree fitness to improve by discovering many rules with small partitions. However, the smaller a rule's partition, the less likely it is that the rule embodies a valid generalization. Therefore, we have implemented the following fitness penalty:

Rules with a corresponding partition with size less than a preselected threshold are considered "illegal", and their contribution to raw fitness is subtracted.[6]

With this fitness penalty in effect, a rule must apply to some minimal number of training cases in order to add to the fitness of the decision tree it is a part of. Therefore, the penalty adds pressure for decision trees to find generalizations by prohibiting decision trees to gather data which is idiosyncratic to the training set.

This fitness adjustment is active in two contexts, both with the same threshold. First, it is used when computing a decision tree's absolute fitness (that is, over the entire training set, uniformly). Second, when computing the competitive (i.e. relative) fitness of a decision tree, it is applied by looking at the partition of the set of training cases selected for that particular fitness measure, which is also a 513 member set, but is not uniform, i.e. it can contain zero copies of some training cases, and more than one copy of other training cases. Therefore, when computing competitive fitness, the penalization is based on an *approximation* of the partition sizes implied by the decision tree. However, since tournament selection is used to select training cases, the distribution of training cases is less skewed than it could be for fitness proportional selection. The penalty is not used when evaluating a decision tree over the test cases.

Note that this strategy for increasing generalization performance is not a change to the evolutionary process, but simply a change to the fitness measure. Also, although this penalty bears similarity to a parsimony factor, it does not directly penalize a tree based on its size.

Table 19.6 shows the results of experiments with particular thresholds. Section 19.8.3 discusses these results, and section 19.9 suggests more sophisticated methods of penalizing raw fitness.

---

[6]Since the rule defining a partition is not necessarily correct for every training case it applies to, the amount it contributes to raw fitness can be different (less) than the size of its partition.

Table 19.6

Results from 4 batches of runs, all with competition. See table 19.1 for the parameters used for these runs. The fourth row shows the performance of the tree in figure 19.1, illustrating the standard deviation which results from the random partitioning of the example data into training and test sets.

Threshold on size of *rule partitions*	Number of runs	Average score over test cases	Standard deviation	Average score over training cases	Standard deviation
0	42	79.12%	1.72	84.09%	1.31
3	58	79.20%	1.71	81.84%	1.22
4	35	78.88%	1.73	80.78%	1.18
N/A *(Tree in fig 19.1)*	100	78.99%	1.23		

19.8 Conclusions

19.8.1 Non-trivial Learning and Generalization Performance

The existence of the decision tree in Figure 19.1, which is small yet achieves a high success rate, adds to the difficulty of this problem domain. It is easy to induce a strategy similar to the one embodied by the small decision tree, even by random search, so every run accomplishes at least that. This weakens the comparisons made between different fitness measures, since the range of possible performance is small. Additionally, the loss in performance over the test cases as compared to the performance over the training cases is just enough that the average test score is comparable to the performance of the small tree in Figure 19.1. This is illustrated in Tables 19.6 and 19.5.

It is important to recognize that a non-trivial task is taking place when evolving a decision tree with a higher success rate than that in Figure 19.1. High scoring decision trees implicitly partition the training and test sets into portions which are mostly non-trivial in size. (See section 19.7 for a description of how decision trees partition the training examples.) For example, the decision tree in Figure 19.3 partitioned the test set into partitions of sizes 12, 130, 4, 5, 1, 2, 11, 17, 32, 97, 21, 6, 14, 5, 1, 31, 15, 94, 1, and 15. Therefore, it would be a mistake to assume that the rules embodied in an evolved decision tree other than the simple rules of the tree in Figure 19.1 are exactly the ones which fail when evaluating the decision tree over the test cases; each rule of an evolved decision tree tends to perform more poorly over the test cases. It is *coincidental* that the average score over the test set is approximate to the score attained by the decision tree in Figure 19.1. In domains without a small, high-scoring tree, evolved decision trees will outperform simple trees over the training set to a greater degree. When this is the case, the relative loss of performance over the test cases will likely not bring performance below that of any simple decision tree.

19.8.2 Competition

The mean training score for the best decision tree found over 500 generations of competitive evolution was significantly different from the mean training score for trials without competition (t=6.760, P<.001). The improvement over test cases with competition is less obvious, however the mean test score for competitive evolution with a threshold of 3 (Table 19.6, second row) was significantly different from the mean test score for trials without competition and with a threshold of 0 (Table 19.5, last row) (P<.0265). The usefulness of competition will prove to be dependent on the domain to which it is applied.

19.8.3 Fitness Penalty

Table 19.6 compares the average train and test performances attained when the threshold on rule partition size is set to 0 (i.e. no fitness penalty), 3 and 4. The usefulness of the fitness penalty for this domain is inconclusive, however the results are informative. With a threshold of 3 or 4, learning is inhibited and the average training score is less than that with a threshold of 0. A higher average training score is expected to correspond to a higher average test score. This can be verified by cross-referencing Tables 19.4 and 19.5. However, the average test score attained with a threshold of 3 is not lower than the average test score attained with a threshold of 0. That is, the difference between average training score and average test score is smaller when the penalty is in use. One way to view this is that the penalty decreases learning potential, but also decreases overlearning. It is possible that with the fitness penalty, many extraneous rules which help a tree's training score, but do not help its test score, are "trimmed".

Since the classification problem in this chapter has only two classes, a rule which has overtuned to the training set (i.e. only applies to a small number of training examples) has as least a 50% chance of correctly classifying a test case it applies to. Therefore, in a classification problem with more than two classes, the negative effects of overlearning will probably prove to be more detrimental. See section 19.9 for variations which could increase the usefulness of a fitness penalty.

19.8.4 Linguistic Data

Evolved decision trees often include rules which provide insightful hints for linguists. For example, the decision tree in Figure 19.3 contains a rule that *and* followed by *in* is of class *discourse*. In looking at the training cases we note that the *in* always prefaces the prepositional phrases *in particular*, *in fact*, and *in a certain respect* when following *and*. These are cases in which *and* is being used to introduce an elaboration. As another example, some decision trees contain the rule that *say* preceded by *to* is of class *sentencial*. This is linguistically viable, since, when preceded by *to*, *say* is most likely a verb, as in, *"That is what I wanted to say."*

19.9 Further Work

There are ways to vary the method by which competition has been implemented for induction over a fixed set of training cases. For one, higher selection pressure for selecting training cases by way of greater than two-member tournament selection or fitness-proportional selection should be evaluated for various domains. Also, it may be beneficial to have the fitness scores of training cases change only at generation boundaries, so that their adaptation is synchronous with the adaptation of the decision trees.

Various fitness penalties should be contrasted for evolving decision trees. In particular, instead of an absolute threshold, a weighted penalty could be implemented by which the bigger the partitions of a decision tree are, the smaller the fitness penalty. The weight would have to be tuned in a domain-specific manner.

The method by which competition is implemented could influence generalization performance. Other parameters which have potential to influence performance over the test set include the fitness penalty weight and the token frequency thresholds mentioned in section 19.4. Schaffer et al. [1990] have used a GA to tune parameters to increase the performance of a neural network over test cases. A similar method could be employed to tune the parameters listed above, i.e. meta-GA.

Ryan's chapter in this book discusses a method by which diversity can be maintained when a parsimony factor is in use. This method could also apply to a penalty based on the partition sizes implied by a decision tree; this penalty bears similarity to a parsimony factor.

Automatically attained statistical data concerning how often words co-occur (e.g. Hatzivassiloglou and McKeown [1993] and Schuetze [1992]) can aid predictive tasks such as word sense disambiguation. [Brown et al. 1991] We intend to evolve disambiguation mechanisms which have access to such data.

Acknowledgments

Thanks to Andy Singleton, Alex Chaffee, David Schaffer and Kathy McKeown for their supportive exchange of ideas. Thanks to Diane Litman for providing the transcript of spoken English used in this work.

Bibliography

Angeline, P. J. and Pollack, J. B., (1993) Competitive Environments Evolve Better Solutions for Complex Tasks, In *Proceedings of the Fifth International Conference on Genetic Algorithms.* San Mateo, CA: Morgan Kaufmann.

Axelrod, R. (1989) Evolution of strategies in the iterated prisoner's dilemma. *Genetic Algorithms and Simulated Annealing*, L. Davis editor, Morgan Kaufmann.

Brown, P. F., DellaPietra, S. A., DellaPietra, V. J., and Mercer, R. L., (1991) Word sense disambiguation using statistical methods, in *Proceedings 29th Annual Meeting of the Association for Computational Linguistics*, (Berkeley, CA), pp. 265-270, June 1991.

Hatzivassiloglou, V. and McKeown, K., (1993) Towards the Automatic Identification of Adjectival Scales: Clustering Adjectives According to Meaning. *Proceedings of the 31st Annual Meeting of the ACL*, Association for Computational Linguistics, Columbus, Ohio, June 1993.

Hillis, D. (1992) Co-evolving Parasites Improves Simulated Evolution as an Optimization Procedure, In *Artificial Life II*, edited by C. Langton, C. Taylor, J. Farmer and S. Rasmussen. Reading, MA: Addison-Wesley Publishing Company, Inc.

Hirschberg, J. and Litman, D., (1993) Empirical Studies on the Disambiguation of Cue Phrases, in *Computational Linguistics*, Vol. 19, No. 3, in press.

Holland, J. (1975) *Adaptation in Natural and Artificial Systems*, Ann Arbor, MI: The University of Michigan Press.

Christie, A. M. (1993) Induction of decision trees from noisy examples, in *AI Expert*, 5(8).

Koza, J. R. (1991) Concept formation and decision tree induction using the genetic programming paradigm. In Schwefel, Hans-Paul, and Maenner, Reinhard (editors), *Parallel Problem Solving from Nature*. Berlin, Germany: Springer-Verlag.

Koza, J. R. (1992a) *Genetic programming:On the programming of computers by mean of natural selection.* Cambridge, MA: MIT press.

Koza, J. R. (1992b) Genetic Evolution and Co-Evolution of Computer Programs. In *Artificial Life II*, edited by C. Langton, C. Taylor, J. Farmer and S. Rasmussen. Reading, MA: Addison-Wesley Publishing Company, Inc.

Quinlan, J.R. (1986) Induction of decision trees. *Machine Learning* 1(1), New York: Kluwer Academic Publishers, 1986, pp. 81-106.

Schaffer, D., Caruana, R. A. and Eshelman, L. J., (1990) Using Genetic Search to Exploit the Emergent Behavior of Neural Networks. Physica D 42, p. 244-248.

Schuetze, H. (1992) Dimensions of meaning. In *Proceedings of Supercomputing '91*.

Tackett, W. A. (1993) Genetic Programming for Feature Discovery and Image Discrimination. In *Proceedings of the Fifth International Conference on Genetic Algorithms*. San Mateo, CA: Morgan Kaufmann.

20 Cracking and Co-Evolving Randomizers

Jan Jannink

Although **pseudo-random number generators** or **randomizers** are of great importance in the domain of simulating real world phenomena, it is difficult to construct functions which satisfy the many criteria, such as uniform distribution, which 'good' randomizers possess. It is computationally expensive to perform the statistical analysis required to establish their quality. Moreover, no current method of analysis can guarantee quality, since even the question of what constitutes the set of criteria defining randomness remains open.

This paper discusses two experiments designed to test the applicability of genetic programming to the analysis and the unconstrained construction of randomizers. The first experiment attempts to unravel the structure of a number of generators recommended in the literature by guessing their output, given previous output data. The second examines the possibility that co-evolving populations of programs in a competitive environment can produce randomizers which conform to the criteria without explicitly testing for them. In other words, the requisite properties must emerge from the experiments' nature.

The experiments have a convenient representation as a guessing game, closely resembling the two player penny matching problem, in which each player choses heads or tails, and one player hopes to outwit the other in order to prevent a match between the pennies, while the other tries to guess what the first will to do and force a match.

Such competition between programs in the genetic programming framework should breed functions whose output sequence is difficult to reproduce. Through this simple mechanism we strive for a further result, a functional approach to randomness, rather than its statistical description. This would have the advantage of being similar to the definition as put forth in information theory.

Finally, in addition to introducing novel fitness measurement techniques, these simulations aim to show that the genetic programming paradigm is suitable for building structured models of randomness from limited information, without using the exhaustive traditional tests of randomness.

20.1 Background

The themes which pervade the discussion that follows are randomness, game theory, co-evolution, information theory and competition, with genetic programming as a binding force linking them together.

Random or stochastic phenomena appear commonly in nature, as well as in human activities such as economics, therefore computer modeling and simulation typically require functions producing randomized sequences of numbers.

These so-called pseudo-random sequences present many of the same characteristics as truly random numbers, if their generators are well designed. Among the important criteria for randomness are uniform distribution of values throughout their range, as well as uniformity of k-tuples of the values. This is defined by considering the k-tuples as coordinates in k-dimensional space, and noting the distance between the hyperplanes they form in the k-space [L'Ecuyer 1988]. Another often used term for this property is high entropy, to indicate that there is little clustering of values.

Randomizer-like functions with this property have been produced with genetic programming [Koza 1992], and it is noteworthy that the GP environment used for these experiments contains an implementation of the "minimal standard" generator [Park, Miller 1988].

That randomness is a non-trivial matter is readily apparent from the considerable literature devoted to the question of what constitutes a 'good' or 'bad' generator [L'Ecuyer 1990], and the well documented examples of poor randomizers sold with many commercial computer systems [Park, Miller 1988].

Most tests for randomness require large amounts of computer time, and are sensitive to the initial conditions of the tests [L'Ecuyer 1988]. One goal in the study of randomizers is to reduce the time required to test them. Another goal is to derive a functional description of the randomness they attain, or in other words to give a measure of how difficult it is to differentiate the pseudo-random sequences from truly random numbers.

Game theory provides a possible method of attaining these goals in the simple guise of the two player zero-sum penny matching game which was presented in Von Neumann and Morgenstern's seminal work [Von Neumann 1944]. There is no deterministic strategy for penny matching which will result in success for either player over the long run [Selfridge 1989].

In order for fully randomized behavior to emerge from the penny matching game, an adaptive technique must be devised to take advantage of its format. The competition between the player trying to force a match, and the player trying to avoid it, provides a perfect arena for the use of co-evolution [Hillis 1992].

Indeed, we may find it extremely time-consuming to test whether a sequence is random, but it is relatively easy to check whether another sequence, a guessing strategy developed for the game, matches it closely. This method of testing gives easily calibrated results which depend on the quality of the sequences produced while playing [Angeline, Pollack 1993].

The assumption is that the programs will adapt, and must evolve successfully to match the sequence they are made to reproduce, or else be condemned never to reach their goal. In the latter case we can assume either that the problem is ill defined, or that the tested sequence is simply not amenable to extrapolation (and therefore equivalent in some sense to a random sequence).

The next two sections, **20.2-20.3**, cover the reasons for pursuing this problem. After those, **20.4-20.5** define the methodology of the research, followed by a description of the experiments and their results, **20.6-20.7**, and closing with some thoughts about future work in the area in **20.8**.

Table 20.1
Physical, mathematical and computational phenomena by phase complexity

phase	constant	periodic	*complex*	chaotic	stochastic
Material	molecule	crystal	*liquid*	gas	
Function	linear	repeating	*fractal*	pseudo-random	random
Computation	regular	recursive	*universal*	divergent	

20.2 Motivation

Randomizers are essentially a juggling act between the exigencies of compact representation, rapid evaluation and high entropy. According to Kolmogorov's information theoretic definition of randomness, a random sequence can not be written in a more compact form, such as a program, than the enumeration of its elements [Kolmogorov 1965]. Randomizers absolutely contradict this tenet. Their most desirable property therefore is to produce sequences which for all intents and purposes mimic the characteristics of a random sequence.

In order to situate the randomizer problem in terms of genetic programming, table 20.1 presents an adaptation of a list of phenomena ordered by phase complexity [Rucker 1993] suggesting a representation of chaos in terms of pseudo-randomness. Literature in the field of artificial life generally focuses on the complex phase as the locus of 'interesting' phenomena, such as universal computation, or by analogy genetic programming. Assuming that the table's categorizations are valid, the experiments presented below are unusual in that they attempt to make chaotic phase behavior emerge out of complex phase processes. In anticipation of difficulties related to this difference, we introduce special fitness testing techniques to deal with the problems' atypical data.

Genetic programming has been shown capable of producing functions that have the property of high entropy [Koza 1992]. Unfortunately, the properties of randomizers go beyond simple entropy, and their interactions have not, and may never be, fully described. Therefore a different, non-explicit method is the only hope at the present time for the study of stochastic patterns.

It appears that there have been no prior attempts to induce and test for randomizing behavior through a competitive mechanism, nor any algorithmic methods to infer the quality of a randomizer through a functional description. The focus has been on learning locally successful strategies [Selfridge 1989] and exhaustive testing for randomness [L'Ecuyer 1988]. These experiments aim to bridge that gap.

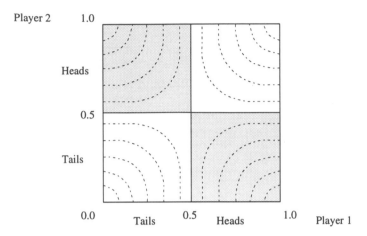

Figure 20.1
Expected gain for players of penny matching game

20.3   Arguments for Success

In order for randomizing behavior to emerge from the experiments there must be reasons why randomness should be a desirable trait to preserve.

20.3.1   Two Player Penny Matching Game

It has been shown that in the penny matching game a random choice of heads or tails is the best strategy for both players, because it minimizes expected loss all the while maximizing expected gain. Penny matching is called the 'natural device' to produce even 50% probabilities [Von Neumann 1944]. Therefore the best programs to solve penny matching should be the ones which exhibit the most random behavior.

Figure 20.1 graphs the expected gain for penny matching players [Selfridge 1989]. The axes give the probabilities of each player choosing heads. The saddle point of the graph is at the center, where both players have an even chance of playing heads. The expected gain is positive in the shaded area for the player trying to prevent the match, negative in the other regions, and vice-versa for the other player.

20.3.2   Uniform Distribution

The most compelling reason to hope for the emergence of a good randomizer out of a penny matching competition is that a distribution of plays which isn't fully uniform is amenable to reverse engineering. If any set of numbers is more likely to appear in the sequences, then it will be targeted and exploited by the guessing programs, and the fitness of the generators would suffer.

20.4 Models

In the descriptions below there are terms which will be used in a particular manner.

player : a program generating or guessing a number sequence

population : a group of players evolving together

program : a parse tree with functions as internal nodes and terminals as leaves

seed : a number which is at the head of a sequence of numbers

20.4.1 Two Player Multi-Penny Matching Game

The idea governing the programs' function is a two person zero-sum game. This game is described as follows:

1. A sequence generator offers up a seed number

2. A guesser program tries to extrapolate the sequence from it

3. The generator's and guesser's sequences are compared

4. The guesser is successful if its extrapolation was accurate

Ordinary penny matching is a similar game in which the number is replaced by a binary (heads or tails) value. The guesser keeps both pennies if its penny matches the generator's, otherwise the generator keeps them.

The only substantive difference in the multi-penny game is that an integer (multi-bit) number is produced, a sequence of which are compared. The information content of the message between the sequence generator and the guesser is therefore much larger. In effect, it corresponds to matching multiple pennies at a time.

The higher level of information exchange provides the potential for better regression, and simplifies to some extent the measurement of the programs' fitness.

20.4.2 Fitness Measure

Two measures of accuracy for the guesser are immediately discernible. The first is the difference between the guess and the target value of each number in the sequence. The second is Hamming distance or the number of bits per guess which differ in the sequences. The resulting value is scaled in a non-linear fashion, to favor more accurate guesses, and to normalize the score to a value between one and zero. The generator's fitness is measured relative to the guesser's, that is: $1 - fitness(guesser)$.

For simplicity, the sequence on which fitness is initially evaluated is a seed and a set of numbers generated immediately following. This appears to be a somewhat weak test because the generator's non-monotonicity implies that a sequence of guesses may be locally accurate, but divergent in the long term.

However, it is vital to see that for fitness to be a useful measure, the test sequences must be presented to programs in a 'natural' order such as: $f(seed), f(f(seed)), f(f(f(seed)))...$ where $f()$ is the randomizer. The choice of a sequence to test is not a trivial matter because given the sequence corresponding to $f(1), f(2), f(3)...$, the problem becomes one of function regression, and guessers very quickly converge to locally perfect solutions, without any ability to generalize.

In order to better judge the value of the results, the best of generation individuals undergo a validation test on a sequence which is not tested for by the GP system during fitness evaluation. If the results of this test are good for some program, its generality is better ensured.

A further refinement to this measure increases progressively the number of fitness cases presented to programs during the course of a GP run. This method is discussed in detail in **20.5.1 Sampling** below.

When a complete fitness measure entails comparing the performance of each program against every other one, as is the norm for co-evolution, it becomes impractical to do so as the problem scales up in size. A partial fitness test in the form of a tournament over the whole population provides a reasonably accurate measure of fitness in a more scalable fashion. This technique is described below in **20.5.2 Fitness Tournament**.

20.4.3   Single Generator

In the simplest version of the multi-penny matching game there is only one sequence generator, and all of the programs attempt to extrapolate its sequence. The generator does not change over time, which may allow the population of guessers to determine its internal structure. This is the the focus of the first experiment outlined in **20.6.1 Tested Randomizers**. The sequence generators chosen for the experiment are widely used, and are presented in the literature as superior in quality.

20.4.4   Separate Generators and Guessers

Two separate populations of evolving programs compete. The fitness of generators is determined by the failure of guessers to come up with the correct sequences. This will evolve parallel sets of programs which for the most part perform the same function, that is, take a seed and generate a sequence of numbers. It may seem to be the most natural way to implement the problem, but it actually gives a considerable fitness advantage to the generators, which are more likely to compete against incorrect guessers, than guessers against similar generators. This argument expresses the fact that in the space of possible programs, most populate the valleys of any problem's fitness landscape. The mechanism below is a compromise that circumvents this weakness.

20.4.5   Sexing Populations

Because the functions of generating sequences and guessing them are so complementary, it is natural to consider merging them into a single population of programs which both generate and guess sequences. However, it may be counterproductive for a program to guess exactly the same sequence as the one it generates. Therefore it is convenient to introduce a tag which differentiates the functionality of generating and guessing. Such a tag could cause its subtrees to execute only if the program is guessing, for example. The complementarity of *generator* and *guesser* functions leads to their description as 'sexes'.

Having a tag in the root node of the program can force the program to act as a generator or guesser exclusively. By allowing fixed and flexible 'sexes' in the same population the highest level of functionality is attained. A more detailed description of the actual construct used in the experiments appears below in **20.6.3 Functions and Terminals**.

20.5   New Techniques

In order to tackle the randomizer problem new approaches to fitness measurement became necessary. Dynamic sampling addresses the issue of overfitting and non-convergence that may occur when too little or conversely too much fitness data is present. The fitness tournament minimizes scalability problems occurring when program fitness must be measured relative to other programs in a population.

20.5.1   Dynamic Sampling

Randomizers are atypical functions in that they are by design non-monotonic. This presents a serious problem to any method of pattern matching attempting to, as it were, perform curve fitting on data produced by a randomizer. If too little data is presented then the pattern matcher is likely to arrive at a locally correct solution only. Moreover, the data appears contradictory and patternless if too much of it is available. To achieve any form of generality there must be a mechanism to allow initially some convergence, and thereafter increase the generality of the solution.

Generally, in genetic programming, the fitness cases correspond to a sampling of the data which comprise a solution to a problem. These are defined at the beginning of the run, and remain unchanged throughout the run. Dynamic sampling proposes to change or add fitness cases throughout the course of the run. This can be on a constant schedule, or a randomized one with a distribution around a given range, or at an increasing rate.

The advantage of this approach for difficult problems is that some initial convergence becomes feasible while the number of fitness cases is small, and that by modifying them or by adding supplementary fitness cases on the fly, a route towards more general solutions becomes a possibility.

20.5.2 Fitness Tournament

This fitness measurement technique overcomes scalability issues when fitness must be calculated relative to the performance of other members of the population, as occurs in problems using co-evolution. It allows the problem size to increase without compromising the accuracy of the fitness measure obtained. It is not to be confused with tournament selection, which finds programs for reproduction and crossover using previously calculated fitness values.

In order to obtain a valid measure of a program's fitness, the whole population must be evaluated in some way. As described above fitness is measured relative to the performance of an opponent. In the case of learning strategies for penny matching when there are only two players, they both evolve towards locally successful strategies for the game, and become caught in chaotic cycles, as one tries to elucidate the other's strategy and still remain unpredictable [Selfridge 1989].

To avoid such cyclic behavior, tournaments between all the players determine an overall best fitness. In the first round all programs compete, and all subsequent rounds consist of smaller tournaments on the winning and losing halves of the previous round. After the tournaments are completed the most successful and weakest strategies have sorted themselves out.

This method's drawback is that there may be classes of strategies which are approximately equally fit, in that they perform well against a similar number of strategies, but never the same ones. This will occur, as a rule, because there is at best a partial ordering of game strategies. By playing multiple tournaments the likelihood of this drawback becoming significant is reduced.

Since every program participates in the same number of tournaments, it is still perfectly acceptable to sort the population by fitness, which remedies the above weakness by retrieving programs which lost early on in the tournament, but performed well otherwise.

The fitness tournament retains most of the accuracy of a complete fitness test with added scalability.

20.6   Experiments

This section describes some implementation details of the experiments, starting with the randomizers examined in the first experiment. The functions, terminals and other parameters constituting the runs are also covered.

20.6.1   Tested Randomizers

The first experiment involves a single randomizer and a population of programs attempting to reproduce its sequence. In order to tune up the system we performed a number of preliminary tests on the trivial randomizing function: $x * 17$ mod 269. The parameters derived as a result of the tuning are listed in table 20.2 below. The randomizers described in this section are promoted in the literature as having good characteristics. Each description comes with a name, to identify it in subsequent discussion.

Two basic types of randomizers are considered here. The multiplicative linear congruential generator introduced by D. H. Lehmer is probably the most widely used method for obtaining pseudo random sequences.

The trivial randomizer above belongs to this category, as well as the Park-Miller randomizer (r0) which corresponds to: $(x * 7^5)$ mod $(2^{31} - 1)$. Its main advantage is that it is very easy to start up, since only a single seed value (x) is necessary, and it can be efficiently implemented on small computers. However, its period, the number of numbers generated before its sequence repeats is limited by the size of the modulus (the last number in the equations above).

A few further generators derived from the above model are included in the tests as well. First among these, integer division by four on the Park-Miller randomizer produces a sequence minus two low order bits (r1), which are considered less random than the others. Others recommended in [L'Ecuyer 1988] combine the input of two or three generators as shown in the pseudo-code below.

Figure 20.2
Randomizer with two seeds: (r2)

```
seed1 = seed1 * 40014 mod 2147483563
seed2 = seed2 * 40692 mod 2147483399
output = seed1 - seed2
if (output < 1)
  output = output + 2147483562
```

Figure 20.3
Randomizer with three seeds: (r3)

```
seed1 = seed1 * 157 mod 32363
seed2 = seed2 * 146 mod 31727
seed3 = seed3 * 142 mod 31657
output = seed1 - seed2 + seed3
if (seed1 - 705 > seed2)
   output = output - 32362
if (output < 1)
   output = output + 32362
```

The above have a longer period than simple generators of the same size while still being quick to initialize. Also, unlike the simple generators they do not exhibit a lattice structure when consecutive outputs from the randomizer are taken as Cartesian coordinates and plotted [L'Ecuyer 1988].

Lagged Fibonacci generators form the second major group of randomizers. These have the advantage of having long period lengths, but require more initialization. An example of this type is a randomizer proposed by G. Marsaglia with 97 seed values and essentially calculates: $x_n = (x_{n-97} - x_{n-33}) \bmod 2^{24}$ (r4).

Finally, an extended version of the Fibonacci generator with 24 seed values calculating: $x_n = (x_{n-24} - x_{n-9} + c) \bmod 2^{24}$, where c represents a 'carry' value set to 1 if in the previous iteration $x_{n-24} < x_{n-9}$ and 0 otherwise (r5). It has a period length of approximately 10^{170}, within which it is simple to generate long disjoint sequences, and would therefore appear to be ideal for heavy users of random numbers [James 1990].

There are other types of randomizers, which do not have as strong a claim to success, a number of which are described and analyzed in [Knuth 1969].

20.6.2 Tableau

In keeping with the format presented in [Koza 1992], table 20.2 gives the initial parameters of the runs. As many results are reported, mnemonic symbols are included to simplify the description of the experimental conditions. The • symbol indicates a default value of the sequence guessing experiment, whereas a ◁ or ▷ refers to a co-evolution experiment. Others appear in the figures or in the text below.

20.6.3 Functions and Terminals

The terminal set is kept as simple as possible. It allows for a variable x (in two tests a second variable y is also used), and the set of random integer constants in the range $[0 - 255]$. The variable x is the output produced by the randomizer in its previous iteration (y is from the iteration prior to x). The purpose of the integer size restriction is twofold. First, it

Table 20.2
Genetic programming parameters associated with a mnemonic

Objective:	●	deduce structure of various randomizers
	◁	co-evolve population(s) of generators and guessers
Terminal set:	●◁	X, random-constant
	two	X, Y, random-constant
	old	..., Y, random-constant (Y is the older of two prior sequence outputs)
Function set:	●◁	+, -, *, MOD, NOT, AND, OR, XOR
	▷	+, -, *, MOD, NOT, AND, OR, XOR, IFGEN
	mem	+, -, *, MOD, NOT, AND, OR, XOR, >>, <<, READ, WRITE
Fitness cases:	●	15 output pairs from successive randomizer iterations
	chg	15 output pairs changing dynamically, with an average rate of change set to once per 14 generations
	frq	15 output pairs changing dynamically, with an average rate of change increasing from once per 12 generations to once every generation
	inc	15-100 output pairs; new pairs added dynamically with an increasing average rate of change
Validation:	●	10 output pairs from successive randomizer iterations
Fitness score:	●◁	sum over all fitness cases of the difference between randomizer output and tested individuals' output
	bit	sum over all fitness cases of hammimg distance between randomizer output and tested individuals' output
Parameters:	●	1 population of 2048 sequence guessers, 100 generations, 25 runs, tournament selection size 6, crossover 83.2%, reproduction 16.6%
	◁	2 separate populations, 512 guessers, 512 generators, 200 generations, 50 runs, tournament selection size 5, crossover 83.2%, reproduction 16.6%
	▷	1 combined population of 1024 guesser/generators, 200 generations, 50 runs, tournament selection size 5, crossover 83.2%, reproduction 16.6%
Termination predicate:	●◁	reach final generation of run

prevents an explosion in the size of the numbers used in the function, second, it fits in with the instruction set which builds up large numbers more simply than it reduces them.

One function set exists in two versions, the second differing only in the presence of the IFGEN 'macro' function which executes its left subtree if it is a *generator*, and its right subtree if it is a *guesser*. The first is used for simulations involving a population of separate program types only, while the second is appropriate for a co-evolving population combining both program types in the same population.

The MOD operator is a protected modulo, which returns the first argument if the second is null. The DIV operator was left out of the function sets, although it could arguably be included.

A few further functions were added to one set of runs. The first two are functions performing an arithmetic shift to the binary values of the data. The first argument for these functions is the one to be shifted, the second the number of bits it must shift. Here, assuming 32 bit sized data, the second value is restricted to the range $[0 - 31]$.

A second pair of additional functions are READ and WRITE, as defined in [Teller 1994], which access an array of 16 values of the same type as the function arguments. The address argument of both functions is restricted to the range $[0 - 15]$. Before each fitness evaluation in runs using these operators, the array is initialized so that each array element contains a value equal to its index plus one. This provides a well defined set of initial data to the program, and also facilitates the possible use of the array for indirect indexing. A detailed discussion of these operators is not within the scope of this paper and appears in chapter 9.

All functions in these experiments take two arguments except for NOT and READ which are unary.

20.6.4   GP Shell Modifications

All results described below were derived using the SGPC system [Tackett 1993], compiled with minimal modifications, most notably on its handling of random number generation, running on SUN sparc10 workstations.

To achieve co-evolution of populations several coding changes were necessary. The new evaluate_fitness_of_populations function goes through the population matching up pairs of programs in multiple tournaments that simultaneously build an ordering of the population by number of wins.

Also, fitness-cases are handled in a slightly different fashion. The functionality of define_fitness_cases is split into two distinct phases. Its new namesake just creates the data structure to contain the fitness values, and sets only a few of them. The function evaluate_fitness_of_populations is now also involved in the creation of new fitness cases. Its added functionality is applied during runs, after the generation of new programs is complete.

20.7   Results

The results of the experiments were sensitive to the settings of the global variables controlling the behavior of the GP system. The parameters indicated under the heading of the same name in the tableau were chosen heuristically, on the basis of tests using the trivial randomizer described above. The fittest programs produced by the tests reported here are on the order of 200 lines of code, even when simplified, therefore none are reproduced here.

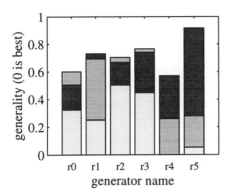 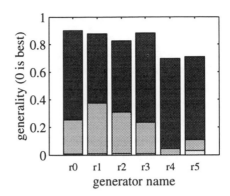

Figure 20.4
Comparison, by validation then fitness, of six sequence guessers by relative improvement during the tests, using as a benchmark value the average score of the fittest generation 0 individuals over all runs. Light shading represents the best individual of the test, dark the best individual of the first generation, and medium the average of the best found in each run.

The following discussion attempts to make up for this deficiency by being as informative and descriptive as possible.

Figure 20.4.a and figure 20.4.b plot values taken from 25 runs, which we call a test, for simplicity, on each of the six randomizers presented above. Referring to the tableau's nomenclature, the default settings indicated by • are active for these tests. All tests are initially identical except for the randomizer producing the sequences the population must reproduce.

From the raw output data, we extract information on the best individuals of each run, and process that down into eight essential values. These values, calculated at the end of each randomizer's test are as follows:

1. average validation score of fittest generation 0 individuals in each run

2. best validation score of fittest generation 0 individuals in each run

3. average validation score of fittest individuals of each run

4. best validation score of the fittest individuals during entire test

5. average fitness score of fittest generation 0 individuals in each run

6. best fitness score of any generation 0 individual

7. average fitness score of fittest individuals of each run

8. best fitness score during entire test

Note that the value in 1. above, typically is not the validation score of the fittest individual of the test in the first generation, but is that of one of the other fittest individuals of the runs.

Fitness and validation scores are not necessarily tightly coupled, as is readily observed in the above charts.

Figure 20.4.a, figure 20.4.b and figure 20.5 measure generality, in other words, the extent to which improvement in fitness, as tested through the fitness cases, reflects fitness changes in an untested data set. Generality is a crucial benchmark in experiments such as these where only a small fraction of the possible fitness cases are used. GP's natural convergence to solutions ensured that performance relative to the tested fitness cases improved by orders of magnitude in each experiment pictured here, a phenomenon which clearly did not always occur with the validation cases.

The plot is normalized so the average validation score of the first generation is equal to 1.0 for each randomizer. The assumption made here is that this score is rather uniformly mediocre, for any simulation. All improvement that GP achieves is in reference to this benchmark value. Direct comparison between randomizers, using this scheme, is prone to some inaccuracy, due to their variation in range, but it is adequate for our purposes.

Each bar in the graph is composed of two or three shaded regions. Those in figure 20.4.a and figure 20.5, correspond to the ratios of the second, third and fourth values from the above list, to the first list value. Figure 20.4.b, on the other hand, displays ratios of the sixth, seventh and eighth list items to the fifth. These ratios are shaded dark, medium and light respectively. In bars having only two shadings, such as $r4$, the missing value was too low to be displayed in the figure.

The equation below calculates the generality value x of the individual, among the fittest of the test, which had the lowest validation fitness score, which appear as the light bars in figure 20.4.a. Here n refers to the number of runs in the test, r a single complete run, $gen0(r)$ the first generation of a run, and $validation(r)$, the validation score of the fittest individual of r. Other generality values are derived in analogous fashion.

$$x = \frac{\min(validation(r_i))}{\sum_{i \in n} validation(gen0(r_i))/n}$$

As is plain from figure 20.4, the randomizers above are extremely diverse in their ability to withstand the test of genetic programming. Although in every case the validation fitness score improved beyond the tested fitness of the fittest individuals of generation 0, indicating a basic level of generality, the actual amount of change served as a good indicator of the randomizers' performance. The lagged Fibonacci randomizers in particular are weak. The normalized score of the fittest program up against $r4$ is a vanishing 0.00004. In other words the program had a very high level of generality, since it not only solved the fitness cases accurately, but also passed the validation test with flying colors. The scale of figure 20.4.a

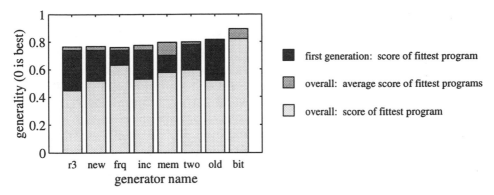

Figure 20.5
Comparison of eight versions of the same sequence guesser, differing in the details of fitness measurement, function set, and terminal set.

is such that the bar corresponding to this result is not visible. Tests with both the defaults and the *inc* sampling method (not shown here) accomplished this validation result, on runs with different starting conditions. This result may actually be due to the initial state of the seeds, to which these randomizers are highly sensitive.

The randomizer with the best performance overall is r3. We investigated whether better results are possible for this randomizer and collect the results in figure 20.5 below.

The different tests on r3 are as follows:

1. initial conditions as in figure 20.4; tableau default: •

2. changing fitness cases with a set average rate; *chg*

3. changing fitness cases with increasing frequency; *frq*

4. increasing the number of fitness cases; *inc*

5. extended function set using a memory array; *mem*

6. two previous sequence outputs as terminals; *two*

7. only the older of two outputs as a terminal; *old*

8. the bitwise hamming distance of the sequences; *bit*

The values graphed for *chg*, are the best of a set of ten tests not shown here, which evaluated changing fitness cases at a fixed rate of once every 6, 8, 10, 12, 14 generations, as well as changing them at a variable rate averaging once in 6, 8, 10, 12 or 14 generations. A slight improvement in average validation score was achieved over the default, at the expense of a less fit overall best individual.

Increasing the frequency of fitness case change, as in *frq*, appears uninteresting from the perspective of absolute results, but there is a mitigating factor. More of the individuals with good validation scores occurred in later generations of the runs, so while their scores were not exceptional, there was more continued progress.

When increasing the number of fitness cases *inc*, not only did overall best scores remain close to the default runs, but the fit individuals continued to evolve longer. The best individual appeared in generation 94, and maintained its high validation score, whereas in the default test the population overfitted to the fitness cases, and the best individual which appeared in generation 20 disappeared immediately, and nothing approaching its score ever returned. Tests on the other randomizers, not included here, provide more impressive evidence of the power of this method, as they all produced better validation scores than any other technique.

Of all the tests, the one using the indexed array memory model *mem*, achieved the best average validation score, although the hoped for score improvements were not. We earlier reported excellent results for this model, and soon discovered that it was the result of a fascinating bug, which in itself opens up vast areas of potential discoveries in genetic programming. Inadvertently, the memory array was not correctly initialized except at the start of each generation. As a result, information was continuously added to the array in the fitness evaluations of the entire population. This property of 'sharing' information, as it were, allowed the population as a whole to evolve towards solutions in a more concerted way, and allowed its best individuals to go beyond what was achieved by any other method. The implications of this bug appear to be countless in the areas of models of communication, cooperation, etc., and offer another direction to explore co-evolution.

It is a disappointment that the test with two terminals, consisting of the two previous outputs of the randomizer *two*, did not result in an improvement. In absolute terms, the best validation score was more than double that of the default test. It appears that the two randomized values are sufficiently uncorrelated to impede the evolution of fit individuals. In this test the best individuals, without exception, made use of only a single terminal, and did not survive long.

Likewise, the test using one terminal which corresponded to the older of the two most recent outputs *old*, did not produce any notable results. Its best individuals achieved a score approximately 40% worse, in absolute terms, than the corresponding test using the more recent output of the randomizer. This result is interesting however, in that it gives an indication of how much information of the randomizer's outputs is retained in the subsequent outputs.

Using hamming distance as a fitness test, *bit*, proved unfruitful, as there were no improvements to the best individual, and only minimal changes in average performance. The

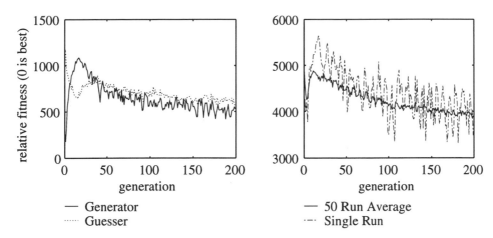

Figure 20.6
Average fitness results over 50 runs of 200 generations each, with co-evolving populations, the first with separate generating and guessing populations, the second with a combined generator/guesser population, showing for contrast an example of a typical single run.

best individual of the test actually appeared in generation 0, which is why in figure 20.5 the dark shading is obscured.

Figure 20.6.a depicts the evolution of fitness for separate populations of guessers and generators ($\lhd$ in table20.2), while figure 20.6.b shows the performance of a population of combined guessers and generators ($\rhd$). The graphed data corresponds to the average over all runs of the performance of the fittest individual of each run at each generation. As fitness scores in the populations are the result of an internal competition with equal numbers of winners and losers, average fitness of the populations is constant throughout the runs.

In both tests we see an early worsening of fitness. At this stage the guessers learn how to decode the generators' sequences. A plateau is reached when generators begin to modify their techniques. The lines in figure 20.6.a, show how the coupling of generator and guesser result in this pattern. Once this plateau is reached, both populations become fitter in tandem, which is surprising until one refers back to figure 20.1, and realizes that the populations are seeking out the point of maximum gain in their competitions, which for both is the saddle point of the graph.

The bold line in figure 20.6.b demonstrates that the average fitness of these fittest programs follows a pattern of steady decrease, whereas the dashed line indicates the variability of fitness in a single run. The comparison of this data provides a further clue as to what is taking place during the course of a run.

This evidence indicates a phenomenon of discovery by the generators of new methods of encoding data, followed by the corresponding decipherment of this strategy by guessers in succeeding generations. Overall, each new discovery in sequence generation improves the best of generation fitness, although this is masked by the variability of the individual runs.

This tendency for the sequence generation phase to contribute more highly to fitness is borne out by results from runs with separate generators and guessers. When the populations are separate, their coupling is weaker, and there is much greater variability within runs. Therefore a larger population size is necessary to achieve similar results.

20.8   Conclusion

Program evolution simulations establish the counter-intuitive result that it is feasible, up to a certain point, to build increasingly accurate representations of pseudo-random number generators, and thereby obtain a relative measure of their quality.

The mechanism of competition is used to co-evolve populations of programs, which by playing a simple game strive to attain a functional definition of randomness. These experiments show that genetic programming is able to build structured models of randomness from limited information, without using exhaustive traditional tests of randomness.

Fitness measurement techniques, such as progressively increasing the number of fitness cases, promote continued evolution by reducing overfitting, and enable the exploration of problems with extremely noisy data, by initially allowing patterns to be recognized within a minimal set of fitness cases.

Further testing of co-evolved randomizers will show how well they measure up to hand-crafted ones, when subjected to statistical analysis. The concept of a shared memory pool for a population is fascinating for its implications regarding population-wide evolution, and merits its own study.

It is worthwhile to note in passing that Kolmogorov considered, in 1965 already, the concept of algorithmic quantification of the density of information in genetic material. Additional relevant material on randomness and complexity of computer programs, and the mathematical meaning of life is available in [Chaitin 1987].

Acknowledgements

Thanks are due to Patrik D'haeseleer and Jason Bluming for late night inspiration and humor while developing this project. Thanks also to professor Bala Ramesh at the Naval Postgraduate School in Monterey, for providing those most valuable of resources, time and processor cycles. The use of Walter Tackett's SGPC source code is also much appreciated.

Bibliography

Angeline, P. J. (1993) and J. B. Pollack, "Competitive Environments Evolve Better Solutions for Complex Tasks," in *Proceedings of the Fifth International Conference on Genetic Algorithms*, S. Forrest Ed. San Mateo, CA: Morgan Kaufmann.

Hillis, D. (1992) "Co-evolving Parasites Improves Simulated Evolution as an Optimization Procedure," in *Artificial Life II*, C. Langton, C. Taylor, J. Farmer and S. Rasmussen, Ed. Reading, MA: Addison-Wesley Publishing Co.

Chaitin, G. J. (1987) *Information, Randomness & Incompleteness*. Singapore: World Press.

James, F. (1990) "A Review of Pseudorandom Number Generators," in *Computer Physics Communications, vol. 60*, Amsterdam: North-Holland Publishing Co.

Knuth, D. E. (1969) *The Art of Computer Programming, Vol. 2 / Seminumerical Algorithms*. Reading, MA: Addison-Wesely Publishing Co.

Kolmogorov, A. (1965) "Three Approaches to the Quantitative Definition of Information," in *Problems of Information Transmission, vol. 1 no. 1*, New York, NY: Faraday Press.

Koza, J. R. (1992) *Genetic Programming: on the Programming of Computers by Means of Natural Selection*. Cambridge, MA: MIT Press.

L'Ecuyer, P. (1988) "Efficient and Portable Combined Random Number Generators," in *Communications of the ACM, vol. 31, no. 6*, New York, NY: ACM press.

L'Ecuyer, P. (1990) "Random Numbers for Simulation," in *Communications of the ACM, vol. 33, no. 10*, New York, NY: ACM press.

Park, S. K. (1988) and K. W. Miller, "Random Number Generators: Good Ones are Hard to Find," in *Communications of the ACM, vol. 31, no. 10*, New York, NY: ACM press.

Rucker, R. (1993) *Presentation on visualizing complexity and 'Bug Land' software, Feb. 2, 1993*, Stanford, CA: Stanford University.

Selfridge, O. G. (1989) "Adaptive Strategies of Learning: A Study of Two-person Zero-sum Competition," in *Proceedings of the 6th International Workshop on Machine Learning 1989*, A. M. Segre Ed. San Mateo, CA: Morgan Kaufmann.

Tackett, W. A. (1993) *SGPC source code*, available via anonymous ftp: `ftp.cc.utexas.edu` under directory `/pub/genetic-programming/code`.

Teller, A. (1994), "The Evolution of Mental Models," in *Advances in Genetic Programming*, K. E. Kinnear, Jr., Ed. Cambridge, MA: MIT Press.

Von Neumann, J. (1944) and O. Morgenstern, *Theory of Games and Economic Behavior*. Princeton, NJ: Princeton U. Press.

21 Optimizing Confidence of Text Classification by Evolution of Symbolic Expressions

Brij Masand

This paper reports some experiments in applying genetic algorithms for assessing the confidence of automatically assigned multiple keywords for news stories. Using Memory Based Reasoning (MBR) (a k-nearest neighbor method) to classify the stories, we would like to assign a confidence score per news story, that allows one to refer stories with low classification confidence to a human coder. Using Genetic Programming (GP) as used for program evolution by [Koza 1992], we discover and evolve symbolic expressions to compute confidence scores for news stories that allow a higher performance on subsets of the database while referring some stories to human editors. We have earlier reported recall and precision of 81% and 72%, if 100% of the stories are coded automatically [Masand, Linoff and Waltz 1992]. Using the evolved confidence measures to refer some stories for manual coding, we can achieve about 80% recall and 80% precision for 92% of the stories. This compares favorably with manually specified confidence functions that could classify 76% of the database with an 80-80 % recall-precision requirement.

21.1 Introduction

The availability of massively parallel machines with large memory and computational power have made it easier to use large example databases with relatively simple similarity metrics for the purpose of information retrieval. The same machines can also be used to reduce the amount of programming required to optimize the performance of information retrieval through techniques such as genetic algorithms.

Since John Holland's introduction of genetic techniques in [Holland 1975], GAs have been applied to a variety of optimization problems such as numeric optimization, discovering optimal routes, schedules etc. Koza has extended this approach for evolving simple computer programs to solve specific problems in various domains, without directly programming the form of the solution itself (see chapter 2 and [Koza 1992]).

We have previously reported the results for using a Connection Machine (CM)-based IR system for classification of news stories [Masand, Linoff and Waltz1992] and the use of a large training database of Census returns for automatically classifying new Census returns [Creecy *et al* 1992]. In this paper, we address the problem of referring difficult cases for manual coding using a confidence measure for classification. Following work described in [Koza 1992] on evolving symbolic expressions, we explore the evolution of formulae using genetic algorithms to compute confidence scores for the keywords assigned to news stories. This allows the choice of optimal subsets of the database for high levels of classification performance.

Sections 2 to 5 introduce the coding and the referral problems. Section 6 gives an overview of genetic algorithms, Section 7 illustrates some representative formulae and associated classification results. Section 8 describes the details of the genetic algorithm as it's applied to this domain. Section 9 presents the best evolved formulae to date. We conclude with a discussion of results and future directions.

21.2   The News Story Classification Problem

Editors at Dow Jones assign codes daily to hundreds of stories originating from diverse sources such as newspapers, magazines, newswires, and press releases. Each editor must master the 350 or so distinct codes (table below shows example codes). Due to the high volume of stories, typically several thousand per day, manually coding all stories consistently and with high recall in a timely manner is impractical.

Code	Name	Code Type
R/CA	California	Region
M/FIN	Financial	Market Sector
N/ERN	Earnings	Subject
I/CPR	Computers	Industry
I/BNK	All Banks	Industry
P/PCR	Personal Computers	Product
G/FDA	Food and Drug Admin.	Government

21.3   Automated keyword assignment using MBR

The coding task consists of assigning one or more codes to a text document, from a possible set of about 350 codes. Fig. 21.1 shows the text of a typical story with codes. The codes appearing in the header are the ones assigned by the editors and the codes following "Suggested Codes" are those suggested by the automated system. Each code has a score in the second column. In this particular case the system suggests 11 of the 14 codes assigned by the editors (overlap marked by *) and assigns three extra codes.

By varying the score threshold, we can trade-off recall and precision. S1, S2 refer to the scores for the corresponding codes. While these scores can be used as a measure of cer-

0023000PR PR 910820
I/AUT I/CPR I/ELQ M/CYC M/IDU M/TEC
R/EU R/FE R/GE R/JA R/MI R/PRM R/TX R/WEU
Suggested Codes:

S1	*	3991	R/FE	Far East
S2	*	3991	M/IDU	Industrial
S3	*	3991	I/ELQ	Electrical Components & Equipment
S4	*	3067	R/JA	Japan
S5	*	2813	M/TEC	Technology
S6	*	2813	M/CYC	Consumer, Cyclical
S7	*	2813	I/CPR	Computers
S8	*	2813	I/AUT	Automobile Manufacturers
S9		2460	P/MCR	Mainframes
S10		1555	R/CA	California
S11		1495	M/UTI	Utilities
S12	*	1285	R/MI	Michigan
S13	*	1178	R/PRM	Pacific Rim
S14	*	1175	R/EU	Europe

"DAIMLER-BENZ UNIT SIGNS $11,000,000 AGREEMENT FOR HITATCHI DATA SYSTEMS DISK DRIVES"

SANTA CLARA, Calif.--(BUSINESS WIRE)--Debis Systemhaus GmbH, a 100 percent subsidiary of Daimler-Benz, has signed a contract to purchase approximately $11 million (U.S.) of 7390 Disk Storage Subsystems. The 7390s will be installed in debis' data centers throughout Germany over the next 6 months.

Daimler-Benz is a diversified manufacturing and services company whose corporate units include Mercedes-Benz, AEG, Deutsche Aerospace and debis. Debis provides computing, communications and financial services along with insurance, trading and marketing services. The 7390 Disk Storage Subsystems are HDS' most advanced high-capacity storage subsystems capable of storing up to 22.7 gigabytes of data per cabinet. 22 gigabytes is the equivalent of approximately 15.7 million double-spaced typewritten pages. First shipped in October of 1990, the 7390s are used in conjunction with high-performance mainframe computers in a wide variety of businesses and enterprises.

Hitachi Data Systems is a joint venture company owned by Hitachi, Ltd. and Electronic Data Systems (EDS). The company markets a broad range of mainframe systems, peripheral products and services. Headquartered in Santa Clara, HDS employees 2,600 people with products installed in more than 30 countries worldwide. e

Figure 21.1
Sample News Story and Codes

tainty to accept or reject a particular code, they need to be combined to have an aggregate measure for the classification certainty of the entire story (all the codes taken together).

Score	Size	Headline
1000	2k	Daimler-Benz unit signs $11,000,000 agreement for Hitatchi Data
924	2k	MCI signs agreement for Hitachi Data Systems disk drives
654	2k	Delta Air Lines takes delivery of industry's first...
631	2k	Crowley Maritime Corp. installs HDS EX
607	2k	HDS announces 15 percent performance boost for EX Series processors
604	2k	L.M. Ericsson installs two Hitachi Data Systems 420 mainframes
571	2k	Gaz de France installs HDS EX 420 mainframe
568	5k	Hitachi Data Systems announces two new models of EX Series mainframes

21.4   The Coding Algorithm

Memory Based Reasoning (MBR) consists of variations on nearest-neighbor techniques [Dasrathy 1991] where the solution for a new case is derived from solutions corresponding to earlier cases similar to the new one, stored in a training database. Following the general approach of MBR [Waltz 1990], close matches to a new story are found using an already coded training database of about 87,000 stories from the Dow Jones Press Release News Wire, and a Connection-Machine Document Retrieval system (CMDRS, [Stanfill and Kahle 1986]) that supports full text queries, as the underlying match engine. The table above shows some of the near matches for the sample story. By combining the codes already assigned to the near matches, new stories are coded with a recall of about 82% and precision of about 71%. Recall and Precision were computed by regarding the codes previously assigned by the editors as "correct" and using cross validation by excluding the story to be classified from training database. Recall measures the proportion of codes assigned by the system that were also assigned by the editors, while precision measures the ratio of the "correct" codes to the total number of assigned codes. Thus in order to have 100% recall and 100% precision an automatic system would assign all the correct codes and no other codes. Details of the coding algorithm and the match engine are described in [Masand, Linoff and Waltz 1992]. Related automated coding approaches can be found in

[Dasrathy 1991], [Hayes and Weinstein 1991], [Creecy *et al* 1992] and [Rau, Jacobs and Paul 1991].The performance of coding depends on the number of near neighbors used and the score threshold used to accept or reject codes and the size of the training database [Masand 1993]. Increasing the number of near neighbors increases recall and decreases precision. Increasing threshold has the opposite effect.

21.5   The Referral Problem

While automated coding systems may have respectable *average* performance it may be still be necessary to detect cases with poor classifications. Using an automated classification system in a practical setting, it may be desirable to have the system automatically code only those documents with respect to which a high classification confidence can be assigned, referring the "difficult" stories to be coded manually. Another way to view this problem is to consider increasing classification performance on a subset of the database, for example by only classifying 80 to, 90% of the database automatically. One might ask: if a recall and precision of 80 and 70 is achievable on classifying 100% of the stories automatically, what fraction of the stories might be coded if the recall and precision are increased to 80-80, or 90-90?

21.5.1   Referral with single keyword per document

An example of this can be seen in the Census Classification task [Creecy *et al* 1992] where an automated system was used to classify Census returns with a high degree of accuracy (90%) on a subset of the entire database (about 60%). The average accuracy for the entire database was about 70%. Each Census return was assigned only one code, along with a certainty score. For a single assigned keyword per document, the task of referral can be attempted by selecting a constant threshold which determines whether a document is classified automatically or manually. Such a set of thresholds (per code) were in fact used for the Census classification task.

21.5.2   Referral with multiple keywords

In the case of news stories where multiple keywords are assigned to each document, one needs to combine the individual certainty scores for all the keywords and then set appropriate thresholds. While a simple approach such as averaging all the scores or taking the minimum score and thresholding on that might work, in general it's not clear how to best combine the individual scores for a composite confidence score for the document. The problem can be stated as:

Given a set of keywords K1, K2, K3...Kn, for each document and their corresponding scores, S1, S2...Sn, where
S1 > S2 >.., Sn, find a function **f** such that a confidence **c** can be computed

$$c = f (K1, K2...Kn, S1, S2....Sn) \tag{21.1}$$

for which a threshold T can then be set such that **c** > T causes documents to be classified automatically. The performance of the automatically classified subset of the test set is used as a measure of recall and precision.

In order to find the form of f in (21.1) using GAs, we simplify the problem by only considering the scores S1, S2 (ignoring the keywords themselves). We begin with an initial population of randomly generated formulae of S1, S2...Sn and then evolve formulae by combining the better formulae (as measured by their "fitness") over several generations.

21.6   Brief Overview of Genetic Algorithms

For a detailed introduction to GAs see [Goldberg 1989].The basic idea of a genetic algorithm is to computationally evolve structures that when interpreted are useful as solutions to some problems. Examples of such structures might be a list of cities for the travelling salesman problem, or a symbolic expression which could be interpreted as a finite state machine. A random population of the structures to be evolved is created as an initial population. In analogy with natural evolution, each individual is evaluated for "fitness" in an appropriate computational "environment" and the best individuals are selected for reproduction and "mating" -- producing new offsprings by combining random elements of parent structures. Mutation is simulated by causing random changes in the structures with a very low probability. The next generation is then evaluated similarly and the process is continued for a specified number of generations or when the best individuals reach a certain specified fitness. Even before the process terminates, the best individuals from each generation may provide reasonable solutions.

The underlying assumption is that such a selective scheme of generating variations relates structure to fitness. The evolved solutions may not be globally optimal, but nevertheless useful if no other easy means exists to find them. For this experiment the individual structures consisted of symbolic expressions as explained in section 7. As a search technique GAs are most suitable for domains where the problem has high dimensionality, and subparts of the problem can be solved independently. For a problem with low dimensionality pure random search in combination with hill-climbing may work as well. See chapter 2 for an introduction to Genetic Programming, an extension of the genetic algorithms approach for evolving programs.

21.7 Example formulae

f1 and f2 represent two random formulae from the initial population:

f1: sqrt (sqrt (S3) + sqrt(S7)) / 5.0

which only considers the third and seventh keyword scores.

f2: (s4 + (S3 + 8.0)* 9.0 +S1 + S6)

this one attempts an average of a few scores.

21.7.0.1 example results
The following functions represent some successful individuals from several runs:

f3: sqrt (S7 + S2) + 6.0

this formula only considers the scores of the first three keywords, resulting in a recall and precision of 75-87 on 84% of the database.

f4: S2/(S1- 4)

this one computes a ratio, resulting in a recall and precision of 76-81 on 88% of the database

f5: sqrt (S1)/9.0 + 5.0

this yields a recall and precision of 76 and 80% on 92% of the database.

f6:sqrt (S2*(sqrt (4.0*S2 + S1)) + sqrt (S3 + S6 +sqrt (S2)))

this formula yields a recall precision of 77-80% on 90% of the database.

Next we describe the representation of the formulae and the details of the GA process for this experiment.

21.7.1 Representation of evolved formulae

A convenient form of representation is important for modifying the structures for the purpose of cross-over and mutation. We use symbolic expressions in the LISP language,

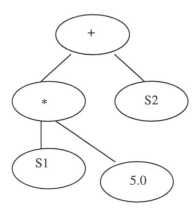

Figure 21.2
Tree Representation of a formula

which allows easy manipulation of a hierarchic symbolic expression as first class data objects and also allows easy evaluation of newly generated structures in an interpreted environment. Figure 21.2 shows the following formula:

`(5.0*S1 + S2)`

which computes the confidence as a weighted average of the first and the second keyword score.

The formula is shown as a binary tree corresponding to the prefix form:

`(+ (* 5.0 S1) S2)`

The intermediate nodes of the tree are functions and the terminal nodes can be variables or constants. Using a set of functions suitable for the problem and a set of variables and constants a random tree such as that in figure 5 can be generated in such a way that each tree is a valid formula.

Starting with the root node of the tree a function and two operands for it are randomly chosen. If both the operands are terminals (constants or variables) the formula is complete, if not, at least one of the nodes is a function for which we find two more operands etc. The depth of such a tree can be controlled to generate functions of a desired complexity.

Table 21.1
Tableau for News Story Classification/Referral

Objective	Evolve a referral formula to compute the confidence for an automatically coded news story, that will be used to accept or reject the codes for the story.
Terminal Set	S1,S2,S3,S4,S5,S6,S7,1,2,3,4,5
Function set	+,-,*,divide,sqrt
Fitness Cases	Test set of 200 randomly selected stories
Raw Fitness	Product of recall, Precision and proportion of test set classified
Standard Fitness	Same as raw fitness
Hits	Same as raw fitness
Wrapper	None
Parameters	M = 100, G = 20
Success Predicate	Number of generations

21.8   The environment for Genetic evolution

The following sections describe some details of the genetic programming environment.

21.8.1   Generation of the initial random populations

The initial function set and terminal set must be such that a random formula generated from such a set has sufficient expressive power to represent a possible solution. For this experiment we used the set of functions

```
(+, -, *, divide, sqrt)
```

where the divide was modified to return a finite value in the case of a zero divisor and the sqrt was modified to take an absolute value of the argument.

The terminal set consisted of

```
(S1,S2,S3,S4,S5,S6,S7,1.0, 2.0,3.0,4.0,5.0,6.0,7.0,8.0, 9.0)
```

this allows the first seven keyword scores to be used in the formula. (the average number of assigned keywords is about 8).The keywords themselves were ignored for this initial experiment.

21.8.2   Evaluating fitness

Each individual in the population is evaluated by using it as the confidence formula to accept or refer already classified news stories from a training set of 200 random news stories. This constitutes the "environment" for the population being evolved. The scores of the keywords in the story are assigned to variables S1, S2 etc. and then the formula is evaluated to compute a confidence number. The number is normalized over all the test documents and a fixed fraction of the range (80%) is used as a threshold to accept or refer the story. The recall and precision of the accepted stories are computed and the fitness of the formula is computed as the product of recall, precision and the fraction of the test set that is classified:

$$\text{fitness} = \text{recall*precision*fraction-of-stories-classified} \qquad (21.2)$$

Fitness can thus range from 0.0 to 1.0. Although most of the individuals in the initial population have low fitness, some of the individuals may perform well enough to be selected for further variation.

21.8.3   Fitness proportionate reproduction

Based on the fitness scores a certain fraction of the population (typically the best 10% or 20%) is included in the next generation without any variation.

21.8.4   Cross-over

The remaining population (about 80%) is generated by picking random pairs of parent structures and producing offspring structures by combining random elements of each parent. For example a random sub-tree of a formula of one parent can be inserted in place of a random sub-tree of the other parent and vice-versa, producing two offsprings. The pairs are chosen with replacement and the probability of being picked is proportional to the fitness of the individual.

21.8.5   Mutation

This experiment used mutation at a low rate (0.05) to modify those terminals of the tree that are numeric.

21.8.6   Population size

Population sizes of 50, 100, and 300 were tried. There was little improvement in results beyond the size 100 population.

21.8.7   Number of generations

Most of the interesting formulae emerged within 10 generations. While the absolute fitness improved from generation to generation, different individuals for the same fitness evolved different trade-offs between recall, precision and the fraction of the test set classified. Usually, after 15 or so generations, a few genotypes tend to dominate the population. Genetic operators can be used to preserve the genetic diversity of the population (by reducing the fraction of the dominating genotypes for example).

21.9   Results

The table below shows the performance for the best functions that were evolved in 20 different runs. Formulae f11 and f12 were specified manually for the purpose of comparison.

function	keyword scores used	Recall	Precision	% coded
f3	S7, S2	75	87	84
f4	S1, S2	76	81	88
f5	S1	78	80	92
f6	S1, S2, S3, S6	77	80	90
f7	S1, S4	78	76	92
f8	S2, S3	76	80	90
f9	S1, S4	79	77	92
f10	S3, S1	84	84	80
*f11	S1,S7	78	80	80
*f12	S1,S2,S3,S4,S5, S6,S7	80	80	76

```
f7:   (14.0 - S4/S1)

f8:   (S2- S3)

f9:   (S1 + S4)

f10:  (S3 + 5.0)*S1

*f11:(S1-S7)/7.0

*f12:(S1 + S2 +S3+ S4+S5+S6+S7)/7.0
```

The results compare favorably (about 14% better) with those obtained by the manually specified formulae such as f11, which computes the range, and f12, which computes the average.

21.9.1   The test environment

For training purposes we used a randomly selected training set of 200 news stories from a larger set of stories each of which had been previously coded by an editor. The recall and precision were computed by treating the keywords originally assigned by the editors as correct. For validating the results on an independent test set, the best evolved functions were tested on an independent test set consisting of 500 stories. Except for minor differences in the trade-offs between recall and precision, the evolved functions were quite robust (with respect to the recall-precision product) indicating that overfitting did not occur with respect to the training data.

21.10   Discussion of results

The best of the evolved formulae have already significantly exceeded the earlier results obtained manually (for example by classifying an additional 14% of the test set for an 80-80% recall and precision). It might have been possible to manually arrive at formulae comparable to the best ones evolved. However the ability to discover them with an automatic process is still useful. While it may also be possible to use a neural network for this purpose, the explicit form of the evolved solution may provide insight into the relationships between the variables under consideration.

   Some formulae with few variables but with non-linear dependencies perform better than formulas that take more keyword scores into account (such as computing an average).

It is also interesting that most of the successful formulae are a function of the first 3 or 4 keyword scores.

If the form of the solution were known we could do more specific parametric optimization (e.g. find the weights for a given formula).

Evolved solutions are probably specific to the distribution of the certainty score generated by the underlying method of classification. Since the process is automatic we believe we could evolve appropriate functions as easily even if the method of classification was modified.

21.11   Conclusions

We have successfully extended the Genetic Programming work by Koza [Koza 1992] to the IR domain for improving classification-referral performance by evolution of symbolic expressions.This suggests that GP could be applied to evolve symbolic expressions for other IR applications especially where the form of the relationship between variables of interest is not known beforehand. Thus there is a potential for discovering interesting relationships unbiased by our preconceptions.

21.12   Continuing and Future Work

More work needs to be done in analyzing and studying the different forms that evolve and correlating them with the different genetic operators, function and terminal sets.

One obvious next step is to apply the technique to optimize the underlying coding step itself by discovering better ways to combine the evidence from several near matches.

Co-evolution in the form of changing training sets could be used to improve performance even more over a wider sample of random news stories.

How large a random search would solve this problem to a comparable level? It will be useful to generate a large random population (maybe 10,000 individuals) and compare the best formulae from that population to those from a smaller population evolved for several generations.

The function set can be expanded e.g. by adding min, max, and comparison functions.

We could also generate formulae that take the actual keywords into account (using logical and string comparison functions for instance).

Acknowledgments

I would like to thank Dave Waltz, Steve Smith and Kurt Thearling from Thinking Machines for help with this project.

References

Creecy, R. H., Masand B., Smith S, Waltz D. (1992) "Trading MIPS and Memory for Knowledge Engineering: Classifying Census Returns on the Connection Machine." The *Comm. ACM* Aug (1992).

Dasrathy B. V. (1991) *Nearest Neighbor (NN) Norms: NN Pattern Classification Techniques*. IEEE Computer Society Press, Los Alamitos, California (1991).

Goldberg, D. E. (1989) "Genetic Algorithms in Search, Optimization and Machine learning" Addison-Wesley, 1989.

Hayes, P. J. and Weinstein, S.P. (1991) "CONSTRUE/TIS: A System for Content-based Indexing of a Database of News Stories." *Innovative Applications of Artificial Intelligence 2*. The AAAI press/The MIT Press, Cambridge, MA, pp. 49-64, 1991.

Hayes P.J. *et al* (1990) "A Shell for Content-based Text Categorization." 6th IEEE AI Applications Conference, Santa Monica, CA, March 1990.

Holland , J. H (1975) "Adaptation in Natural and Artificial Systems". University of Michigan Press, Ann Arbor, MI 1975

Koza , J. R. (1992) "Genetic Programming: On the programming of computers by natural selection",The MIT Press, Cambridge, Mass. 1992.

Lewis, David D. (1992) "An Evaluation of Phrasal and Clustered Representation on a Text Categorization Task." *Proceedings, SIGIR '92*, Copenhagen,Denmark.

Masand, Brij M., Linoff, Gordon S.,and Waltz, David L. (1992) "Classifying News Stories on the Connection Machine using Memory Based Reasoning" *Proceedings, SIGIR '92*, Copenhagen, Denmark.

Masand, Brij M. (1993) "Effects of Query and Training Database size on Classification of News Stories using Memory Based Reasoning", To appear in the proceedings of the AAAI Spring Symposium on Case-Based Reasoning and Information Retrieval, 1993.

Rau, Lisa F. and Jacobs, Paul S. (1991) "Creating Segmented Databases From Free Text for Text Retrieval." *Proceedings, SIGIR 1991* (Chicago, Illinois).

Stanfill, C. and Kahle, B (1986) "Parallel Free-Text Search on the Connection Machine System." *Comm. ACM* 29 *12* (December 1986), pp. 1229-1239.

Waltz, D. L. (1990) "Memory-Based Reasoning." In M.A. Arbib and J.A. Robinson (eds), *Natural and Artificial Parallel Computation*, The MIT Press, Cambridge, MA. (1990), pp. 251-276.

22 Evolvable 3D Modeling for Model-Based Object Recognition Systems

Thang Nguyen, Thomas Huang

This paper presents a system that evolves 3D models over time, eventually producing novel models that are more desirable than initial models. The algorithm starts with some crude models given by the user, or randomly-generated models from a given model-grammar with generic design rules and loose constraints. The underlying philosophy here is of gradually evolving the initial models into better models over many generations. There is a close analog in the evolution of species where better-fit species gradually emerge and form specialized niches, a highly efficient process of complex structural and functional optimization. Our simulation results for 3D jet aircraft model design illustrate that this approach to model design and refinement is feasible and effective. The intended application domain is for automatic object recognition systems, though the model fitness criteria is currently determined by user interactive selection.

22.1 Introduction

Since we intend our evolvable modeling eventually for use in a 3D recognition system, let us mention some observations about model-based 3D object recognition systems, which could potentially benefit from a model evolution module, as follows:

1. There have been existing vision systems (e.g. [Goad 1983], [Ikeuchi 1988], and others) that allow us to generate recognition strategies/programs, once we have an appropriate 3D model of the target. These are highly automatic 3D model-based recognition systems. And there are many other model-based recognition systems ([Brooks 1981], [Lowe 1991], etc.) with various degrees of sophistication. Still, a bottleneck in the development of these systems is the high cost of good model base design, indexing and coding.

2. Real-world objects are very diverse. For cars, planes, chairs, etc. there is much greater variability/parametrization than could be captured by hand-crafting of models. Vision system developers most often do not have at hand the design blueprints for their objects of interest. Furthermore, the number of object designs can be expected to grow exponentially over time. We observe that even man-made objects such as real airplanes exhibit an evolutionary trend towards higher complexity and specialization. Thus if a system can automatically evolve its model base then it can achieve a higher degree of autonomy in dealing with new-object cases, with lower requirements for human "teaching" or tedious hand-crafting.

In general, there are four functional components for 3D model-based recognition [Brooks 1981], [Goad 1983], [Ikeuchi 1988]:

- a model knowledge base that allows bidirectional inference from image features to object structures, and vice versa.

- a library of feature detection and grouping routines that can extract reliable and discriminating features from given images. The extracted and grouped features will subsequently be used to form initial object hypotheses.

• a hypothesis generator, which will construct initial hypotheses about the possible objects that could have given rise to those image features found by the feature detection and grouping stage. Frequently, integral to hypothesis generation is a precompilation of all the view aspects of the object models, effectively decomposing the more general 3D recognition problem into a hierarchy of 2D recognition/matching.

• a hypothesis resolver/verifier to compute the degree of match between a given hypothesis and an object model in the system's model base. Usually, object hypothesis verification is a sequential, iterative process that gradually accumulates supporting and/or refuting evidence for all currently active hypotheses, keeping only those with high plausibilities for the next iteration. The process terminates with either: a few best hypotheses, one best hypothesis, or none remaining.

One can see that the model knowledge base and the hypothesis generator (or case retriever) can be enhanced with an evolvable modeling component: a self-evolving model base can derive new models to deal with new objects for which there are no existing models in the model base. GA/GP, or genetic algorithms/programming, has been applied successfully to a few vision and graphic applications, such as: segmentation of intensity images by Bhanu et al [Bhanu 1990], segmentation of range images by Meygret et al [Meygret 1992], pattern recognition [Calloway 1987], and automatic generation of abstract color art [Sims 1991]. Baker [Baker 1993] used GP to evolve line drawings of faces with user interactive selection. Watabe and Okino [Watabe 1993] used GA to evolve free-form-deformation of 2D lattices. Todd and Latham [Todd 1992] evolve highly complex, biologically-inspired, surrealistic 3D "creatures". Notably, Hill et al [Hill 1992] applied GA to generate flexible 2D templates for recognition of left ventricles in echocardiograms. However, we are unaware of any published work on GP application to 3D model evolution for model-based recognition.

The organization of this chapter is as follows. In section 22.2 we discuss the overall framework of a hybrid GP/GA for the evolution of 3D hierarchical structures. In section 22.3 we discuss the implementation details and some results in evolving simple 3D jet aircraft models. Section 22.4 explores related works and future extensions and applications of this work.

22.2 Evolvable 3D Modeling with Multi-level GP/GA

It should be kept in mind that this work is, first and foremost, about an evolvable modeling platform for 3D models. In reminding you this, we hope to avoid possible confusions in subsequent sections where there is the question of determining the *fitness* of a 3D model. Some subtle confusions can arise because 3D model-based recognition, the intended application domain, is, in many respects, an inverse problem of 3D computer graphics. While computer graphics aims to synthesize realistic images of objects given their structural descriptions and other properties, computer vision/recognition aims at inferring structures and properties of objects in a given image [Brooks 1981], [Goad 1983], [Ikeuchi 1988]. Unless otherwise stated in this paper, a 3D model's *fitness* is evaluated only from a graphics perspective, that is, by the user's visual perception. When this evolvable modeling platform is integrated in an object recognition system, however, a model's fitness is ultimately determined quantitatively by how accurately it can match the 3D shape of the target object.

Our approach is to define 3D geometric models in such a way that one can crossover (or "mate"), in the spirit of GA/GP, between two "parent" models and get two new models (offsprings), most often with inherited features, but sometimes with new characteristics not possessed by either parents:

1. Grammar-based, structured object modeling: a plane is defined structurally as composed of a body, wings, engines, stabilizer tails and vertical rudder. Important configuration variations like engine and tail placement (tip-of-rudder or tail-body) are also incorporated into the grammar. We are thus able to "crossover" structural elements among the symbolic designs.

2. Parametrization of numerical model feature values, feature relation coefficients, and inter-feature constraints. We define many numerical parameters in terms of other parameters. The numerical parameters and coefficients are coded into binary strings so a pair of them can be crossed and give offsprings of binary strings.

3. Performance feedback contributes to shape the path of progress: models that are more similar to target models have more chance of surviving and evolving further. Simulated evolution processes have been shown to be a robust approach to optimization [Goldberg 1989], [Koza 1992].

With respect to crossover and mutation, each of our models has two levels (figure 22.1):

• structural level: with a production grammar and symbolic parameters. A model design expressed at this level has the form of a tree. A crossover at this level is performed by exchanging two subtrees of the same functionality: e.g., two Wings subtrees.

• binary string level: for the numerical values of the parameters, all coded in binary strings. At this level, a crossover between two models can be performed by crossing two binary strings, both of which refer to some common parameter, e.g. WingHalfSpan.

In practice, some substructure is picked randomly, from among all substructures of a symbolic design, for crossover or mutation. This has the same effect of hierarchical sampling of both structures (GP style) and parameters (GA style).

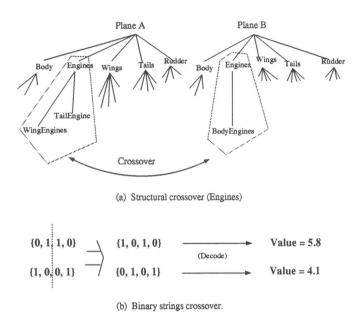

(a) Structural crossover (Engines)

$\{0, 1, 1, 0\}$ $\{1, 0, 1, 0\}$ $\xrightarrow{\text{(Decode)}}$ Value = 5.8

$\{1, 0, 0, 1\}$ $\{0, 1, 0, 1\}$ $\xrightarrow{\hspace{2cm}}$ Value = 4.1

(b) Binary strings crossover.

Figure 22.1
Crossover between two jet aircraft models at the structural level (Engines), and binary
string level for numerical parameter values.

22.3 Evolvable modeling of jetliners — implementation

Next we discuss some details of the implementation for interactive evolution of 3D
jetliner models. With appropriate augmentations, the system here can be embedded
as a hypothesis generator and model acquisition module within an automatic object
recognition system.

22.3.1 The Data Structure of the Evolvable 3D Jet Models

Each 3D jet model is a data structure describing how a set of geometric primitives, such
as polygons, cylinders, cones, and ellipsoids of different size, orientation and relative
position, are put together to give an appearance of a jetliner. There must be discrete,
structural descriptors concerning, for example, the number and configuration of the
engines, where its horizontal stabilizer tail is mounted (high at the tip of the rudder,
or at the tail body's level), etc. And there need be many numerical shape descriptors
concerning, for example, the wing span, the width of the wing's base, and many others.

The need to enable a 3D model to crossover with another 3D model *and* give
meaningful results requires that the structural descriptors and the numerical shape

descriptors be suitable for meaningful "genetic" manipulation. For example, consider the crossing of the horizontal tail structures between two jets, and suppose one horizontal tail is mounted high at the rudder tip, and one is mounted at the tail-end level. (Note that fuselage diameters can be different.) One must ensure that the mounting heights of the resulting horizontal tails are meaningful, i.e., either at the respective rudder tip's height, or the tail-end's height, and not somewhere in midair, without any physical support. This illustrates that we have to keep a symbolic variable, say, tail-mounting-height, separate from its numerical value, say, 3 meters. Keeping them separate also enable the system to evolve 3D models either purely in the sense of GA, or by swapping functionally equivalent susbstructures as in GP, or a mixture of both, which we did for the results here.

The data structure of an individual jet model is a list of six components:

1. the birth record, keeping track of the parents of the individual (if crossover or mutation birth), or noting that the individual is generated randomly. This is helpful in tracking the statistical behavior of the evolutionary process.

2. the display position of the rendered 3D model in the population.

3. the structural design of the model, *planeStructure*, with symbolic parameters not yet instantiated. These structures are suitable for GP-style genetic operations.

4. *planeParameters*, the list of numerical parameters coded into binary strings for GA-style genetic operations.

5. a list of the most recent design changes incurred by either crossover or mutation. For newly random individuals, this slot is an empty list.

6. the score, or *fitness*, assigned to the individual model, either by the user's judgement or by an object's recognition system's evaluation.

The *planeStructure*, similar to those in figure 22.1(a), is constructed according to a *planeGrammar*, whose rules are as follows:

- Plane —> DesignPlane[Body, AllEngines, Wings, Tails, Rudder]
- Body —> SketchBody[midbodyLength, tailLength, bodyDiameter]
- AllEngines —> EnginesPick[NumOfEngines[Random[ ]]]
- WEngines —> WEnginesAt[wingEngineFromBase]
- BEngines —> BEnginesAt[bodyEnginesFromTailEnd]
- Wings —> SketchWings[wingHSpan, wingbaseWidth, wingExtbaseWidth,
                     wingbaseleadPointX, wingElbowK, wingSweep, wingtipWidth]
- Tails —> SketchTails[tailHSpan, tailbaseWidth, tailtipWidth, tailSweep, tailsMount]
- Rudder —> SketchRudder[rudderbaseWidth, ruddertipWidth,
                     ruddertipHeight, rudderSweep]

EnginesPick is a function that randomly select an engine configuration out of five possible configurations possible given a maximum number of three engines. The syntax of EnginesPick means: select the sublist of EngineConfigs that have the first numerical slot equal to the number of engines given, then pick randomly among the list of possible configurations (2nd item in the sublist selected) to get one. For example, there is only one configuration with one engine, that is, { TEngine }.

Likewise, TailsMount evaluates automatically to a random sampling between two possible tails-mounting configurations: RudderTailsMount or BodyTailsMount. As we will see later, each of these in turn evaluates to a 3D point in space, specifying the position of the leading point of the tails, when properly mounted.

- EnginesPick[numOfEngines] := RandomPick[Select[EngineConfigs,
                                            (#[[1]] == numOfEngines)&][[1, 2]]];

- EnginesConfigs = { { 1, { { TEngine } } },
                     { 2, { { WEngines }, { BEngines } } },
                     { 3, { { WEngines, TEngine }, { BEngines, TEngine } }

- TailsMount := RandomPick[{ RudderTailsMount, BodyTailsMount }];

SketchBody, SketchWings, WEnginesAt, BEnginesAt, SketchTails, and *SketchRudder* are all dummy functors: they only serve to hold the parameters in place and do not evaluate them since we want to keep the *planeStructure* un-instantiated, all symbolic, for GP-style swapping of functional subtrees. When a graphical rendering is called for, all parameter values are first computed from their strings, parameter symbols are instantiated, then a substitution of all the dummy functors by the corresponding functors will let the instantiated *planeStructure* automatically expands into a list of all 3D geometric primitives of the model.

22.3.2 The Parameter Constraints and Parametric Relations

Each numerical parameters is allowed to take on a value within some constraint interval, into which its binary string will be mapped and interpreted. There are nineteen parameters which participate in GA-style genetic operations. Some of these parameters, like *wingbaseleadPointX*, are dependent on other parameters. The list of all parameter constraints is as follows:

- {midbodyLength, IsWithin, {6.0, 9.0}}
- {bodyDiameter, IsWithin, {0.8, 1.2}}
- {wingHSpan, IsWithin, {4.5, 6.8}}
- {wingbaseWidth, IsWithin, {1.8, 3}}
- {wingExtbaseWidth, IsWithin, {0.7, 1.2}}

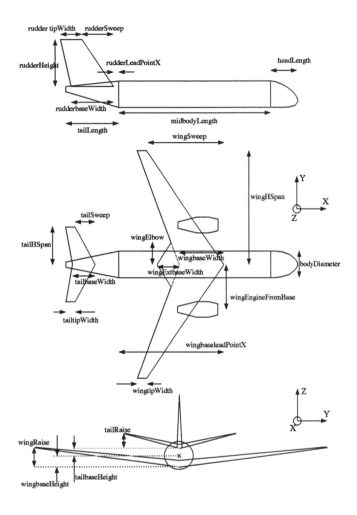

Figure 22.2
Three simplified orthographic views of a generic 3D model design framework of a jet plane, showing some basic parameter definitions.

- {wingbaseleadPointX, IsWithin, {0.52, 0.65}, midbodyLength}
- {wingElbowK, IsWithin, {0.15, 0.35}, wingHSpan}
- {wingSweep, IsWithin, {2.0, 5.0}}
- {wingtipWidth, IsWithin, {0.4, 0.7}}
- {wingEngineFromBase, IsWithin, {0.25, 0.5}, wingHSpan}

- {bodyEngineFromTailEnd, IsWithin, {-0.15, 0.05}, midbodyLength}
- {rudderbaseWidth, IsWithin, {1.6, 1.9}}
- {ruddertipHeight, IsWithin, {2, 2.4}}
- {ruddertipWidth, IsWithin, {0.8, 1.2}}
- {rudderSweep, IsWithin, {0.8, 1.6}}
- {tailHSpan, IsWithin, {1.8, 2.7}}
- {tailbaseWidth, IsWithin, {0.9, 1.4}}
- {tailSweep, IsWithin, {0.8, 1.8}}
- {tailtipWidth, IsWithin, {0.25, 0.5}}

A constraint interval by itself does not give a parameter value directly, but given an N-bit binary string S for any independent parameter and its corresponding parameter constraint interval *{lowerpoint, upperpoint}*, the parameter value *ParamValue* is determined by a linear mapping of the binary string S into the interval, including the lower end point:

$$ParamValue = lowerpoint + \frac{(upperpoint - lowerpoint) \sum_{k=0}^{N-1} 2^k * S_k}{2^N - 1} \qquad (22.1)$$

where S_k is the kth bit's value (0 or 1). For dependent parameters, the binary string is interpreted as a coefficient. For example, suppose we have a string 0110 for *wingEngineFromBase*, and suppose the value of *wingHSpan* is 4.96, one first computes the string's value to be 0.35, and then one uses that 0.35 as a coefficient to compute the value for *wingEngineFromBase* to be 0.35 * 4.96, which is 1.736.

Beside the above nineteen basic parameters, there are also constant-valued parameters and parametric relations, which are actually equations with evolvable variables (those parameters coded into GA binary strings), two of them being 3D vector-valued functions:

- rudderbaseHeight —> tailLength * sin[0.1] + 0.02
- rudderleadPointX —> rudderbaseWidth – tailLength + 0.05
- tailLength —> 2.5
- tailRaise —> 0.15
- BodyTailsMount —> { tailbaseWidth – tailLength + 0.05,
                              0, rudderbaseHeight }
- RudderTailsMount —> { rudderleadPointX – rudderSweep + 0.05,
                              0, ruddertipHeight – 0.1 }
- wingRaise —> 0.5

- wingbaseHeight —> -0.4 * bodyDiameter
- enginemaxR —> 0.3
- enginewingSpace —> -1.5 * enginemaxR
- TEnginebodySpace —> -0.1 + enginemaxR + bodyDiameter / 2
- enginewingLead —> -0.6
- enginebodySpace —> 0.8 * (bodyDiameter + enginemaxR)

HiddenParameters do not directly participate in any genetic operations. However, those that depend on the nineteen "evolvable" parameters have indirectly "evolvable" values. The two parameters *BodyTailsMount* and *RudderTailsMount* are rather special: they are actually 3D points in space (a list of three components) specifying the foremost point of the horizontal tails when at the appropriate mounting position.

22.3.3 The Population Structure and Population Transition

As with other GA/GP applications, the basis for diversity is the population of individuals and the genetic operations that create a new population from among the individuals of the current population. For simplicity of the presentation and discussion of our results, each generation consists of only four jetliner models, probably the barest minimal size for any kind of simulated evolution. For such a small population size, we employed a simple probabilistic, rank-based selection for both the proportionate reproduction and the mutation operators, but we use deterministic selection for the only crossover per generation. Because the runs reported here require interactive fitness evaluation, deterministic selection for crossover might not be as "unnatural" as it seems.

Further more, we keep a simple format of the model population for quick comprehension of the process, and to keep track of the birth origin of each model individual. See Figure 22.3: there are 4 populations, from the 0th generation to the 3rd generation of the evolution process. In each population (except the 0th generation), models at positions #1 and #2 are children of a crossover operation (sexual production) from the previous generation; the model at position #3 is a reproduced offspring, a perfect copy of its ancestor; and the model at position #4 is either a mutation or a new randomly generated model (with even probability of either case.) For the run shown in Figure 22.3, the objective, hence fitness criteria, is to find a jet model as similar to a Boeing 777 as possible. (There is a scanned photograph of a Boeing 777 in Figure 22.5, also for discussion of automatic object recognition processes.) Figure 22.4 shows four different models, including the best-of-run, or best-so-far, model, #4 of generation 3 from Figure 22.3, shown at top left. Two other models shown in Figure 22.4 are among those randomly generated, and one from a different run with a fitness criteria geared towards aethetics rather than any particular recognition. When examining the best-so-far B777-lookalike in Figures 22.4 and 22.3, one should keep in mind that the

Table 22.1
Tableau for 3D jetliner model evolution.

Objective:	Find and display some 3D jetliner model that is as similar as possible to what the user has in mind. (In an automatic object recognition system, the objective would be for the system to come up with an accurate 3D model of a previously unknown jetliner, given some of its images.)
Terminal set:	- Lowest structural terminals are geometric primitives: polygons, cylinders, cones, ellipsoids, and combinations of them. - Parameters and spatial relation between these geometric primitives as object's subparts are coded in binary strings in GA style.
Function set:	DesignBody, DesignWings, DesignTails, DesignRudder, WingEnginesAt, BodyEnginesAt.
Fitness cases:	User dependent
Raw fitness:	- Assigned by user, in the manner of interactive-evolution: the better a model matches the target he wants, the higher raw fitness value he assigns to it. - In model-based recognition: fitness can be computed by various error metric between a 3D model and the detected image features, with or without 3D backprojection.
Standardized fitness:	Same as raw fitness for this problem.
Hits:	Not defined in this problem.
Wrapper:	Graphical interpretation of geometric primitives, rendering and display.
Population size:	Very modest by GA/GP standard, typically under 10. (Actually, it is questionable whether a human user can consistently evaluate fitness values for a population of, say, thousands of 3D models, especially over many generations. However, if used in a 3D model-based recognition system, there is no fundamental limitation except the memory and processing time constraints.)
Termination:	- User-preset threshold of termination fitness value. - User-preset maximum number of generations, typically under 50 (user's patience is a limiting factor, since the graphical rendering for each population of 4 models can take 4 minutes on a Sun Sparc II).

"evolution" has taken place for a mere total of sixteen individual models, with too small a population size, all just for a demonstration of the general principles. In fact, less than ten runs total have been tried to get all the jet models shown in this paper.

The overall flow of the evolution process for these 3D jet models can be summarized as follows:

1. The first population of models consists entirely of randomly generated models from the model grammar, model primitives, parametric relations and constraints.

2. In every generation, the 3D jet models are displayed after they have been structurally and parametrically defined (either by random generation or by crossover or mutation.) Each jetliner model is then given a score, judged by the user (or by some internal performance measure, if within an object recognition system).

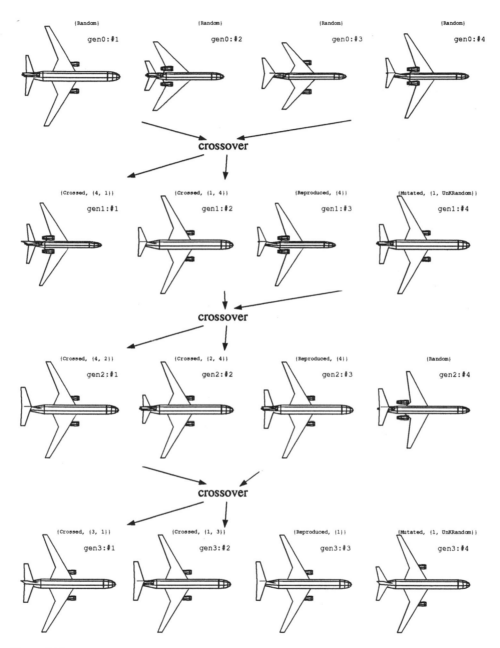

Figure 22.3
A typical run of 4 generations, from gen0 to gen3, in search for a Boeing 777 lookalike.

3. The two models with the highest scores are then selected for crossover, or "mating". One model will reproduce perfectly. One new model will come from either a mutation or a random creation. The probabilities of both mutation and random-creation here are set to 0.5*(1/4). This process creates the new generation of four models.

4. The above loop in (b) and (c) is continued until at least one model is given a score higher than a threshold of acceptability (say, any model to be accepted has score > 100.) Alternatively, a maximum number of generations can be set beforehand, and many runs can be tried.

Incidentally, we treat a mutation as a crossover with some random unknown model, which is appropriate in our framework. (A random unknown model is one that is

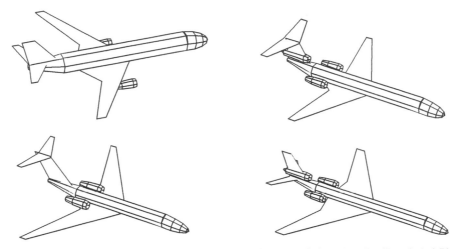

{{3, 4, Mutated, {1, UnKRandom}}, {49.5, 0., 0.}, DesignPlane[SketchBody[midbodyLength, tailLength, bodyDiameter], {WEnginesAt[wingEngineFromBase]}, SketchWings[wingHSpan, wingbaseWidth, wingExbaseWidth, wingbaseleadPointX, wingElbowK, wingSweep, wingtipWidth], SketchTails[tailHSpan, tailbaseWidth, tailtipWidth, tailSweep, BodyTailsMount], SketchRudder[rudderbaseWidth, ruddertipWidth, ruddertipHeight, rudderSweep]], {{bodyDiameter, 1.3, {1, 1, 1, 1}}, {bodyEngineFromTail, -0.0033, {1, 0, 1, 1}}, {midbodyLength, 8.6, {1, 1, 0, 1}}, {rudderbaseWidth, 1.9467, {1, 1, 0, 1}}, {rudderSweep, 1.4933, {1, 1, 0, 1}}, {ruddertipHeight, 2.133, {0, 1, 0, 1}}, {ruddertipWidth, 0.9, {0, 1, 1, 0}}, {tailbaseWidth, 1.32, {0, 1, 1, 0}}, {tailHSpan, 2.34, {1, 0, 0, 1}}, {tailSweep, 1.3933, {1, 0, 0, 0}}, {tailtipWidth, 0.4400, {0, 1, 1, 1}}, {wingbaseleadPointX, 0.55467, {0, 1, 0, 0}}, {wingbaseWidth, 1.8733, {0, 0, 1, 0}}, {wingElbowK, 0.34733, {1, 1, 0, 1}}, {wingEngineFromBase, 0.3833, {1, 0, 0, 0}}, {wingExbaseWidth, 2., {1, 1, 1, 1}}, {wingHSpan, 6.8, {1, 1, 1, 1}}, {wingSweep, 4., {1, 0, 1, 0}}, {wingtipWidth, 0.6, {1, 1, 1, 1}}}, {CrossEngines, CrossTailMounts, {TailsMountPoint, {newVal, BodyTailsMount}, {oldVal, RudderTailsMount}}, {bodyEngineFromTail, {newVal, -0.0033}, {oldVal, -0.0833}, {newString, {1, 0, 1, 1}}, {oldString, {0, 1, 0, 1}}}, {tailHSpan, {newVal, 2.34}, {oldVal, 2.7}, {newString, {1, 0, 0, 1}}, {oldString, {1, 1, 1, 1}}}, {tailSweep, {newVal, 1.3933}, {oldVal, 0.8733}, {newString, {1, 0, 0, 0}}, {oldString, {0, 0, 1, 0}}}, {tailtipWidth, {newVal, 0.440}, {oldVal, 0.560}, {newString, {0, 1, 1, 1}}, {oldString, {1, 1, 0, 1}}}, {wingEngineFromBase, {newVal, 0.3833}, {oldVal, 0.25}, {newString, {1, 0, 0, 0}}, {oldString, {0, 0, 0, 0}}}}, {ScoreGivenIs, 105}}

Figure 22.4
Some of the jet models at a closer view. Top left is the best-so-far B777-lookalike jet model, individual #4 in generation 3, from Figure 22.3. The full design expression of the B777-lookalike is listed. Top right is a random model, #4 of generation 0. Bottom left is another random model, #4 of generation 2. Bottom right is another best-so-far model from a different run where the fitness criteria from the user is more of aethestic nature.

randomly generated and participates in a crossover-mutation, but its full design structure and parameters are not kept in the population records). Thus mutation and crossover in this context do not differ much in any fundamental sense. We view evolution in nature more as a guide than a dogma, and the usefulness of the results are more of concern than "rigorous" theoretical justifications. The results seem encouraging. In the last row of Figure 22.3, there are two models that are more similar to a Boeing 777, while the other two models also have much relevant similarity. One can readily see from the course of the population evolution that the chance of finding a model more similar to a Boeing 777 has been gradually enhanced. If one looks more closely into the relevant individual records, one can also trace the inherited features of the evolved models over time. For example, between the best-of-generation in generation 0, and the best-of-run (the best-of-generation in generation 3), out of a total of nineteen numerical parameters and four configuration variables (three for engines and one for tails-mounting configuration), the best-of-run has inherited fourteen numerical parameters, one set of engines (wing-mounted), and the tail-mounting. On the average, the feature inheritance dynamics in the run is such that there are about 3-5 feature changes per offspring individual per generation, with the rest being inherited unchanged from the dominant parent. A natural consequence of the model tree structure and the hierarchical sampling for crossover, a feature of GP, is that one parent will be the dominant parent, whose features are more likely to appear in the offspring.

22.4 Related Works, Extensions, and Applications

To put our work in perspective, let us mention some of the more relevant works on the application of GA/GP to design problems. Of more interest to our discussion are those works by Bramlette and Cusic [Bramlette 1989], Powell and Skolnick [Powell 1993], and [Baker 1993]. Baker used GP to evolve line drawings of faces, with user interactive selection. Bramlette and Cusic did a comparative evaluation of search methods applied to parametric design of fighter aircraft. Among fifty defining parameters, they chose eight parameters to study with five search methods: random search, biased random walk, iterated genetic algorithm (GA), GA with biased random walk, and iterated simulated annealing. The last two methods equally produced the best answers. Powell and Skolnick compared GA and Num Opt (among the better numerical optimizers) on ten difficult engineering design optimization problems with non-linear constraints, and found GA to be quite competitive, finding the global optimum point or region in many cases and getting stuck only once more often than Num Opt. Notably, the GA search even improved on the best known solution in one problem. The design problem set in [Powell 1993] includes a five-speed gearbox, a chemical reactor, and mathematical programming models of three- and five-stage membrane separation processes. These authors have also been very successful in real-world design applications with EnGENEous, a hybrid of expert system and genetic algorithms that has improved significantly designs of cooling fans, molecular electronic structures, and aircraft engine turbines [Powell 1989].

Perhaps with the exception of Baker's work, which employed GP in a comparatively small problem domain, the works above are fairly extensive and complex systems, each being an effort by many, but they have not had the advantage of being able to evolve hierarchical designs directly with the GA implementations, except with more programming from the users. In contrast, the hierarchical structure of our design lends itself to more modular development, also with the added degree of freedom in GP-style evolution of tree structures, suitable to a broader problem spectrum. If one is to evolve designs with GP, say, to optimize aircraft, the richer combinatorial design spaces could potentially lead to previously unexplored high-performance configurations.

Certainly, the evolvable modeling framework we have built so far is fairly crude, with many limitations that suggest further significant improvements. One obvious refinement is to speedup the graphical rendering and display so that much larger populations with more flexible selection schemes could be run at interactive rate. This would enable the user to observe many essential evolutionary behaviors of GP/GA more clearly for future optimization. Fortunately though, even with a small population size of four, many interesting and useful evolutionary behaviors of GP/GA have been observed. One can see that the chance becomes better and better for some new jet model(s) to come close to what the user is looking for. Another improvement would be to allow for more structural and parametric variations in the 3D jet models. For example, there could be more shapes of engines, more shape primitives for the fuselage and the nose section, and more detailed structures like control surfaces (flaps) in the wings, tails and rudder. Also, front stabilizers (canard) could be allowed as a possible structural configuration.

The necessary augmentation for model-based recognition would involve more substantial extensions to the current platform, due to the inverse-problem nature of recognition. First, there must be a flexible module for bi-directional inferential links between 3D object models and detectable image features. For example, there must be some inferential rules that suggests the possible existence of either a wing, tails or a rudder, when a tapered quadrilateral region is detected [Goad 1983], [Brooks 1981], [Ikeuchi 1988]. And there should also be efficient and flexible rendering mechanisms so one can simulate the image feature of any structural component of the object (say, the length of a wing segment) under various conditions. Despite the availability of high-performance, photo-realism of current graphics and modeling platforms, the specific needs of model-based object recognition has not been well met. Model-based vision researchers most often have to develop their own modeling environments, among them Vantage [Ikeuchi 1988] and Geomod [Brooks 1981] being somewhat better known. Besides the above essential inference modules, there must also be efficient feature detection and feature grouping mechanisms for low-level image analysis, and the hypothesis constructor and verifier modules. We will not delve more deeply into these issues which, together with spatial inferential modules, have been the topics of intensive research by the vision community. Among the many published works which are of more interest to our work, there are [Hill 1992], [Meygret 1992], [Biederman 1987], [Burns 1992], [Dickinson 1992], [Grimson 1991], [Lowe 1991], besides those rather well-known works [Goad 1983], [Brooks 1981], and [Ikeuchi 1988]. Figure 22.5 illustrates some important steps in model-based recognition, with the assumed goal of recognizing the presence of a Boeing 777 in a scanned photograph (top left). The grouping of line and curve segments into parts of a complete object hypothesis is one of the more difficult problems. For illustration purposes, the idealized groupings of line and curve segments in Figure 22.5 are delineated symbolically by ellipses. The final recognition/discrimination stage is highly complex. Due to space and scope limitation here, the interested readers could consult publications such as the collection of papers in [Flynn 1993], for a more thorough treatment on 3D object recognition.

22.5 Comments and Conclusions

We have demonstrated that it is feasible to evolve 3D jet models to search for better models, starting from random models. There are many ways to apply our evolvable modeling system. This is one method to let a system learn to acquire a new model with only implicit and partial, non-specific information (human users or recognition modules need only give some "fuzzy" score to the quality of any model proposed.) In many realistic situations, the evaluation criteria for a model cannot be stated very concretely, especially with objects unknown to even the system designer.

This system of 3D model evolution can be useful for model-based recognition. For clarity of the figures, we had simplified some graphical rendering parameters to avoid clutters by too many details of the 3D models at small scales. We could render the models at greater detail, or incorporate shading, but the simple graphic rendering helps keep the model features more transparent for discussion. Many simulations of this nature have been run and the results found to be interesting, with some novel and pleasing designs. On the other hand, conventional jetliner models are just as conveniently generated. The best feature of our system is its high autonomy, thanks to the underlying philosophy of evolutionary processes. Currently we are also working on extensions for facial model evolution and incorporation of low-level image analysis modules for object recognition.

Acknowledgments

Comments and suggestions from Kim Kinnear have been the source of invaluable help and cheering support for getting the materials in shape. Thanks are also due to James Rice, Craig Reynolds, and Simon Handley for much of constructive criticism. This work has been supported in part by the National Science Foundation, grant NSF IRI-89-02728.

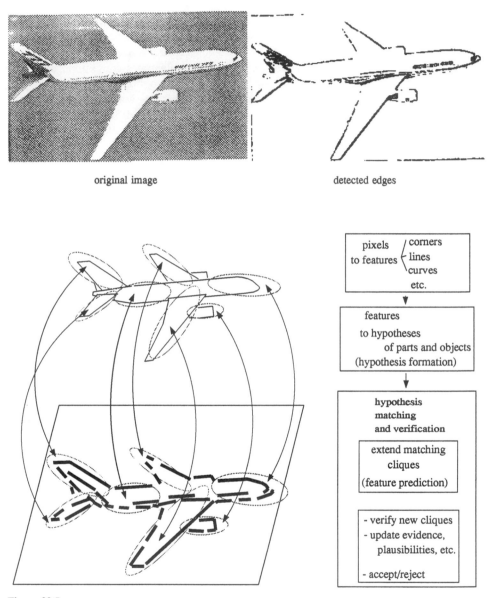

original image   detected edges

Figure 22.5
An image of a Boeing 777 (top left), its detected edge pixels (top right), and an
illustration of the matching/recognition process (bottom).

Bibliography

Baker, E. (1993) "Evolving Line Drawings", *in Proceedings of the Fifth International Conference on Genetic Algorithms*, S. Forrest, Ed. San Mateo, CA: Morgan Kaufmann.

Bhanu, B. (1990), Lee, S. and J. Ming, "Self-optimizing control system for adaptive image segmentation", *in Proceedings of the Image Understanding Workshop*, pp. 583–596.

Biederman, I., (1987) "Matching Image Edges to Object Memory," *in Proceedings of the 1st International Conference on Computer Vision*, London, UK, 1987, pp. 384–392.

Bramlette, M. A. (1989) and R. Cusic, "A Comparative Evaluation of Search Methods Applied to Parametric Design of Aircraft," *in Proceedings of the Third International Conference on Genetic Algorithms*, J. D. Schaefer, Ed. San Mateo, CA: Morgan Kaufmann.

Brooks, R. A. (1981) "Symbolic Reasoning Among 3-D Models and 2-D Images," *Artificial Intelligence* 17.

Burns, J. B. (1992) and E. M. Riseman "Matching Complex Images to Multiple 3D Objects using View Description Networks," *Proceedings of the Image Understanding Workshop*, pp. 675–682.

Calloway, D. (1987) "Using a genetic algorithm to design binary phase-only filters for pattern recognition," *Proceedings of the 2nd Intl. Conf. on Genetic Algorithms*, pp. 422–429.

Dickinson, S. (1992), Pentland, A., and A. Rosenfeld "3-D Shape Recovery Using Distributed Aspect Matching," *IEEE Transactions on Pattern Analysis and Machine Intelligence* 14(2), February 1992, pp. 174–198.

Flynn, P. J. (1993) and A. K. Jain (Eds.) *Three-Dimensional Object Recognition Systems.* Amsterdam, The Netherlands: Elsevier Science Publishers.

Goad, C. (1983) "Special Purpose Automatic Programming for 3D Model-based Vision," *Proceedings of the Image Understanding Workshop*, Virginia, pp. 94–104.

Goldberg, D. E. (1989) *Genetic Algorithms in search, optimization and machine learning.* MA: Addison-Wesley.

Grimson, W. E. L. (1991) and D. P. Huttenlocher, "On the Verification of Hypothesized Matches in Model-Based Recognition," *IEEE Transactions on Pattern Analysis and Machine Intelligence* 13(12), December 1991, pp. 1201–1213.

Hill, A. (1992) and C. J. Taylor "Model-based image interpretation using genetic algorithms," *Image and Vision Computing* 10(5) June 1992, pp. 295–300.

Ikeuchi, K. (1988) and T. Kanade "Automatic Generation of Object Recognition Programs," *Proceedings of the IEEE* 76(8), August 1988, pp. 1016–1035.

Koza, J. (1992) *Genetic Programming.* Cambridge, MA: MIT Press.

Lowe, D. G. (1991) "Fitting Parametrized Three-Dimensional Models to Images," *IEEE Transactions on Pattern Analysis and Machine Intelligence* 13(5), May 1991, pp. 441–450.

Meygret, A. (1992), Levine, M. D. and G. Roth "Robust Primitive Extraction in a Range Image," *Proceedings of the 11th Intl. Conf. on Pattern Recognition*, 1992, vol. III, pp. 193–196.

Powell, D. J. (1989), Tong S. S. and M. M. Skolnick "EnGENEous, domain independent machine learning for design optimization," *in Proceedings of the Third International Conference on Genetic Algorithms*, J. D. Schaefer, Ed. San Mateo, CA: Morgan Kaufmann.

Powell, D. J. (1993) and M. M. Skolnick "Using genetic algorithms in engineering design optimization with non-linear constraints," *in Proceedings of the Fifth International Conference on Genetic Algorithms*, S. Forrest, Ed. San Mateo, CA: Morgan Kaufmann.

Sims, K. (1991) "Artificial Evolution for Computer Graphics," *Computer Graphics* 25(4), July 1991, pp. 319–328.

Todd, S. (1992) and W. Latham *Evolutionary Art and Computers.* London, UK: Academic Press.

Watabe, H. (1993) and N. Okino "A Study on Genetic Shape Design," *in Proceedings of the Fifth International Conference on Genetic Algorithms*, S. Forrest, Ed. San Mateo, CA: Morgan Kaufmann.

23   Automatically Defined Features: The Simultaneous Evolution of 2-Dimensional Feature Detectors and an Algorithm for Using Them

David Andre

Although automatically defined functions (ADFcts) with genetic programming (GP) appear to have great utility in a wide variety of domains, their application to the automatic discovery of 2-dimensional features has been only moderately successful [Koza 1993]. Boolean functions of pixel inputs, although very general, may not be the best representation for 2-dimensional features. This chapter describes a method for the simultaneous evolution of 2-dimensional hit-miss matrices and an algorithm to use these matrices in pattern recognition. Hit-miss matrices are templates that can be moved over part of an input pattern to check for a 'match'. These matrices are evolved using a 2-dimensional genetic algorithm, while the algorithms controlling the templates are evolved using GP. The approach is applied to the problem of digit recognition, and is found to be successful at discovering individuals which can recognize very low resolution digits. Possibilities for expansion into a full-size character recognition system are discussed.

23.1   Introduction and Overview

The problem of determining which feature set to use has always been somewhat of a dilemma in the pattern recognition field. Not only must a feature set have features that can be reliably extracted from patterns, but the feature set must maintain the linear separability of the input patterns. These factors cannot always be predetermined for a suggested feature set, and sometimes many months of work can be rendered useless when a researcher realizes that the feature set for which he or she had been developing a classification scheme does not separate the pattern space linearly--i.e., two patterns that should have different classifications have identical features. One method for dealing with this problem is to develop a learning mechanism that learns both the feature set as well as a mechanism that distinguishes between different combinations of features.

John Koza [1993b] has successfully developed such a paradigm using automatically defined functions (ADFcts) with genetic programming (GP - for more information on GP, see [Koza 1992a]). His method of evolving individual programs that both automatically define functions (ADFcts) to solve subproblems and learn a method for using them to solve the main problem has shown to be successful in several domains [Koza 1992b], including the problem of learning a minimum path to traverse an array, i.e., finding the shortest path for a lawnmower [Koza 1993b], and the problem of recognizing an L and an I from other noisy patterns [Koza 1993a]. Koza used Boolean functions of a 3x3 matrix of pixels as feature detectors for a 'turtle' that could travel around the 6x4 pattern, calling ADFcts to examine the area around it. The method was somewhat successful--several successful individuals were evolved, but only after many unsuccessful runs with a population size of 8,000 individual programs.

There are several reasons why Koza's system needs further work before it can be used for developing a character recognition system using GP. First, reducing an NxN pattern to a set of N^2 Boolean values does not retain the possibly very useful 2-dimensional information contained in the pattern. For example, given only the Boolean inputs, it is very difficult to determine which inputs are 'next' to each other. Throwing this information away puts the recognition system at a great disadvantage. Second, the classification of each pattern was quite complicated. The 'turtle' moved 25 times on a problem where it was only necessary to move 4 times. Koza's 'turtles' were given some redundant information that complicated the trees, such as the nature of the pixel the turtle was on. Also, the function set for the turtle's control code had 12 functions other than ADFcts. These factors probably complicated the search for individuals, and made it more likely that individuals would be inefficient. Koza's 'LI' problem is actually more difficult than the problem of training a recognizer for character recognition, as the usual method in optical character recognition (OCR) is to have recognizers for each character that return 'yes' for that character and 'no' for anything else.

This chapter presents an approach that builds on Koza's results, but is more applicable to the problems inherent in character recognition. Instead of using Boolean functions in ADFcts, it uses hit-miss matrices as automatically defined features (ADF's). Hit-miss matrices are templates that can be moved over part of an input pattern to check for a 'match'. These matrices are evolved using a 2-dimensional genetic algorithm (GA), while the algorithms controlling the templates are evolved using GP. In addition, the function set for the control code is simplified, so that the structure of the control code resembles a decision tree [Quinlan 1986]. The approach is tested on three sets of problems; one is an 'LT' problem similar to Koza's 'LI' problem and the other two are problems where the individual must learn to recognize either a single digit's pattern or a number of very similar patterns. The 'LT' problem was chosen instead of the 'LI' problem because the 'LI' problem is

Section 2 of this chapter discusses hit-miss matrices in greater detail and mentions several motivations for their use in character recognition systems. It also discusses the 2-dimensional GA used to evolve these matrices. Section 3 defines the three different digit recognition problems. Section 4 explains the method for applying GP and the 2-dimensional GA to these problems. Section 5 presents the results on the three different problems. Section 6 discusses the possibilities for expansion into a full-size character recognition system and presents some conclusions.

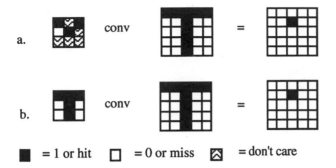

a.

b.

■ = 1 or hit   □ = 0 or miss   ▨ = don't care

Figure 23.1
Two simple examples of convolution: a) Full convolution, b) Hit-Miss Convolution

23.2 Hit-Miss Matrices and the Two-Dimensional Genetic Algorithm

Hit-miss matrices arose in the pattern recognition field as an attempt to solve the following problem: Researchers had been using convolution matrices on entire bitmaps in attempts to recognize various features, but convolution even of a small matrix over an entire bitmap took too long. If the bitmap is NxM, then there are NxMxK array indexing operations per convolution, where K is the number of pixels in the convolution matrix. A possible solution is to convolve a partial matrix--i.e. to ignore many of the components in the convolution matrix. These are hit-miss matrices: matrices where each element can either be a 'hit' (1), a 'miss' (0), or a 'don't care' (?). Convolutions of hit-miss matrices only require those components that are hits or misses to be accessed [Giardina and Dougherty 1988].

  Convolution on binary images is very similar to pattern matching. For example, the 3x3 matrix in Figure 23.1a) can be convolved over the 5x5 'T' matrix by passing the 3x3 matrix over the 5x5 in such a way that the center of the 3x3 crosses every pixel in the 5x5. If the 3x3 matrix matches the elements of the 5x5 beneath it, a '1' is placed into the output array corresponding to the center pixel in the 3x3. The hit-miss convolution is similar, but only the hits and misses need match in order to output a '1'. Examples of convolution are given in Figure 23.1.

23.2.1 Hit-miss Matrices in OCR

Hit-miss matrices have been used in OCR to handle two different types of noise. First, imagine the scenario--a n x n potential 'character' has been snipped out of some large image and passed to the recognition package, with some boundary area around the

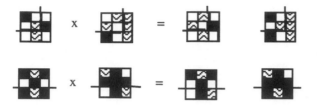

Figure 23.2
Examples of crossover : a) Window size is 2x2; upper-left at (0,0).b)Window size 1x3, upper left at (2,0).

character. If one then compares this snippet to a stored version of a character, it may not match the correct character for several reasons. First, there is always noise in actual documents, and thus the snippet may be slightly different than the stored character. Also, if the character within the snippet is not perfectly centered, then the stored character and the character in the snippet will not exactly overlap, and thus not match. Hit-miss matrices are used to solve this problem by having hits only in the middle of the character, well away from noisy edges, and misses only a distance away from the character. The hit-miss matrix can be convolved on the snippet to avoid the non overlapping problem [Dougherty and Zhao 1992; Giardina and Dougherty 1988].

23.2.2   Hit-Miss Matrices in GP

That hit-miss matrices can be efficient and somewhat noise resistant is encouraging support for their use as 'features' for this system. The paradigm here is slightly different-- a snippet is passed to a GP-evolved recognizer with hit-miss matrices of size NxN as features, and the recognizer's control code moves an NxN window around the snippet, testing to see if the part of the pattern under the window matches a specified feature. The control code uses these matches to traverse a decision tree to determine its output.

23.2.3   Evolutionary Operators for Hit-Miss Matrices

In order to use hit-miss matrices as features with GP, a method is required to evolve the features that utilizes the 2-dimensional nature of the matrices. The standard 'bit-string' GA, [Goldberg 1989], which works well for 1-dimensional patterns, would not work well if directly applied to 2-dimensional matrices by assuming that the end of one row is connected with the start of the next. This 1-dimensional approach would forfeit the knowledge of the 2-dimensional adjacency information available in a matrix. Pixels which were vertically adjacent in the input image would end up a row's width apart in the 1-dimensional representation, thereby creating a more difficult problem. Thus, crossover

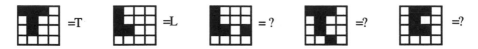

Figure 23.3
Several examples of patterns and classifications in the {L,T} problem

and mutation operations that operate on 2-dimensional portions of the matrix are necessary. Mutation can be easily implemented: following traditional string mutation, each matrix element has a probability of mutation if the pattern is selected to undergo mutation. Crossover between two NxN matrices can be done as follows: 1) choose a random value between 1 and N for each of the width and height of the regions to undergo crossover, 2) randomly choose a row and a column in the matrices representing a valid 'top-left' corner of the 'window' that will undergo crossover, and 3) replace the contents of the 'window' in one matrix with the contents from the other, and vice-versa. Two examples of crossover are shown in Figure 23.2.

It is important that these operators can attain the entire space of hit-miss matrices, because although it may appear simple, the space of all possible hit-miss matrices of size NxN is quite large. The equation is 3^{N^2}. For N=2, this number is 81, for N=3, this number is 19,683, for N=4, it is 4.3 x 10^7. The number of schema in an NxN matrix is 4^{N^2}, because the 'alphabet' size is 3, and there are NxN 'bits' [Goldberg 1989]. The crossover operation has a plethora of ways to occur for a given size of N, as well. For N= 2, crossover can occur in 9 ways, whereas with N=3, crossover can occur in 39 ways. The function for this is $\sum_i \sum_j (i*j)$. It is important to note that these NxN matrices are less general and can express fewer concepts than can a Boolean function over NxN inputs. These matrices can only express conjunctive concepts. However, the control code controlling the movement of the templates can implement disjunctive concepts, as will be discussed in section 23.4.3. The facts that the crossover operation is versatile, powerful, and that it operates in two dimensions render it a good choice for this problem.

23.3   Definition of Three Test Problem Sets

The three sets of test problems were chosen to achieve two very different goals. The first problem set, consisting of only the {L,T} problem, was chosen to provide a metric with which to compare the hit-miss algorithm and Koza's algorithm. Koza's {L,I} problem was not used because of its relative simplicity -- it is not truly a two-character recognition problem, as the 'I' is a subset of the 'L'. The solutions that Koza found utilized this

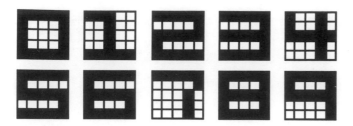

Figure 23.4
Basic Digits for problem sets 2 and 3

feature in their solution [Koza 1993b]. The {L,T} problem was also chosen to investigate whether the hit-miss algorithm could evolve a solution which could recognize two distinct characters. The second problem, a set of single example digit discrimination tasks, was designed to illustrate the capabilities of the method in a situation similar to that of 'full-size' OCR. The goal in this task is to evolve recognizers for each of the 10 digits. The third problem , a set of multiple example digit discrimination tasks, was chosen to explore the robustness of the method and to gain some notion of the difficulty of using this method in a more realistic and noisy environment. Each problem, in any problem set, consists of a set of pairs of bitmaps and output classes. The task is to return the correct output class for each bitmap. The three problem sets are more fully defined below.

23.3.1   Problem Set 1:   The {L,T} problem:

This set consists of a single problem. There are 52 patterns overall. The 'target' patterns are one example of an L, and one example of a T. The other 50 patterns consist of patterns with one pixel missing from an L or T, or one pixel added, and several other patterns, such as all on, or all off. Several examples are given in Figure 23.3.

23.3.2 Problem Set 2: The single example digit discrimination task:

This set consists of 10 separate problems; the goal of each is to recognize one of 10 5x5 digits. For each problem, the correct response is to return a 'yes' for the correct digit, and 'no' for each of the 10 others. For example, problem 2.1 is to recognize only '1' and to reject all of the other digits. Examples of all ten digits are given in Figure 23.4.

Figure 23.5
Five examples of a '0' -- the positive cases for problem 3.0

23.3.3   Problem Set 3: The multiple example digit discrimination task:

This set of problems consists of seven problems: the goal of each is to recognize the five similar examples of the correct digit, and to reject the remaining patterns. The digits tested here are {0,1,2,3,4,7,9}. Results for {5,6,8} are not presented here due to the computational limits caused by the greater complexity of producing a recognizer for these digits. The set of patterns consists of five noisy examples of each digit. Each example is fairly close to the original digit, and does not introduce any ambiguity between digits. An example for 0 is given in Figure 23.5.

23.4   Method for applying GP and GA to digit recognition

There are five main steps in preparing to use GP. One must choose:

I.   the set of terminals,
II.   the set of primitive functions,
III.   the fitness measure,
IV.   the parameters for controlling the run, and
V.   the method of designating a result and the criterion for terminating a run.

23.4.1   Functions, Terminals, and Basic Architecture

Each individual in the population consists of a set of five hit-miss matrices and a LISP S-expression representing the control code for that individual. The terminal set for the {L,T} problem was

T = {(Nada), (Elle),(Tee),(moveE),(moveW),(MoveN),(MoveS)}.

The terminal set for problem sets 2 and 3 was

T = { 0,1, (moveE), (moveW), (moveN), (moveS) }.

Table 23.1
Tableau for Problem Set 1: {L,T}

Objective:	To recognize L and T and to reject all others.
Terminal Set:	Nada,Elle,Tee, MoveE, MoveW, MoveN, MoveS
Function Set:	Progn2, Ifdf0,Ifdf1, Ifdf2, Ifdf3, Ifdf4
Fitness Cases:	52 bitmaps -- 1 'L', 1 'T', 50 'others'
Raw Fitness:	(#of False Pos * 1) + (#of False Neg * 25)
Standardized fitness:	Same as Raw Fitness
Hits:	Number of correct classifications
Wrapper:	None
Parameters:	M = 1200, G = 501, others as in [Koza 1992a]
Success Predicate:	Number of hits equal to zero

Table 23.2
Tableau for Problem Set 2: Single Digit Recognition

Objective:	Recognize only the desired digit in each problem
Terminal Set:	0,1 MoveE, MoveW, MoveN, MoveS
Function Set:	Progn2, Ifdf0, Ifdf1, Ifdf2, Ifdf3, Ifdf4
Fitness Cases:	10 Simple digits
Raw Fitness:	(#of False Pos * 1) + (#of False Neg * 9)
Standardized fitness:	Same as Raw Fitness
Hits:	Number of correct classifications
Wrapper:	None
Parameters:	M = 500 G = 201, others as in [Koza 1992a]
Success Predicate:	Number of hits equal to zero

Table 23.3
Tableau for Problem Set 3: Five examples of each digit with noise

Objective:	Recognize examples of the desired digit; reject all others
Terminal Set:	0, 1, MoveE, MoveW, MoveN, MoveS
Function Set:	Progn2, Ifdf0, Ifdf1, Ifdf2, Ifdf3, Ifdf4
Fitness Cases:	50 images -- 5 of each digit
Raw Fitness:	(#of False Pos * 1) + (#of False Neg * 9)
Standardized fitness:	Same as Raw Fitness
Hits:	Number of correct classifications
Wrapper:	None
Parameters:	M = 800 G = 201, others as in [Koza 1992a]
Success Predicate:	Number of hits equal to zero

The four 'move' terminals are really functions that change only the two state variables, CurX and CurY, as a side effect. These represent the current position of the upper left-hand corner of the 'window' onto which the feature matrices are 'matched'. The world is toriodal, so moving off the edge 'wraps around'. The four move terminals return one of the values possible in the domain, i.e., for problem set 3, MoveE and MoveW returned 0, and MoveN and MoveS returned a 1. The other terminals, 0,1,Nada,Elle, and Tee, represent the classifications possible in each of the two problem types. The 'move' terminals could just as easily been defined as functions.

The control code was theorized to be a decision-tree-type structure, so the function set consists mostly of 'if' macros [Koza 1992a]. The function set for all problems consisted of five 'if' functions to check if the corresponding hit-miss matrix matched the values of the test pattern under the window, and to evaluate their first argument if a match was found, and the second argument otherwise. The only other function was 'progn2', a macro which takes two arguments and evaluates the first, and then the second, and returns the value of the second. The existence of the 'progn' distorts the clean structure of a decision tree, but the resulting structure is still decision-tree-like.

Thus, the function set for all of the problems was

```
F = {progn2, Ifdf0, Ifdf1, Ifdf2, Ifdf3, Ifdf4}.
```

where the Ifdfn functions are the if functions specified above. The control code is an S-expression consisting of primitive functions from the function set and terminals from the terminal set.

23.4.2 Methods of generation and application of the genetic operators

The initial populations of individuals are created by randomly filling the hit-miss matrices of each individual and then creating an S-expression for each individual consisting of a random composition of elements from both the terminal and function sets. The initial S-expression lengths are determined by the ramped half-and-half method [Koza 1992a]. Crossover takes place between individuals by randomly choosing either the S-expression or the hit-miss matrices. If the S-expression is chosen, crossover takes place by exchanging a subtree of one expression for a subtree in another expression. If the hit-miss matrices are chosen, then a matrix is selected at random from each individual to undergo crossover as described in section 2. Mutation takes place in a similar fashion, occurring half the time in the S-expression and half the time in one of the hit-miss matrices.

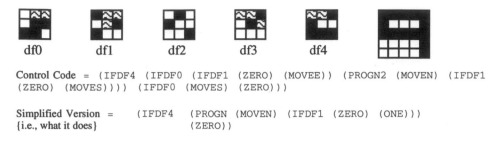

df0   df1   df2   df3   df4

Control Code = (IFDF4 (IFDF0 (IFDF1 (ZERO) (MOVEE)) (PROGN2 (MOVEN) (IFDF1 (ZERO) (MOVES)))) (IFDF0 (MOVES) (ZERO)))

Simplified Version =   (IFDF4   (PROGN (MOVEN) (IFDF1 (ZERO) (ONE)))
{i.e., what it does}   (ZERO))

Figure 23.6
Example of an individual from problem 2.9, generation 126. Its 'goal' was to recognize a '9'.

23.4.3 Example of an individual

As an example of an individual, consider the individual shown in Figure 23.6 that evolved to recognize a single '9' with 100% accuracy, and thus had 0 fitness. This individual first checks if df4 matches when CurX=CurY=0, i.e. when the window is in the upper-left corner. If it does not match, then it cannot be a '9', and the second argument to IFDF4 is evaluated. This second argument checks if df0 matches, which it never does, so this branch returns 0. Now, the only case when the first branch of IFDF4 will be evaluated is if the test character is one of {5,6,8,9}. The first argument to IFDF4 must only discriminate between these four digits. IFDF0 is not true, and so its second argument is evaluated. MoveN is executed, moving the 'window' so that it overlaps the first three columns in the bottom row, and the first three columns in the top two rows. Then, IFDF1 checks to see if there is a pixel in the lower left of the test pattern. (All of the other pixels in df1 must match if the program is at this point in the code.) Since {5,6,8} all have a pixel in the lower left, the individual only returns true in the case that the test character is a 9, as desired.

An interesting behavior of the individual in Figure 23.6 is that it performs a check for the existence of a single pixel. This behavior could be extended, so that the hit-miss matrix contained only the single pixel being checked. This kind of structure could be used to trace out the entire pattern matrix, examining each pixel. Although Koza's Boolean functions [Koza 1992b] are more expressive features than the hit-miss matrices; the individuals in both studies are equally expressive. The lack of generality of the feature detectors in this method may result in slightly larger trees some of the time, but the greater sensitivity to 2-dimensional structure, the smaller feature space to search, and greater simplicity should more than compensate.

23.4.4 The fitness function

The third major step in preparing to use GP is to determine the fitness measure. The fitness cases consist of the set of patterns for the particular problem at hand. Each individual is tested against an environment consisting of the number of fitness cases in the problem being run. There are then several possibilities as to how the individual has classified each instance. First, the individual's classification could be correct, a 'true-positive'. In this case the individual's fitness is not altered. The individual may have mistakenly classified the instance as a positive instance of a character. In this case 1 is added to the fitness score. The last case is that the individual may incorrectly classify what should be a positive example -- a false negative. In this case, a quantity equal to $(N-p) / |p|$ is added to the fitness, where N is the number of fitness cases and p is the number of positive examples. In addition to being penalized for errors, 50 points are added to the fitness score in the case that an individual classifies all examples as belonging to one category. This is done so that there is no initial advantage for focusing on either rejection or acceptance--an individual must do both to score well.

23.4.5 Values of run parameters and the success predicate

The fourth step in preparing to use GP is to determine the values for the various run parameters. New S-expressions were allowed to be only six levels deep. An individual's S-expression was limited to a depth of 20. Fitness proportionate reproduction was used. The fitness proportionate reproduction fraction was 0.1, the crossover at any point fraction was 0.2, and the crossover at function point fraction was 0.7. The population was initially created using the ramped-half-and-half method. For more information about these parameters, see [Koza 1992a]. The population size and maximum number of generations varied for the different problems. For problem 1, the population was chosen to be 1200, the maximum number of generations set to 500. The seemingly high maximum number of generations was to insure that solutions were found as early investigations had indicated that fitness was still improving after 300 generations. For problem set 2, the population size was 500, the maximum number of generations was 200. For problem set 3, the population size was 800, the maximum number of generations was 200. These choices of population size reflected the expected difficult of the problems as well as available computing resources.

The fifth step in preparing to use GP requires specifying termination conditions and choosing a method for designating the result of the run. In this study, runs were terminated if an individual with 0 fitness was found, or after the maximum number of generations. The result of the run is the best-so-far individual--the best individual obtained during the run.

df0   df1   df2   df3   df4

Code: (IFDF1 (PROGN2 (MOVEE) (IFDF3 (MOVEE) (IFDF2 (NADA)   (PROGN2
(PROGN2 (MOVEE) (IFDF3 (MOVEE)   (IFDF2 (NADA)   (PROGN2 (MOVEE)
(MOVEN))))) (IFDF3 (MOVEE)   (IFDF2 (PROGN2 (PROGN2 (MOVEE)   (IFDF3
(MOVEN)   (IFDF2 (NADA)   (PROGN2 (MOVEE) (IFDF3 (MOVEE) (IFDF2 (NADA)
(MOVEN)))))   (IFDF3 (MOVEE)   (IFDF2 (IFDF2 (NADA) (MOVEN)) (PROGN2
(PROGN2 (MOVEE)   (IFDF3 (MOVEE) (IFDF2 (NADA)   (PROGN2 (IFDF3 (MOV
(IFDF2 (MOVEE) (MOVEN)))   (PROGN2 (IFDF2 (MOVEE) (MOVEN) (IFDF3 (MOVEW)
(MOVES)))))))) (IFDF3 (MOVEE)   (IFDF2 (IFDF2 (NADA)   (MOVEE)) (MOVEN)))))))))
(PROGN2 (PROGN2 (MOVEE)   (IFDF3 (MOVEE)   (IFDF2 (NADA) (PROGN2 (IFDF3
(MOVEE) (IFDF2 (IFDF2 (MOVEE) (MOVEN)) (MOVEN)))   (PROGN2 (IFDF2 (MOVEE)
(MOVEN))   (IFDF3 (MOVEW) (MOVES))))))))) (IFDF3 (MOVEE) (IFDF2 (IFDF2
(NADA) (MOVEN))   (MOVEN))))))))))) (IFDF0 (PROGN2 (IFDF2 (MOVEE) (MOVEN))
(IFDF3 (MOVEW) (MOVES))) (MOVEE)))

Figure 23.7
Solution individual for the {L,T} problem, found on Generation 117.

23.5 Results

The runs were all done using a slightly modified version of John Koza's LittleLisp GP
kernel [Koza 1992a].   All runs were done on DEC 5400 and SPARC 2 workstations.
These workstations were all time-shared machines, so running times for a single run
varied from 2 hours to 52 hours.

23.5.1 Results for Problem   Set #1:   { L , T }

The goal in this experiment was to verify that a two-class problem such as the {L,T}
could be solved by this method. Population size was 1,200, maximum generations was
set at 500. In one successful run, an individual scoring 100% emerged. It had 94 nodes,
and was found on generation 117. It is displayed in Figure 23.7.

While the 94 node solution shown in Figure 23.7 is not very elegant, it does get the
job done. In recognizing the 'L', it moves only seven times. It recognizing the 'T', it
moves only twice. The logical structure of the program was somewhat difficult to
follow, but one subtree checked for an 'L', another for a 'T', and the remainder checked for
specific negative cases. The individual had a good deal of code that was specifically
tailored to reject particular negative cases, instead of having a general 'theory' of the
letters. Although not as theoretically desirable as the solution obtained in [Koza 1993a],
it still managed to score 100%, and moved far less often than did Koza's solution to the
easier {L,I} problem. In addition, the individual managed to solve the problem using 2x2

Table 23.4
Results on Problem Set #2: Single example digit discrimination

Digits	# of Runs	# Correct	% Correct	P(M,0)	P(M,Best)	I(M,i,z)
0	198	191	96.46	86.4	*	1000
1	147	147	100	98.6	*	500
2	143	35	24.48	2.87	9.7	67,500
3	132	31	21.27	3.8	*	59,500
4	147	147	100	97.3	*	500
5	267	33	10.86	0.375	2.6	605,500
6	142	12	8.45	2.10	*	108,000
7	47	47	100	97.8	*	500
8	122	28	22.95	0	12.3	490,000
9	185	76	35.14	5.40	13	99,000

feature detectors, rather than 3x3. This is a clear performance advantage, although it is unknown whether using such tiny feature detectors will provide any degree of generalization. In any event, this result indicates that the algorithm presented in this chapter can handle a two-class classification with a large number of distractors.

23.5.2 Problem Set #2: Single example digit discrimination problem

In examining the ten 5x5 digits, it is clear that some digits are easier to recognize than others. In looking only at the initial contents of the window, it is possible to recognize {0,1,4,7}. These characters can be recognized without any movement, and individuals able to recognize these should be very common. Next, {9,2,3} can be recognized with only one move. Finally, {5,6,8} are the hardest to distinguish; they require a minimum of two moves. The apparent simplicity of recognizing these characters is somewhat deceiving, however. The correct 3x3 feature detectors must evolve, and this is not a trivial task given the size of the search space. While distinguishing between these 10 digits is not a exceedingly complicated task, it serves to illustrate some of the potential of the method.

Overall, 1530 runs were done, and 48.82% of them were successful. The population size was 500; maximum generations was set at 200. Table 23.4 summarizes much of the data. I(M,i,z) is the minimum number of individuals that must be processed to obtain a 99% likelihood of solving the problem. P(M,Best) is the probability of finding a successful individual at the generation where I(M,i,z) is attained (asterisk represents that

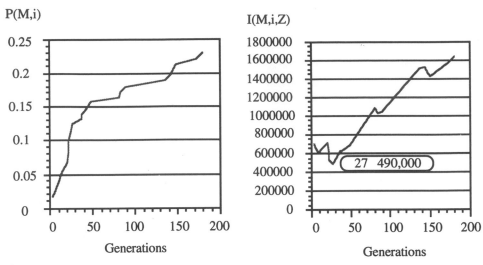

Figure 23.8
Graph of performance on problem 2.8. P(M,i) is the cumulative probability of finding a solution at each
generation. I(M,i,z) is the number of individuals that must be processed to attain a 99% probability of finding
a solution.

I(M,i,z) was found at generation zero, which is indicative that either the problem is quite
easy or that there were too few successful runs to provide reliable statistics). P(M,0) is
the probability that a successful individual is found at generation 0.

The results seem to match the predictions about complexity, at least to a first degree of
approximation. Certainly the easiest to learn are {0,1,4,7}. For some reason, {9} is next,
perhaps because it has a region where only one match need be done. Then, {2, 3} are
next, probably for the reasons discussed above regarding the number of moves necessary
to solve the problem. Clearly more difficult are {5,6,8}. It is important to notice that
although many runs were done for each digit, the total number of successful runs is not
high enough to provide entirely accurate statistics. The 'graphs' are still very rough, but
the strong tendencies represented here are undoubtedly not statistical abnormalities. The
graph for problem 2.8 is given as an example in Figure 23.8.

Although the graphs in Figure 23.8 are noisy and sparsely populated (only 28 points),
they show that the GA/GP algorithm is more effective than random search for this
problem. The minimum number of individuals required to solve the problem with 99%
probability occurs at generation 27, indicating that the effects of evolution are beneficial
when compared with random search. In order to solve this problem with 99%

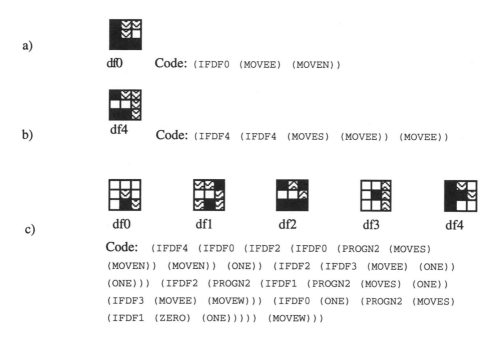

a)   df0   Code: (IFDF0 (MOVEE) (MOVEN))

b)   df4   Code: (IFDF4 (IFDF4 (MOVES) (MOVEE)) (MOVEE))

c)   df0   df1   df2   df3   df4

Code: (IFDF4 (IFDF0 (IFDF2 (IFDF0 (PROGN2 (MOVES)
(MOVEN)) (MOVEN)) (ONE)) (IFDF2 (IFDF3 (MOVEE) (ONE))
(ONE))) (IFDF2 (PROGN2 (IFDF1 (PROGN2 (MOVES) (ONE))
(IFDF3 (MOVEE) (MOVEW))) (IFDF0 (ONE) (PROGN2 (MOVES)
(IFDF1 (ZERO) (ONE))))) (MOVEW)))

Figure 23.9
An analysis of one run on Problem 2.2: a) At generation 0: Fitness 4 = 6 hits.
b) At generation 80: Fitness 1 = 9 hits. c) At generation 158: Fitness 0 = 10 hits

probability using random search, one would have to generate 697,500 individuals, rather than the 490,000 required after twenty-seven generations of evolution.

An analysis of a particular problem by looking at the best-of-generation individuals at the beginning, middle, and end of a run will illustrate how the hit-miss matrices and the control code evolve together. We will look at a case where the '2' recognizer evolves.

The 500 randomly generated individuals in the initial population had an average fitness of 55.74, which indicates that a vast majority of them were classifying all 10 patterns as either 'yes' or 'no'. The best individual had a fitness value of 4, and had 5 points in its control code. It basically checked the initial window as to whether or not it had the pattern in Figure 23.9a. If it did, it responded 'no', if not, it responded 'yes'. While the individual correctly rejects six of the patterns in this fashion, it fails to reject {1,3,0,7}, corresponding to the fitness score of 4. The individuals in the population gradually learned to reject more and more of the other cases, until only one error remained. This

Table 23.5
Results on Problem Set #3: Multiple example digit discrimination with noise

Digits	# of Runs	# Correct	% Correct	P(M,0)	I(M,i,z)
0	31	30	96.7	74	2400
1	39	39	100	71.79	3200
2	21	4	19	0	225,600
3	28	3	10.7	0	812,800
4	29	28	96.6	62	4000
7	29	28	96.6	72.4	3200
9	20	8	40	0.05	72,000

process was mostly a process of finding the correct hit-miss matrix. At generation 80, the individual in Figure 23.9b appeared.

This individual scores a 1, receiving 9 hits. It incorrectly returns a 'yes' for the '3'. At this point, movement is required to solve the problem. The window had to move because 2 and 3 have the same upper-left 3x3 corner in their pattern. Finally, at generation 158 an individual emerged that scored perfectly, displayed in Figure 23.9c.

This 31 point program correctly responds for the '2', as follows. IFDF4 is never true in the starting window, so control goes to the second argument. Then, when IFDF2 returns true for the '2', control goes into the progn2 statement. IFDF1 is also true for the starting window position with a '2'. Then, the window is moved down by one, and then, IFDF0 will not be true, and the window moves down one again. Then, IFDF1 is reached. If the test pattern were a '3', IFDF1 would be true, and this individual would return a '0'. If the test pattern is a '2', then IFDF1 is false, and the individual correctly returns a '1'.

This algorithm's success on this set of problems indicates that GP and the GA can work together effectively to produce feature detectors and a method for using them to produce a character recognizer.

23.5.3 Problem Set #3: Multiple example digit discrimination tasks

The goal of this task was to investigate the ability of the method to handle some variation in the character's structure. It is expected, as before, that {0,1,4,7} will be the easiest, the {9}, and then {2,3}. In this problem, {5,6,8} were not tested. The population size was 800, maximum generations was set at 200. There were a total of 197 runs in this problem set, 111 were successful (56%). It is important to note that this is without doing runs on the three hardest problems so the percentage is not comparable to the percentage in problem 2. Table 23.5 summarizes the data.

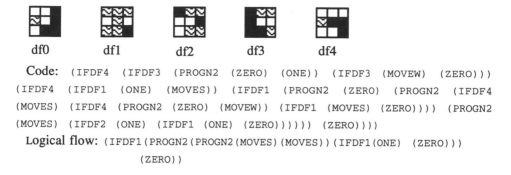

df0   df1   df2   df3   df4

Code: (IFDF4 (IFDF3 (PROGN2 (ZERO) (ONE)) (IFDF3 (MOVEW) (ZERO)))
(IFDF4 (IFDF1 (ONE) (MOVES)) (IFDF1 (PROGN2 (ZERO) (PROGN2 (IFDF4
(MOVES) (IFDF4 (PROGN2 (ZERO) (MOVEW)) (IFDF1 (MOVES) (ZERO)))) (PROGN2
(MOVES) (IFDF2 (ONE) (IFDF1 (ONE) (ZERO)))))) (ZERO))))

Logical flow: (IFDF1(PROGN2(PROGN2(MOVES)(MOVES))(IFDF1(ONE) (ZERO)))
(ZERO))

Figure 23.10
A solution individual for problem 3.3

Solutions were found in all of the tested cases. Although Table 23.5 is based on very few successful runs, it can still allow some generalizations. It seems very clear that learning more examples is a harder task, as the required numbers of processed individuals to have a 99% probability of obtaining a solution are much higher than in problem set #2. The complexity rankings are as predicted. The individual in Figure 23.10 was evolved to recognize a '3' with 100% accuracy. The solution was found on generation 167. The individual has evolved a very robust feature and utilized it to the utmost. It is very certain that a three will have a black pixel in the middle of a 5x5, and will not have black pixels in the left part of the second row. Also, it is nearly certain that there will be a black pixel in the middle of the bottom row, and no black pixels on the left part of the fourth row. The individual has learned something about the shape of the object. Although its concept is neither complete nor sufficient, it is a robust concept that allowed perfect performance on a fairly difficult task given this representation.

23.6 Conclusions and Future work

This chapter has described an approach to using GP and GA to solve pattern recognition problems. The system was found to be successful at solving all three problem domains. First, a solution was found on the {L,T} problem, which was a problem similar to previous work using Boolean feature detectors rather than the hit-miss matrices used here. Second, individuals were evolved that could recognize any 5x5 digit in a non-noisy environment. Third, and perhaps most importantly, the algorithm was found to extend into more noisy domains. The approach was found to perform better than random search on the less trivial problems.

The approach used here could potentially be used on full size characters with 8x8 hit-miss matrices. It should easily scale to larger size characters, and be somewhat resilient to noise. However, there are several problems that would have to be resolved prior to attempting the classification of full-size patterns. First, the current system was very location dependent. The starting position was constant, and character positions were fixed. Future work will attempt to train location-independent pattern recognizers. Another problem with expansion is the speed. As the hit-miss matrices get larger, they will take more time to operate, and learning will be computationally more expensive. Future work will include a slight fitness benefit for terse hit-miss matrices. In addition, future work will attempt to use larger size patterns, larger pattern sets, and multiple 'fonts' to further explore the ability of this paradigm to perform well in realistic conditions.

The most important facet of this research was the use of hit-miss matrices to capture the 2-dimensional information of the patterns. The use of these 2-dimensional structures as features to be used with GP is important not only because it illustrated a method by which GP and GA can be combined, but also because the 'programs' that were evolved were able to access more complicated data structures than simple variables. This represents a very preliminary step down the exciting avenue of adding more complex data structures and thus greater computational power to the programs evolved with GP.

Bibliography

Doughterty, E.R. and D. Zhao (1992). Model-based characterization of statistically optimal design for morphological shape recognition algorithms via the hit-or-miss transform. *Journal of Visual Communication and Image Representation.* vol. 3. no 2. p147-160.

Giardina, C.R., and Doughtery, E.R. (1988). *Morphological Methods in Image and Signal Processing.* Englewood Cliffs, New Jersey: Prentice Hall. p125-132.

Koza, J.R., (1992a). *Genetic programming: on the programming of computers by the means of natural selection,* Cambridge, Mass: MIT Press.

Koza, J.R., (1992b). Hierarchical automatic function definition in genetic programming. In Whitley, Darrell (editor). *Proceedings of Workshop on the Foundations of Genetic Algorithms and Classifier Systems, Vail, Colorado, 1992.* San Mateo, CA: Morgan Kaufmann.

Koza, J.R., (1993a). Simultaneous discovery of detectors and a way of using the detectors via genetic programming. *1993 IEEE International Conference on Neural Networks, San Francisco.* Piscataway, NJ: IEEE 1993. Volume III. p 1794-1801.

Koza, J.R. (1993b) Discovery of a main program and reusable subroutines using genetic programming. In prep.

Quinlan, J.R. (1986). Induction of decision trees. *Machine Learning*, 1, p 81-106.

24 Genetic Micro Programming of Neural Networks

Frédéric Gruau

Cellular Encoding is a method for encoding families of similarly structured boolean neural networks, that can compute scalable boolean functions. Genetic Programming uses the Genetic Algorithm to evolve LISP computer programs. This chapter demonstrates that Cellular Encoding is a micro-programming language of neural networks and that genetic search of neural networks using Cellular Encoding is equivalent to Genetic Micro Programming. The concept of genetic language is defined. Cellular Encoding and LISP are two particular Genetic Programming languages. Other programming languages are proposed. A criterion is put forward to classify genetic languages with increasing complexity. With respect to this criterion, Lisp is more complex than Cellular Encoding. Which language is better for Genetic Programming? We argue that Cellular Encoding is better than LISP for the synthesis of neural networks, and LISP is better for symbolic manipulation. Ultimately, it is possible to evolve the genetic language itself.

24.1 Introduction

Cellular Encoding is a method for encoding families of similarly structured boolean neural networks, that can compute scalable boolean functions. The object that is encoded is a graph grammar. The grammar is encoded as a set of trees (called *grammar trees*) instead of a set of rules. Cellular Encoding can describe networks in an elegant and compact way. Moreover, the representation can be readily recombined by the Genetic Algorithm. Kitano has proposed to encode a matrix grammar in [Kitano 1990]. Although he was the first to propose to encode neural networks with grammar, the use of a matrix grammar do not allow to describe networks in an elegant and compact way and do not suit the GA. We refer the reader to [Gruau 1992a, 1992b] for a comparison between Cellular Encoding and matrix grammar.

Various properties of Cellular Encoding have been formalized and proved by Gruau [Gruau 1992a]. A GA has been used to search the space of grammar-trees. The GA can find a Cellular Encoding that yields both architecture and ± 1 weights specifying a particular neural network for solving the parity, symmetry and decoder boolean function . Furthermore, the grammar trees are recursive encodings that generate whole families of networks that compute parity, symmetry or decoder. In this way, once small parity, symmetry or decoder problems are solved, the grammars typically can generate solutions to parity, symmetry and decoder problems of arbitrary large size [Gruau 1992b, 1993a].

Koza has shown that the tree data structure is an efficient encoding which can be effectively recombined by genetic operators. *Genetic Programming* has been defined by Koza as a way of evolving computer programs with a GA [Koza 1992]. In Koza's Genetic Programming paradigm the individuals in the population are LISP S-expressions which can be depicted graphically as rooted, point-labeled trees with ordered branches. Since the chromosomes (grammar-trees) in Cellular Encoding have exactly the same structure, the same approach is used for generating and recombing grammar trees. This chapter shows that the genetic search of the space of Cellular Encodings is really a Genetic Programming approach, not only because the genetic representation uses the same tree data structure. We first explain why Cellular Encoding can be considered as a micro-programming language.

We present a compiler JaNNeT (Just an Automatic Neural NEtwork Translator). Its input is a Pascal program and its output is a neural net that computes what the Pascal program specifies. Any Pascal program that does not make dynamic memory allocation can be compiled. Recursive function can be compiled, but the depth of recursion must be bounded. JaNNeT uses two rewriting systems: The first system rewrites the parse tree of the Pascal Program into a Cellular Encoding. The second rewriting system is a graph grammar. It develops the desired neural net from the Cellular Encoding.

The concept of genetic programming language is defined. LISP and Cellular Encoding are two distinct genetic programming languages. A criterion to classify genetic languages with increasing complexities is introduced. Using this criterion, LISP turns out to be more complex than Cellular Encoding. We propose a hierarchy of genetic languages that starts from a very simple micro programming language and increases continuously in complexity to reach a high level language LISP equipped with libraries. We then address the question of which genetic language is better. We show that the combination of Genetic Programming with neural networks using Cellular Encoding is totally different from the one using the encapsulation operator previously proposed by Koza and Rice using LISP [Koza and Rice 1991]. The advantages of Cellular Encoding are put forward. With all the results of Koza, LISP is better when the task involves manipulation of symbols. On the other hand, this is not the case when the task is to synthesize neural networks.

A scheme for the evolution of the genetic language itself is also briefly sketched. In this scheme an ideal genetic language appears as a hierarchy of building blocks.

24.2   Review of Cellular Encoding

The genetic generation of neural networks with Cellular Encoding makes more a phenotype/genotype separation. The genotype is the Cellular Encoding, the phenotype is the neural network. This is a difference with most GP application where the program is both the genotype and the phenotype. In this section we describe the genotype to phenotype mapping. That is, we present the details of Cellular Encoding.

The chromosome is represented as a *grammar tree* with ordered branches whose nodes are labeled with character symbols. The reader must not make the confusion between grammar

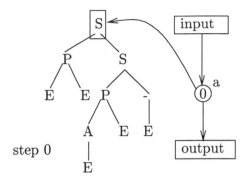

Figure 24.1
Step 0: Development of the neural net for the exclusive-OR (XOR) function in the right half of this figure. The development starts with a single ancestor cell labeled "a" and shown as a circle. The O inside the circle indicates the threshold of the ancestor cell is 0. The ancestor cell is connected to the neural net's input pointer cell (box labeled "input") and the neural net's output pointer cell (Box labeled "output") The Cellular Encoding of the neural net for XOR is shown on the left half of this figure. The arrow between the ancestor cell and the symbol **S** of the graph grammar represents the position of the reading head of the ancestor cell. A continuous line indicates a weight 1, and a heavy line a weight −1. By default all the weights are 1.

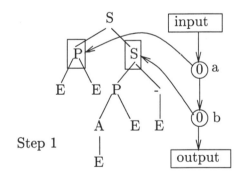

Figure 24.2
The execution at step 1 of the sequential division "S" pointed to by the ancestor cell "a" of figure 24.1 causes the ancestor cell "a" to divide into two neurons "a" and "b". Neuron "a" feeds into neuron "b" with weight +1. The reading head of neuron "a" now points to the left sub-tree of the Cellular Encoding on the left (the box with "**P**") and the reading head of new neuron "b" points to the right subtree (the box at the second level down with "S").

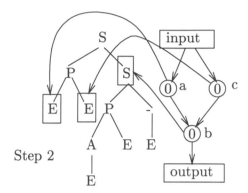

Step 2

Figure 24.3
The execution of the paralle division "**P**" at step 2 causes the creation of neuron "c". Both neuron "a" and "c" inherit the input formerly feeding into "a". Both neurons "a" and "c" output to neuron "b" (The place where "a" formerly sent its output.) The reading head of neuron "a" now points to the left sub-tree ("**E**") and the reading head of the new neuron "c" now points to the right sub-tree (which also has an "**E**").

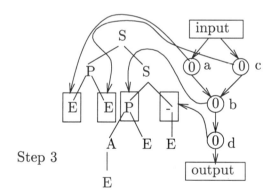

Step 3

Figure 24.4
The execution at step 3 of the sequential division "**S**". pointed to by the neuron "b" of figure 24.3 causes the neuron "b" to divide into two neurons "b" and "d". Neuron "b" feeds into neuron "d" with weight +1. The reading head of neuron "b" now points to the left sub-tree (the box at the third level with "**P**") and the reading head of new neuron "d" points to the right subtree (the box at the third level down with "−").

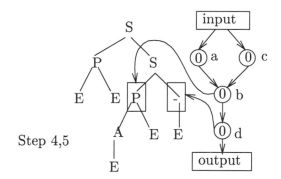

Step 4,5

Figure 24.5
The execution of the two end-program symbols. The **E**'s causes the neural cells "a" and "c" to lose their reading head and become finished neurons. Since there are two **E**'s, it takes two time steps, one time step for each **E**

tree and tree grammar. Grammar tree means a grammar encoded as a tree, whereas tree grammar means a grammar that rewrites trees. A cell is a node of an oriented *network graph* with ordered connections. Each cell carries a duplicate copy of the chromosome (i.e., the grammar tree) and has an internal reading head that reads from the grammar tree. Typically, each cell reads from the grammar tree at a different position. The character symbols represent instructions for cell development that act on the cell or on connections that fan-in to the cell. During a step of the development process, a cell executes the instruction referenced by the symbol it reads, and moves its reading head down in the tree. One can draw an analogy between a cell and a Turing machine. The cell reads from a tree instead of a tape and the cell is capable of duplicating itself; but both execute instructions by moving the reading head in a manner dictated by the symbol that is read. In this section we shall refer to the grammar tree as a *program* and each character as a *program-symbol*.

A cell also manages a set of internal registers, some of which are used during development, while others determine the weights and thresholds of the final neural net. The link register is used to refer to one of possibly several fan-in connections (i.e., links) into a cell.

Consider the problem of finding the neural net for the exclusive OR (XOR) function. Neurons can have thresholds of 0 or 1. Connections can be weighted −1 or +1. If the weighted sum of its input is strictly greater than its threshold, the neuron outputs 1, else it outputs 0. The inputs of the neural net are 0 or 1.

Development of a neural net starts with a single cell called the *ancestor cell* connected to an input pointer cell and an output pointer cell. Consider the starting network on the right half of figure 24.1 and the Cellular Encoding depicted on the left half of figure 24.1. At the starting step 0 the reading head of the ancestor cell is positioned on the root of the tree

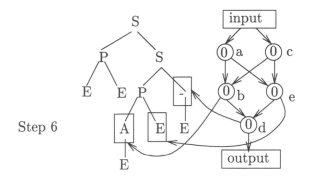

Step 6

Figure 24.6
The execution of the paralle division "**P**" at step 6 causes the creation of neuron "e". Both neuron "b" and "e" inherit the input from neuron "a" and "c" formerly feeding into "b". Both neurons "b" and "e" send the output to neuron "d" (The place where "a" formerly sent its output.) The reading head of neuron "b" now points to the left sub-tree ("**A**") and the reading head of the new neuron "e" now points to the right sub-tree (which has an "**E**").

as shown by the arrow connecting the two. Its registers are initialized with default values. For example, its threshold is set to 0. As this cell repeatedly divides it gives birth to all the other cells that will eventually make up the neural network. The input and output pointer cells to which the ancestor is linked (indicated by empty square boxes in the figure) do not execute any program-symbol. Rather, at the end of the development process, the upper pointer cell is connected to the set of input units, while the lower pointer cell is connected to the set of output units. These input and output units are created during the development, they are not added independently at the end. After development is complete, the pointer cells can be deleted. For example, in figure 24.9, the final decoded neural net has two input units labeled 1 and 2, and one output unit labeled 1.

• A division-program symbol creates two neural cells from one. In a *sequential division* (denoted by **S**) the first neuron child inherits the input links, the second neuron child inherits the output links of the parent cell. The first child connects to the second with weight 1. The link is oriented from the first child to the second child. Since there are two child cells, a division program-symbol must label nodes of arity two. The first child moves its reading head to the left subtree and the second child moves its reading head to the right subtree. This is illustrated in steps 1 and 3.

• Parallel division (denoted by **P**) is a second kind of division program symbol. Both child cells inherit the input and output links from the parent cell (in step 2 and step 6). Finally, when a cell divides, the values of the internal registers of the parent cell are recopied in the child cells.

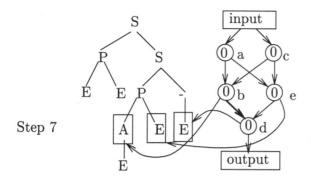

Step 7

Figure 24.7
The neuron "c" executes the value-program symbol "-". The link register is one (the default value) it points the left-most register. The action of "-" is to set the weight of the first link to -1. The heavy line is used to indicate a -1 weight. After the execution of "-" the neuron "c" goes to read the next program symbol which is an "E" located on the fourth level of the tree.

• The *ending-program symbol* denoted **E** causes a neural cell to lose its reading head and become a finished neuron. Since the cell does not read any subtree of the current node, there need not be any subtree to the node labeled by **E**. Therefore **E** labels the leaves of the grammar tree (i.e., nodes of arity 0).

• A value-program symbol modifies the value of an internal register of the cell. The program-symbol **A** increments (and **O** decrements) the threshold of a neuron. The value-program symbol **I** increments (and **D** decrements) the value of the link register, which points to a specific fan-in link or connection. Changing the value of the link register causes it to point to a different fan-in connection. The link register has a default initial value of 1, thus pointing to the leftmost fan-in link. Operations on other connections can be accomplished by first resetting the value of the link register. The program-symbol denoted + sets the weight of the input link pointed by the link register to 1, while - sets the weight to −1 (see step 7). The program-symbols +, and - do not explicitly indicate to which fan-in connection the corresponding instructions are applied. When + or - is executed it is applied to the link pointed to by the link register.

• An unary program-symbol **C** cuts the link pointed by the link register. This operator modifies the topology by removing a link, whereas the reseting of the link register does not modify the topology.

Operators I,D,C,W are not illustrated, they are not required for the development of a neural net for the XOR problem. The sequence in which cells execute program-symbols

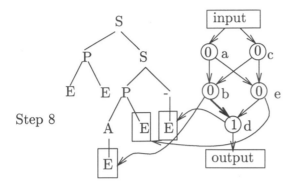

Step 8

Figure 24.8
Neuron "b" executes the value-program symbol "**A**". The action of "**A**" is to increase the threshold of neuron "b" by 1. After execution, the threshold of neuron "b" is 1, and the reading head of neuron "b" points to an "**E**" at the fifth level of the tree.

is determined as follows: once a cell has executed its program-symbol, it enters a First In First Out (FIFO) queue. The next cell to execute is the head of the FIFO queue. If the cell divides, the child which reads the left subtree enters the FIFO queue first. This order of execution tries to model what would happen if cells were active in parallel. It ensures that a cell cannot be active twice while another cell has not been active at all. In some cases, the final configuration of the network depends on the order in which cells execute their corresponding instructions. For example, in the development of the XOR, performing step 7 before step 6 would produce a neural net with an output unit having two negative weights instead of one, as desired. The waiting program-symbol denoted **W** makes the cell wait for its next rewriting step. **W** is necessary for those cases where the development process must be controlled by generating appropriate delays.

Up to this point in our description the grammar tree does not use recursion. (Note that recurrence in the grammar does not imply that there is recurrence in the resulting neural network.) One is able to develop only a single neural network from a grammar tree. But one would like to develop a family of neural networks, which share the same structure, for computing a family of similarly structured problem. This would allow to get a property of scalability. For this purpose, we introduce a recurrent program-symbol denoted **R** which allows a fixed number of loops L. The cell which reads **R** executes the following algorithm:

```
life := life - 1
If (life > 0) reading-head := root of the tree
Else  reading-head := subtree of the current node
```

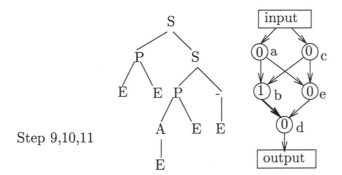

Step 9,10,11

Figure 24.9
The last tree steps consist in executing three end-program symbols. The three **E**'s cause the neural cells "b", "e" and "d" to lose their reading head and become finished neurons. Since there are now only finished neurons, the development is finished. The final neural net has two input units "a" and "c", and one output unit "d". The neuron "a" is the first input unit, because the link from the input pointer cell to "a" is the first output link of the input pointer cell. The neuron "c" is the second input unit, because the link from the input pointer cell to "a" is the second output link of the input pointer cell

where *life* is a register of the cell initialized with L in the ancestor cell.

Thus a grammar develops a family of neural networks parametrized by L. The use of a recurrent-program symbol is illustrated figure 24.10 24.11. The cellular code in this figure is almost the same as the cellular code of a XOR network. the only difference is that an end-program symbol "**E**" has been replaced by a recurrent-program symbol "**R**". The resulting cellular code is now able to develop a neural net for the parity function with an arbitrary large number of inputs, by cascading XOR networks. In figure 24.10 the network for parity of three inputs is shown. This implementation of the recurrence allows precise control of the growth process. The development is not stopped when the network size reach a predetermined limit, but when the code has been read exactly L times through. The number L parametrizes the structure of the neural network. We also introduce a branching program-symbol denoted **B** which labels nodes of arity 2. This program-symbol allows us to encode a family of networks where the first network has a particular structure that may be different than other networks developed from the same grammar tree. A cell that reads **B** executes the following algorithm:

```
If (life > 1) reading-head := left subtree of the current node
Else  reading-head := right subtree  of the current node
```

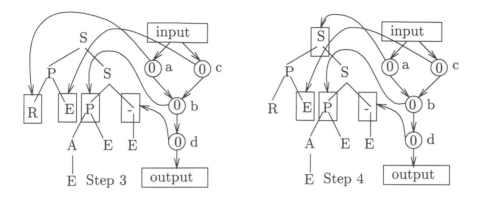

Figure 24.10
A recurrent developmental program for a neural network that computes the even parity for three binary inputs.
The life of the ancestor neuron is 2 The first three steps are the same as those shown in the preceding figure.
During the fourth step, neuron "a" executes a recurrent-program symbol, its reading head points now to the label
"S" on the root of the grammar-tree, as if we were back in the beginning of the development at step 0. The life
of "a" has been decremented and is now equal to 1. Neuron "a" will give birth to a second XOR network, whose
outputs will be sent to the child neurons of neuron "b".

24.3 Genetic Programming

Koza has shown that the tree data structure is an efficient encoding which can be effectively
recombined by genetic operators. *Genetic Programming* has been defined by Koza as a
way of evolving computer programs with a GA [Koza and Rice 1991]. In Koza's Genetic
Programming paradigm the individuals in the population are LISP S-expressions which can
be depicted graphically as rooted, point-labeled trees with ordered branches. Since our
chromosomes (grammar-trees) have exactly the same structure, the same approach is used
for generating and recombining grammar trees. The set of alleles is the set of cell program-
symbols that label the tree, the search space is the hyperspace of all possible labeled trees.
During crossover, a sub-tree is cut from one parent tree and replaces a subtree from the
other parent tree; the result is an offspring tree. The subtrees which are exchanged during
recombination are randomly selected.

There exist two models of parallel GA. In the massive parallel model, individuals are
located on a 2-D grid, and recombination is done between individuals which are near each
other, with respect to the 2-D topology. In the island model, individuals are distributed
in sub-population. Each sub-population is on an island, and islands exchange some of
their individuals from time to time. Our GA is the Mixed Parallel GA described in [Gruau
1993c]. The massive parallel moded is simulated on each processor, whereas the processors
between themselves implement an island model of PGA. Hence advantages of both the

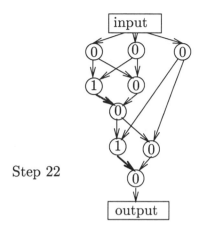

Step 22

Figure 24.11
The final decoded neural network that computes the even parity for three binary inputs. The life of the ancestor neuron is 2. The total number of steps is 22 since the tree is executed 2 times. One can clearly see the two cascaded XOR networks. With an initial life of L we would have developed a neural network for the $L + 1$ input parity problem, with L cascaded XOR networks.

massive parallel model and the island model are combined. The mixed PGA runs on four processors of an IPSC860 parallel machine.

Müehlenbein [Müehlenbein Schomish, and Born 1991] promotes the use of individual hill-climbing. Each individual generated by the GA should try to enhance itself with hill-climbing. Hill-climbing is done by exploring the neighborhood of the individual, with respect to a natural topology given by the problem. In the case of neural network optimization, a learning procedure can be used as a hill-climber. The learning procedure defined in [Gruau 1993b] was implemented. This learning is a cross-over between backpropagation and the bucket brigade used in classifier systems. It learns boolean weights. The learning is applied for each chromosome (solution) produced by the GA, on the first neural network of the family. The learned information is then coded back on the chromosome and can be used to develop the other networks of the family. In [Gruau and Whitley 1993] experimental results suggest that it is the best combination between GA and learning on the phenotype.

Table 1 summarizes the key features of the problem for evolving a Cellular Encoding that generates a family of Neural network for computing the symmetry function of arbitrary large size. In fact we develop the 10 first neural networks. So the solutions are tested for the symmetry problem with up to $2 * 10 = 20$ inputs. Nevertheless, the analysis of the grammar shows that it generates solutions to symmetry of arbitrary large size, because the recurrence has been learned.

Table 24.1
Tableau for genetic search of a neural network for the even parity problem up to 11 inputs, using Cellular Encoding

Objective:	Find the Cellular Encoding of a neural network family that computes the symmetry of an arbitrary large number of bits
Terminal set:	E
Function set:	B,P,S,C,W,I,D,A,O,-,+,R
Fitness cases:	For each developed neural net: If the neural net have less than 6 inputs the whole input space is tested otherwise 32 symmetric and 32 non symmetric random patterns of bits are tested.
Hits:	Fitness cases for which the neural net produces the right output
Raw fitness of one neural net	The number of hits of the neural net:
Standardized fitness of one neural net	raw fitness divided by the total number of fitness cases. It ranges between 0 and 1
Standardized fitness	The sum of standardized fitness of the developed neural net. It ranges between 0 and 10
Parameters:	$M = 1,024, T = 300s, t_m = 0.005, l = 30$
Success Predicate:	The standardized fitness equals 10. That is the 10 first networks are developed and score the maximum number of hits

Not all the 10 neural networks are developed in the early generation, this allows to save time. The following rules decide when to continue the development of other networks: If N_{L_1}, N_{L_2}, ..., N_L have already been developed, then, continue to develop N_{L+1} if N_L correctly processes more than a fraction r of the training set, where r is a given threshold. If $L > L_{max}$ stop the development. In the present experiment, $r = 0.7$ and $L_{max} = 10$.

We used Genetic Programming algorithm but with some differences from Koza. We do not stop the GA after a given number of generations. Instead we stop when elapsed time is greater than a variable T (5 minutes here). We use a mutation rate $t_m = 0.005$, each allele has a probability t_m to be mutated into another allele of the same arity. We found that mutation consistently improves the Genetic search. It may be because all the leaves have the same label "E", and Koza's argument that cross-over on leaves mimics mutation is not valid. Koza consider the depth of the individuals in the initial population. We take into account only the number of nodes and generate an initial population of chromosomes having exactly the same number of nodes which is 30 here. After, when exchange of sub-tree is done during cross-over, the size of the tree can increase (not above 50 nodes)

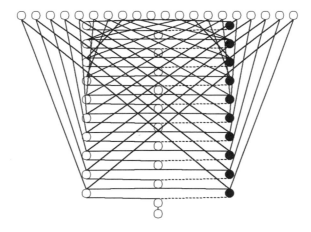

Figure 24.12
A neural network for the symmetry problem of 20 bits. when the threshold is 1, the neuron are black disks, otherwise the empty circle indicated a threshold 0. Plain line represent a weight 1, dashed line represent a threshold -1.

or decrease. Here, we did things differently from normal GP only because at the time of implementation we where not aware of the details of Koza's GP.

The average time over 100 experiment for finding a solution is 62 seconds. An example of chromosome found by the GA is S(S(P(E)(B(P(R(E))(E))(E)))(P(E)(A(I(B(-(E))(R(E)))))))(W(W(W(-(E))))) , The corresponding $10^t h$ neural network that solves the symmetry of 20 inputs is shown in figure 24.12. It is clearly a combination of 20 times the same neuronal group.

The method can be applied to generate neural networks other than boolean neural networks. It is not a problem to encode integer weight and threshold, or even real numbers. Cellular Encoding is efficient if function to be computed by the neural network has regularities. Cellular Encoding is able to encode regularities in a compact way. But, regularity is everywhere in nature.

24.4 JaNNeT: A Neural Compiler of Pascal Program

In this section is described how to compile a Pascal Program into a Neural Network. We begin with some general definitions.

Definition: Genetic Programming language
A genetic programming language $\mathcal{L}_S$ is a set S of symbols of different arities. A genetic

program in such a language is an ordered set of trees labeled using symbols, where at each node, the number of child nodes is the arity of the labeling symbol. Syntactic constraints can limit the range of possible genetic programs.

Koza distinguished between symbols of arity 0 which he calls terminals, and symbols of arity greater than 0 which he calls functions. Here this distinction would be misleading, because a function symbol does not correspond to a real function. The definition of genetic program gives a theoretical skeleton. Its goal is to capture the most useful property of programming which is **MODULARITY**. Each tree t represents a piece of code for a particular function f. This code can be called from other trees, using the number associated to t. With Cellular Encoding, the calling of function f amounts to the inclusion of the neural network that computes f. The calling is implemented using a program-symbol called **JMP** (see [Gruau 1993a]). This program-symbol has an argument. When a cell reads this program-symbol, it places its reading head on the root of the grammar tree which number is the argument.

Instead of using the tree number to refer to the trees, it is possible to use a pattern matching process like the one used by Koza (chapter on spontaneous self replication). Although this referring system can be interesting from a genetic point of view, the general structure of a genetic program is the same, it can always be decomposed into a set of trees. It is only a different implementation. We consider a special kind of grammar defined as follows:

Definition: GEN-grammars
A GEN-grammar is a tree-grammar used to rewrite a genetic program written in a language $\mathcal{L}_S$ into a genetic program written in a language $\mathcal{L}_{S'}$. A rule r of the grammar must check the following condition: The left member is a symbol s of S of arity r. The right member is a tree labeled with symbols of S' and special symbols $T_i, i = 1, \ldots, r$. The right member contains one and exactly one occurrence of each special symbol. When s is rewritten by r, these symbols specify were to insert the subtrees of the node labeled by s.

A GEN-grammar is very simple. In fact it describes a morphism of trees. In rewriting of a genetic program, each node is just replaced by a subtree of symbol from S'. Thus the size of the new program (number of nodes) is bounded by a constant K times the size of the old program, where K can be taken as the maximum size of right members of the grammatical rules.

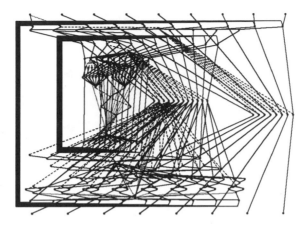

Figure 24.13
A neural network for sorting 8 digits. There are 9 input units (top) and 8 output units (bottom). The activities
flow from top to bottom. The first input unit (the unit in the upper left corner) is the start neuron, which is a
special neuron used to start the network. The last 8 input units are used to input the 8 value to be sorted. When
the network has finished its computation, the output unit activities will be the sorted values. The neuron are more
complex than in the previous section. They have different sort of sigmoïd which allow them to compute arithmetic
and boolean elementary functions.

In [Ratajszcza, Wiber and Gruau 1993] we present a compiler that inputs a Pascal
program and outputs a neural net that computes what the Pascal program specifies. The
compiler starts by computing the parse tree of the Pascal program. The parse tree is more
complex than an usual parse tree used by a compiler, because it includes nodes for definition
of variables. As a result, there is no major differences between this parse tree and a LISP
program, modulo a reduction of the LISP instruction set. The second step of the compiler
is to rewrite this parse tree into a Cellular Encoding using a GEN-grammar. The third step
of the compiler is to develop the Cellular Encoding having chosen an initial value of the life
L, this will produce a neural net that computes what the Pascal program specifies, provided
that no more than L encapsulated recursive calls are done during the execution. Apart
from the development module which is a separate program, the software of the compiler is
merely the 27 rewriting rules of the GEN-grammar that rewrites a parse tree into a Cellular
Encoding.

We now show an example of compilation. The input units of the compiled neural
network correspond to the value read during the execution of the Pascal program. If in the
Pascal program, the instruction *read(a)* appears, there will be one input neuron created.
By setting the initial value of this neuron to v, the value of a will be initialized to v. The

output units correspond to the value written during the execution of the Pascal program. If in the Pascal program, the instruction *write*(*a*) appears, there will be one output neuron created. Whenever the instruction *write*(*a*) is executed, the value of this output neuron will be the value of *a*. There is one particular input unit which is called the *start cell*, which is used to start the network. In order to run the network, one must initialize the input units with the desired values, and then relax the network with a dynamic between parallel and sequential. A neuron fires only if all its neighbors have already fired. The network has finished its computation when no any neurons fire.

Figure 24.13 shows the neural net compiled from the following program:

```
program p;
type tab = array [0..7] of integer;
var t : tab; i, j, max, aux : integer;
begin
  read (t);
  while i < 5 do
  begin
    j := i + 1;max := i;
    while j <= 5 do
    begin
      if t[max] < t[j] then
          max := j
     .fi;
      j := j + 1;
    end;
    aux := t[i];t[i] := t[max];
    t[max] := aux;i := i + 1;
  end;
  write (t);
end.
```

The neural net has 9 input units: a start cell which is used to run the network, plus 6 values to be sorted. There is 8 output units that corresponds to the instruction `write(t)`. Clearly, we can see the two encapsulated loops, that are mapped on two sets of recurrent links. The recurrent links are represented by links which have three segments. Neurons inside the inner loops compute the indice of the max element within the remaining elements to be sorted, neuron inside the outer loops are used to permute element in the array.

24.5   A Hierarchy of Genetic Languages

We propose a classification of Genetic Language based on this ordering relation.

Definition: Level of complexity

Given two genetic languages $\mathcal{L}_1$ and $\mathcal{L}_2$, $\mathcal{L}_1$ is higher level than $\mathcal{L}_2$ ($\mathcal{L}_1 \succ \mathcal{L}_2$) if there exists a GEN-grammar able to rewrite any genetic program P written in $\mathcal{L}_1$ into a genetic program P' written in $\mathcal{L}_2$, that does the same computation.

In the the preceding section we show that Pascal-Parse-Tree $\succ$ Cellular-Encoding. Using a LISP with a reduced instruction set, We have also LISP $\succ$ Cellular Encoding, because our Pascal Parse Trees are similar to LISP S-Expressions, since they include the definition of variables. We now propose other genetic languages, and classify them using $\succ$.

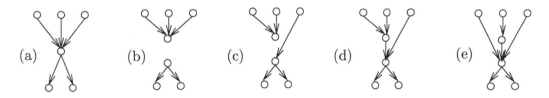

Figure 24.14
Microcoding of the program-symbol ADL 2. (Add one Link), 2 is an argument. This program symbol adds a cell on the link number 2. The microcode is DIV(m(<)(s(m(>)) (2) (a) before the division ADL, (b) division in two children. (c) the segment "m>" is analysed,. (d) the segment "s" is anlysed (e) the segment "m<" is analysed.

In order to make the compiler produce a more compact Cellular Code, we had to introduce many other division program-symbols, as well as program-symbols that modify cell registers, or that make local topological transformations. Each new program-symbol may have an integer argument that specifies a particular link or a particular register, this favors compact representation. For clarity, we have renamed the program-symbols with more explicit name than a single letter, and have microcoded them. A microcode is composed of two parts. The first part specifies the category of the operator. It uses three capital letters. DIV indicates a division, TOP a modification of the topology, CEL, SIT, LNK a modification of respectively a cell register, a site register or a link register. A cell possesses an ordered set of input sites for the links that fan in, and an ordered set of output sites for the link that fan out. EXE indicates a program-symbol used to manage the order of execution, HOM refers to a division in more than 2 children, AFF denotes operators used only for the graph display. The program-symbols are listed in table 24.2. The second part of the microcode is an operand. For CEL, SIT, LNK the operand is the name of the register, for EXE it is the name of the particular order of execution encoded. Among EXE operators, the test operators test a condition on the registers of the cells or on the local

```
******************Local Topological Transformation**************************
CUTL     TOPm-<m->      Cuts the input links to the left
CUTI     TOPm<m>        Cuts an input link
CUTO     -TOPm<m>       Cuts an output link
MRG      TOPm<d^*m>     Merge the link of the input neighbour
CPFO     -TOPm<d^#m>=   Copy with first inputs
CHOPI    TOPm=          Cut all the input links except one
CHOPO    -TOPm=         Cut all the output links except one
PUTF     TOPm=m<m>      Makes a link permutation with the first link
SWITCH   TOPm<m$m>$m=   Makes a link permutation with the last link
PUTL     TOPm<m>m=      Makes a link permutation with the last link
KILL     TOP            Delete the neurone
CYC      TOPs           Create a recurrent link
********Cell division into two children which will execute a separate   code******
PAR      DIVd*R*        Parallel
SEQ      DIVs           Sequential
ADL      DIVm<sm>       Add on Link
AD       DIVm_<sm_>     Add on Link using the link register
IPAR     DIVm<sd=m>     Input Par
SPLFI    DIVd<=s        Split with first inputs
SPLFIL   DIVd-<=s       Split with last inputs
XSPL     DIVm<=s        Exclusive split
******Cell division into more than two children which will execute the same code***
CLO      HOMc           Clone into n children, n is the argument
SPLIT    HOMs           Split into m children, nis computed from the topology
TAB      HOMt           Same has split, chil i as its bias set to i
***********************Modification of  register********************************
SBIAS    CELb           Set the bias
SSIGMO   CELs           Set the type of sigmo\"{\i}
SDYN     CELd           Set the dynamic
EVER     CELe           Set the level of simplifiability
LR       CELl           Set the link register
SITE     SITd           Sets the divisibility of the site
DELSITE  SITm           Merges all the site
VAL      LNKv           Sets the value of the weight
**********************Managing of the execution order*********************
WAIT     EXEw           Wait the specified number of step
JMP      EXEj           Include a subnetwork
END      EXEe           Become a finished neuron
BLOC     EXEq           Wait for the neighbor to loose their reading head
BLIFE    EXEf           Test the value of the life register
BUN      EXEu           Test wether the neighbor has the specified number of links
BPN      EXEp           Test wether the neighbor is a pointer cell
BLR      EXEl           Test the value of the link register
```

Table 24.2
Microcode of the operators used for the compilation. For clarity, microcodes are represented as strings

topology, if it is true, the reading head is placed on the left sub-tree else it is placed on the right sub-tree.

An example is the operator **B** already defined in section 2 which has been renamed as BLIFE. For DIV, the operand is as list of segments. Each segment is composed of one letter and a possibly empty list of arithmetic signs. The letter 's' means the output site of the first child is connected to the input site of the second child, it needs no arithmetic sign. The letter 'm' or 'M' means move a set of links, the letter 'd' or 'D' means duplicate a set of links. the letter 'r' or 'R' is the same as 'd' or 'D' except that the added links are inserted in the reverse order. If the letter is a capital, the links are duplicated or moved from the output site of the second child cell to the input site of the first child cell else, the links are duplicated or moved from the input site of the first child cell to the input site of the second child cell. Arithmetic signs specify the set of links to be moved or duplicated. '$<$', '$>$', '$=$', '$*$' specifies the links lower than, equal, or greater than the argument, '#' and '$' refers to the first and the last link, ' $\cdot$' refers to links from the site of the neighboring cell, '$-$' is used when it is necessary to count the links in decreasing order, rather than in increasing order as usual, '$-$' placed before the three capital letter exchanges the role of input site and output site, '$\_$' is used to replace the argument by the value of the link register. When the division takes place, the first child inherits the input links, the second child inherit the output links. Then the links are redistributed, by analyzing each segment of the microcode in the reverse order. Figure 24.14 gives an example of a DIV microcode analysis. A similar microcoding is used for the operand of TOP, that locally modifies the topology , by cutting or reordering links. In this case, the segments specify which of the input links must be kept. By replacing each program-symbols by its microcode, we get a Cellular Micro Encoding. This replacement can be done using a Gen-Grammar. The left member of a rule is the operator name, the right member is the microcode encoded as a tree.

During the software development of JaNNeT, we defined grammar trees of neural building blocks that are called by the compiled code. These grammar trees form a library and can be considered as macros written in Cellular Encoding. The genetic language that includes Cellular Encoding plus this library can be called Cellular Macro Encoding. It is possible to derive a Cellular Encoding from a Macro Cellular Encoding using a Gen-Grammar. The left member is the name of the macro, the right member is the code of the macro. Finally, if one uses a Lisp Library, the resulting language can be called a Macro LISP. We can summerize the results with the formula:

Macro Lisp $\succ$ LISP $\succ$ Macro Cellular Encoding $\succ$ Cellular Encoding $\succ$ Micro Cellular Encoding.

We now present a concrete example illustrating this hierarchy. Consider the Pascal Program : program p; var a: integer; begin write (a) end. Since variables are set to zero by default, this program will be compiled into a single output neuron with bias 0, that outputs 0. The program has a parse tree which is:

PROGRAM(DECL("A")(TYPE-SIMPLE)(WRITE(IDF-LEC("A"))))

The translation into Macro Cellular Encoding done by the compiler follows Note that the name of the variables must be replaced with a number encoding the moment of their appearance. Therefore the parse tree should contain variable number instead of variable names. But this is an implementation technical detail of JaNNeT, it does not put into question our hierarchy. It is possible to modify Cellular Encoding in order to keep variable names.:

```
IPAR(1)(START-)(SEQ(CPFO(1(SWITCH(1)(IPAR(2)(VAR(LR(1)(VAL(0)(EVER(2))))
(PUTL (2)(BLOC(PAR(CHOPO(1)(SEQ(CPFO(1)(CUTO(1)(CHOP(3)(BPN(1)(END)(MRG
(PN-(2))))))) )(BLOC(WRITE-MACRO))))))))(CUT(1)(CUTO(1)))))))))(BLOC(MRG
(1)(CUTO(1))))).
```

The translation into Cellular Encoding involves replacing each call to a Macro (here WRITE-MACRO) by the cellular code of the macro, it produces:

```
IPAR(1)(START-)(SEQ(CPFO(1(SWITCH(1)(IPAR(2)(VAR(LR(1)(VAL(0)(EVER(2))))
(PUTL(2)(BLOC(PAR(CHOPO(1)(SEQ(CPFO(1)(CUTO(1)(CHOP(3)(BPN(1)(END)(MRG
(PN-(2)))))))(BLOC((LABEL(-29)(BPN(1)(END)(MRG(1)(SPLIT(JMP(-29))))))))))
(CUT(1)(CUTO(1)))))))))(BLOC(MRG(1)(CUTO(1)))))
```

Because we took a simple program the tree do not increase a lot between Macro Cellular Encoding and Cellular Encoding. The translation into Micro Cellular Encoding is done by replacing each program-symbol by its microcode, it produces:

```
DIV(m(<)(s(d(=)(m(>)))))(1)(EXE(e)(0))(DIV(s)(0)(-(TOP(m(<)(d(^(#))(m(>=))))
(1))(TOP(m(<)(m($)(m((>$)(m(=)))))))(1)(DIV(m(<)(s(d(=)(m(>)))))(2)(AFF(var)
(0)(CEL1(1)(LNK(v)(0)(CEL(e)(2)))))(TOP(m(<)(m(>)(m(=))))(2)(EXE(q)(0)(DIV(
d(*)(R(*)))(0)(-(TOP(m(=))(1))(DIV(s)(0)(-(TOP(m(<)(d(^(#))(m(>=))))(1)(-(
TOP(m(<)(m(>)))(1))(TOP(m(=))(3)(EXE(p)(1)(EXE(e)(0))(TOP(m(<)(d(^(*))(m(>))
))(1)(CEL(e)(2))))))(EXE(q)(0)(EXE(y)(-29)(EXE(p)(1)(EXE(e)(0))(TOP(m(<)(d(^
(*))(m(>))))(1)(HOM(0)(0)(EXE(j)(-29)))))))))(TOP(m(<)(m(>)))(1)(TOP(m(<)(m(>
)))(1))))))))))(EXE(q)(0)(TOP(m(<)(d(^*)(m(>))))(1)(-(TOP(m(<)(m(>)))(1)))))
```

We can observe that the size the final tree is much bigger, but the necessary number of label has diminished. It is possible to code the above tree using only 8 different labels.

24.6 Cellular Encoding Versus LISP

Koza has successfully used LISP-like language as a genetic language in a broad range of problems. However, he got impressive results mainly on problems that involve manipulation of symbols. We do not mean mathematical symbols, but entities with a high level meaning. This success may be due to the fact that LISP is a language adapted to symbolic manipulation. On the other hand, when Koza and Rice [Koza and Rice 1991] attempt to search the space of neural nets using LISP, the results are not impressive. They found a neural net for the XOR and for the one-bit adder.

The combination of Genetic Programming with neural network using Cellular Encoding is totally different from the one proposed by Koza and Rice. Here, it is not a program that simulates the neural network, but the neural networks that behave in a program-like manner, since we have shown that a program can be compiled towards a neural net. Modulo this new kind of combination, Genetic Programming of neural networks allows to combine a symbolic search led by the GA, with a connectionist search obtained through learning. We have indeed already combined the genetic search with a learning[Gruau 1993 b, Gruau and Whitley 1993]. It is a natural and elegant solution to the fundamental problem of building neural networks that can process symbols, without an **EXPLICIT MAPPING** from symbols to neurons or neuron-states. Here, the method exploits all the magic of Genetic Programming that builds a different sort of program without using an **EXPLICIT MAPPING** of a target problem to an algorithm. Another interest of the method is to incorporate modularity into neural networks. Modularity is the property of programming. When considered with neural network encoding it can be defined as:

Definition: Modularity
Consider a network N_1 that includes at many different places, a copy of the same subnetwork N_2. The code of N_1 is **modular** *if it includes the code of N_2, a single time.*

In [Gruau 1993a] we show that Cellular Encoding has the property of modularity, and conduct an experiment where modularity is used to solve a difficult problem. Another advantage of Cellular Encoding over LISP is that Cellular Encoding can infer a recurrence where the number of inputs grows. For example, the GA could find a genetic code for a neural network family that computes the parity of n inputs, n arbitrarily large. It seems uneasy to do that with LISP in a natural way, because the inputs must be specified as terminals, which are in finite number.

The general conclusion of this section is that LISP is a better genetic language to do symbol manipulations, and Cellular Encoding is more adapted for neural network synthesis. We believe that neural network synthesis is a more fundamental problem than symbol manipulation. The root problem of Artificial Intelligence is the symbol grounding problem. It is hard to bridge a symbol to the real world. There is a gap between symbols and the real world which is noisy, fuzzy and very large. If the genetic search is made at the level of symbol, it cannot solve the symbol grounding problem. Genetic search with Cellular Encoding allows to search at lower level of complexities, as we have shown in the previous section, and eventually to define structures like retina or cortical column using macro of Cellular Encoding. Such structures are more in touch with the real world and not directly connected with symbols. A concrete example is the neural network found by Beer and Gallagher that solves a six leg locomotion problem in [Beer and Gallagher 1993]. They encoded a large network with a few bits, using symmetries. Here the structure is represented by the symmetries, and symmetries can be captured by a Cellular Encoding, using a modular code. If our final aim is to breed complex cognitive structures like a brain, rather than problem solving based on symbol, we believe that it is better to start with a lower level language than LISP.

24.7 Conclusion

Cellular Encoding is a method for encoding families of similarly structured boolean neural networks, that can compute scalable boolean functions. In this chapter, we demonstrate that genetic search of neural networks using Cellular Encoding is equivalent to Genetic Micro Programming and that Cellular Encoding is a micro-programming language of neural-networks. For this purpose a Pascal neural compiler has been introduced. In order to build a theoretical frame that allows generalization, we propose a definition of a Genetic Programming language, together with a criterion to classify their range of complexity. This criterion uses a special kind of very simple tree-grammars called GEN grammars. Five genetic languages have been classified: Macro Lisp $\succ$ LISP $\succ$ Macro Cellular Encoding $\succ$ Cellular Encoding $\succ$ Micro Cellular Encoding. This classification shows that a decomposition is possible from a high level and abstract description based on an efficient programming language down to very simple micro-micro-microcode that specifies simple and local topological modifications of a graph. These modification are duplication or movement of a set of contiguous links. There is no proof as to which of these five languages is the most appropriate for genetic search. We argue that high level genetic languages (LISP) are more adapted to symbolic processing and low level genetic language (Cellular Encoding) to neural network synthesis. In fact the right genetic language itself can be found by genetic search. The initial genetic program must be built using the most simple genetic language which is Micro Cellular Encoding. A building block is a piece of code that on average increases the fitness of a structure. With this in mind, we can view each symbol of a given language at level l as a building block with respect to the language at level $l - 1$. In this scheme, evolution is a process whereby efficient genetic language of higher complexity are acquired, and stored using a GEN-grammar. The rules of the GEN grammar are the building blocks; they are coded in trees within the chromosome which is an ordered set of trees. The scheme needs an addressing mechanism, a mechanism to decide when to increase the set of trees from n to $n + 1$, and a mechanism for initializing the $n + 1$ component. The first mechanism can be an addressing by template where the complementary template is searched only on trees of lower order than the calling tree. The second mechanism can increase the number of trees whenever a change in the environment occurs, reflected by a change in the average fitness of the population. The initialization of the $n + 1$ component can be simply random.

A genetic language found with this method would be a hierarchy of building blocks. If the initial structures are built with a higher level languages than micro-Cellular Encoding, the scheme is a model for the evolution of modular neural networks, where each neural building block of level $l - 1$ is a symbol at level l.

Bibliography

Randall Beer and John Gallagher (1993), Evolving dynamical neural networks for adaptive behavior. *Adaptive Behavior*, 1:92–122.

F. Gruau (1992a), Cellular encoding of genetic neural network. Technical report 92.21, Laboratoire de l'Informatique du Parallélisme, Ecole Normale Supérieure de Lyon.

F. Gruau (1992b), Genetic synthesis of boolean neural networks with a cell rewriting developmental process. In Darell Whitley and David Schaffer *Combination of Genetic Algorithms and Neural Networks*.

F. Gruau (1993a), Genetic synthesis of modular neural networks. In Stephanie Forrest, editor, *5th International conference on Genetic Algorithm*.

F. Gruau (1993b), A learning and pruning algorithm for genetic neural networks. In Michel Verleysen *European Symposium on Artificial Neural Network*.

F. Gruau (1993c), The mixed parallel genetic algorithm. In Denis Tristram, editor, *ParCo 93 (Parallel Computing)*.

F. Gruau and D. Whitley (1993), Adding learning to the the cellular developmental process: a comparative study. *Evolutionary Computation*. to appear V1N3.

H. Kitano (1990), Designing neural network using genetic algorithm with graph generation system. *Complex Systems*, 4:461–476.

J. R. Koza and J. P. Rice (1991), Genetic generation of both the weights and architecture for a neural network. In *IJCNN91, International Joint Conference on Neural Networks*, Seattle pages 397–404.

John R. Koza (1992), *Genetic programming:On the programming of computers by mean of natural selection*. MIT press.

H. Müehlenbein, M. Schomish, and J.Born (1991), The parallel genetic algorithm as function optimizer. *Parallel Computing*, 17:619–632, 1991.

Jean-Yves Ratajszcak, Gilles Wiber, and Frédéric Gruau (1993), Compilateur pascal de rèseaux de neurones. Rapport de stage de projet de fin d'année, Ecole Nationale Supérieure d'Informatique et de mathématiques appliquée de Grenoble.

Author Index